AF342548

Operator Theory, Analysis and Mathematical Physics

Jan Janas
Pavel Kurasov
Ari Laptev
Sergei Naboko
Günter Stolz
Editors

Birkhäuser
Basel · Boston · Berlin

Editors:

Jan Janas
Institut of Mathematics
Polish Academy of Sciences
ul. Sw. Tomasza 30/7
31-027 Krakow
Poland
e-mail: najanas@cyf-kr.edu.pl

Pavel Kurasov
Department of Mathematics
Lund Institute of Technology
221 00 Lund
Sweden
e-mail: kurasov@maths.lth.se

Sergei Naboko
Department of Mathematical Physics
St. Petersburg State University
Ulyanovskaya 1
198904 St. Petersburg
Russia
e-mail: naboko@snoopy.phys.spbu.ru

Ari Laptev
Dept. of Mathematics, Room 675
South Kensington Campus
Imperial College London
SW7 2AZ London, UK
e-mail:a.laptev@imperial.ac.uk

and

Department of Mathematics
Royal Institute of Technology
100 44 Stockholm
Sweden
e-mail: laptev@math.kth.se

Günter Stolz
Department of Mathematics
University of Alabama
452 Campbell Hall
Birmingham, AL 35294-1170
USA
e-mail: stolz@math.uab.edu

2000 Mathematics Subject Classification 35J10, 35P05, 37K15, 39A11, 39A12, 47A10, 47A40, 47A45, 47A48, 47B36, 47F05, 81V99, 82D10, 99Z99

Library of Congress Control Number: 2007922252

Bibliographic information published by Die Deutsche Bibliothek
Die Deutsche Bibliothek lists this publication in the Deutsche Nationalbibliografie; detailed bibliographic data is available in the Internet at <http://dnb.ddb.de>.

ISBN 978-3-7643-8134-9 Birkhäuser Verlag, Basel – Boston – Berlin

Contents

Introduction

This volume contains mainly the lectures delivered by the participants of the International Conference: Operator Theory and its Applications in Mathematical Physics – OTAMP 2004, held at Mathematical Research and Conference Center in Bedlewo near Poznan. The idea behind these lectures was to present interesting ramifications of operator methods in current research of mathematical physics. The topics of these Proceedings are primarily concerned with: functional models of non-selfadjoint operators, spectral properties of Dirac and Jacobi matrices, Dirichlet-to-Neumann techniques, Lyapunov exponents methods and inverse spectral problems for quantum graphs.

All papers of the volume contain original material and were refereed by acknowledged experts.

The Editors thank all the referees whose critical remarks helped to improve the quality of this volume.

The Organizing Committee of the conference would like to thank all session organizers for taking care about the scientific programm and all participants for making warm and friendly atmosphere during the meeting.

We are particulary grateful to the organizers of **SPECT**, without whose financial support the OTAMP 2004 would never been so successful. We also acknowledcdge financial support of young Polish participants by Stefan Banach International Mathematical Center and thank the staff of the Conference Center at Bedlewo for their great support which helped to run the conference smoothly.

Finally, we thank the Editorial Board and especially Professor I. Gohberg for including this volume into the series **Operator Theory: Advances and Applications** and to Birkhäuser-Verlag for help in preparation of the volume.

<table>
<tr><td>Birmingham – Krakow – Lund</td><td>August 2006</td></tr>
<tr><td>St. Petersburg – Stockholm</td><td>The Editors</td></tr>
</table>

Operator Theory:
Advances and Applications, Vol. 174, 1–7

Finiteness of Eigenvalues of the Perturbed Dirac Operator

Petru A. Cojuhari

Abstract. Finiteness criteria are established for the point spectrum of the perturbed Dirac operator. The results are obtained by applying the direct methods of the perturbation theory of linear operators. The particular case of the Hamiltonian of a Dirac particle in an electromagnetic field is also considered.

Mathematics Subject Classification (2000). Primary 35P05, 47F05; Secondary 47A55, 47A75.

Keywords. Dirac operators, spectral theory, relatively compact perturbation.

1. Introduction

The present paper is concerned with a spectral problem for the perturbed Dirac operator of the form

$$H = \sum_{k=1}^{n} \alpha_k D_k + \alpha_{n+1} + Q, \tag{1.1}$$

where $D_k = i\frac{\partial}{\partial x_k}$ $(k = 1, \ldots, n)$, α_k $(k = 1, \ldots, n+1)$ are $m \times m$ Hermitian matrices which satisfy the anticommutation relations (or, so-called Clifford's relations)

$$\alpha_j \alpha_k + \alpha_k \alpha_j = 2\delta_{jk} \ (j, k = 1, \ldots, n+1), \tag{1.2}$$

$m = 2^{\frac{n}{2}}$ for n even and $m = 2^{\frac{n+1}{2}}$ for n odd. Q is considered as a perturbation of the free Dirac operator

$$H_0 = \sum_{k=1}^{n} \alpha_k D_k + \alpha_{n+1} \tag{1.3}$$

and represents the operator of multiplication by a given $m \times m$ Hermitian matrix-valued function $Q(x), x \in \mathbb{R}^n$. In accordance with our interests we assume that the elements $q_{jk}(x)$ $(j, k = 1, \ldots, m)$ of the matrix $Q(x)$ are measurable functions from the space $L_\infty(\mathbb{R}^n)$. The operators H_0 and H are considered in the space $L_2(\mathbb{R}^n; \mathbb{C}^m)$ with their maximal domains of definition. Namely, it is considered that the domain of the operator H_0 is the Sobolev space $W_2^1(\mathbb{R}^n; \mathbb{C}^m)$ and, because Q is a bounded operator, the perturbed Dirac operator H is defined on the same domain $W_2^1(\mathbb{R}^n; \mathbb{C}^m)$ as well. The Dirac operators H_0 and H are selfadjoint on this domain. For the free Dirac operator H_0 is true the following algebraic relations

$$H_0^2 = \sum_{k=1}^{n} \alpha_k^2 D_k^2 + \sum_{j \neq k} (\alpha_j \alpha_k + \alpha_k \alpha_j) D_j D_k + \sum_{k=1}^{n} (\alpha_{n+1} \alpha_k + \alpha_k \alpha_{n+1}) D_k + \alpha_{n+1}^2$$

$$= \sum_{k=1}^{n} D_k^2 + E_m = (-\Delta + I) E_m,$$

so that

$$H_0^2 = (-\Delta + I) E_m. \tag{1.4}$$

Here Δ denotes the Laplace operator on $\mathbb{R}^n$ and E_m the $m \times m$ identity matrix. It follows from (1.4) that the spectrum of the operator H_0^2 covers the interval $[1, \infty]$ and, since the spectrum of the operator H_0 is a symmetric set with respect to the origin, it results that its spectrum coincides with the set $\sigma(H) = (-\infty, -1] \cup [1, +\infty)$. We note that the symmetry of the spectrum of H_0 can be shown easily by invoking, for instance, another matrix β which together with α_k ($k = 1, \ldots, n+1$) the anticommutation conditions (1.2) are satisfied. Then

$$(H_0 + \lambda)\beta = -\beta(H_0 - \lambda)$$

for each scalar λ, and so the property of the symmetry of $\sigma(H_0)$ becomes to be clear. The unperturbed operator H_0 has no eigenvalues (in fact the spectrum of H_0 is only absolutely continuous). If the entries of the matrix-valued function $Q(x)$ vanish at the infinite, the continuous spectrum of the perturbed Dirac operator H coincides with $\sigma(H_0)$ and the perturbation Q can provoke a non-trivial point spectrum. Our problem is to study the point spectrum of the perturbed Dirac operator H. This problem has been studied by many researchers in connection with various problems (note that the most of the results were concerned with the case $n = 3$ and $m = 4$). A good deal of background material on the development and perspectives of the problem can be found in [1], [2], [3], [5], [7], [10], [12], [13], [14]. Apart from the already mentioned works, we refer to the [15] and the references given therein for a partial list.

In this paper, we give conditions on $Q(x)$ under which the point spectrum of H (if any) has ± 1 as the only possible accumulation points. Specifically, we assume that $Q(x)$ satisfies the following assumption.

(A) *$Q(x) = [q_{jk}(x)]$, $x \in \mathbb{R}^n$, is an $m \times m$ Hermitian matrix-valued function the entries of which are elements from the space $L_\infty(\mathbb{R}^n)$ and*

$$\lim_{|x| \to \infty} |x| \, q_{jk}(x) = 0 \ (j, k = 1, \ldots, m).$$

The main results are obtained by applying the abstract results from [6] (see also its refinement results made in [9]). Below, we cite the corresponding result.

Let $\mathcal{H}$ be a Hilbert space. Denote by $\mathbb{B}(\mathcal{H})$ the space of all bounded operators on $\mathcal{H}$ and by $\mathbb{B}_\infty(\mathcal{H})$ the subspace of $\mathbb{B}(\mathcal{H})$ consisting of all compact operators in $\mathcal{H}$. The domain and the range of an operator A are denoted by $\mathrm{Dom}(A)$ and $\mathrm{Ran}(A)$, respectively.

Theorem 1.1. [9] *Let A and B be symmetric operators in a space $\mathcal{H}$ and let the operator A has no eigenvalues on a closed interval Λ of the real axis. Suppose that there exists an operator-valued function $T(\lambda)$ defined on the interval Λ having the properties that*

(i) *$T(\lambda) \in \mathbb{B}_\infty(\mathcal{H})$ $(\lambda \in \Lambda)$,*
(ii) *$T(\lambda)$ is continuous on Λ in the uniform norm topology, and*
(iii) *for each $\lambda \in \Lambda$ and for each $u \in \mathrm{Dom}(B)$ such that $Bu \in \mathrm{Ran}(A - \lambda I)$ there holds the following inequality*

$$\| (A - \lambda I)^{-1} Bu \| \leq \| T(\lambda)u \| . \tag{1.5}$$

Then the point spectrum of the perturbed operator $A + B$ on the interval Λ consists only of finite number of eigenvalues of finity multiplicity.

Remark 1.2. *The assertion of Theorem 1.1 remains true if in place of (1.5) it is required the following one*

$$\| (A - \lambda I)^{-1} Bu \| \leq \sum_{k=1}^{N} \| T_k(\lambda)u \|, \tag{1.6}$$

where the operator-valued functions $T_k(\lambda)$ $(k = 1, \ldots, N)$ satisfy the conditions (i) and (ii).

As we already mentioned we will apply Theorem 1.1 to the study of the problem of the discreteness of the set of eigenvalues of the perturbed Dirac operator H. The main results are presented in the next section.

2. Main results

Let H be the Dirac operator defined by (1.1) in which the matrix-valued function satisfies the assumption (A). The unperturbed Dirac operator H_0 represents a matrix differential operator (of the dimension $m \times m$) of order 1. The symbol of the operator H_0 is a matrix-valued function which we denote by $h_0(\xi), \xi \in \mathbb{R}^n$. Note that by applying the Fourier transformation to the elements of the space $L_2(\mathbb{R}^n; \mathbb{C}^m)$ the operator H_0 is transformed (in the momentum space) into a multiplication operator by the matrix $h_0(\xi)$. The Fourier transformation is defined by the formula

$$\hat{u}(\xi) = (Fu)(\xi) = \frac{1}{(2\pi)^{\frac{n}{2}}} \int u(x) e^{i<x,\xi>} dx \ (u \in L_2(\mathbb{R}^n))$$

in which $< x, \xi >$ designates the scalar product of the elements $x, \xi \in \mathbb{R}^n$ (here and in what follows $\int : \int_{\mathbb{R}^n}$). The corresponding norm in $\mathbb{R}^n$ (or $\mathbb{C}^m$) will be denoted as usually by $| \, . \, |$. The operator norm of $m \times m$ matrices corresponding to the norm $| \, . \, |$ in $\mathbb{C}^m$ will be denoted by $| \, . \, |$, as well.

Our main result is the following

Theorem 2.1. *Let H be the perturbed Dirac operator defined by (1.1) for which the assumption* (A) *is satisfied. Then the point spectrum of the operator H has only ± 1 as accumulation points. Each eigenvalue can be only of a finite multiplicity.*

Proof. That the spectrum in the spectral gap $(-1, 1)$ is only discrete without any accumulation points in the interior of this interval follows at once due to Weyl type theorems. Let Λ be an closed interval contained in the set $(-\infty, -1) \cup (1, +\infty)$ and let λ be an arbitrary point belonging to Λ. It will be shown that under assumed conditions the operators H_0 and H verify all of hypotheses of Theorem 1.1. To this end, we estimate the norm of the element $(H_0 - \lambda I)^{-1} Qu$ for each $u \in L_2(\mathbb{R}^n; \mathbb{C}^m)$ such that $Qu \in \mathrm{Ran}(H_0 - \lambda I)$. Let $\widehat{Qu}$ be the Fourier transform of Qu, and denote

$$\widehat{v}(\xi) := (h_0(\xi) + \lambda)\widehat{Qu}(\xi), \ \xi \in \mathbb{R}^n.$$

According to (1.4), we may write

$$\| (H_0 - \lambda I)^{-1} Qu \|^2 = \int |(h_0(\xi) - \lambda)^{-1}\widehat{Qu}(\xi)|^2 d\xi$$

$$= \int |(|\xi|^2 - r(\lambda)^2)^{-1}\widehat{v}(\xi)|^2 d\xi, \tag{2.1}$$

where $r(\lambda) := \sqrt{\lambda^2 - 1}$.

Next, we let

$$\Omega(\Lambda) = \cup_{\lambda \in \Lambda} \{\xi \in \mathbb{R}^n : |\xi| = r(\lambda)\}$$

and we choose a sphere U of radius R with center of the origin such that $U \supset \Omega(\Lambda)$ and let $V = \mathbb{R}^n \setminus U$. Then passing to spherical coordinates $\xi = |\xi|\omega, \rho = |\xi|$ (we write ds_ω for the area element of hypersurface S_{n-1} of the unit sphere S in $\mathbb{R}^n$), and denoting

$$\hat{f}(\rho, \omega) = \frac{\rho^{\frac{n-1}{2}} \hat{v}(\rho\omega)}{\rho + r(\lambda)} \ (0 \leq \rho < \infty, \ \omega \in S_{n-1}),$$

we have

$$\int_U |(h_0(\xi) - \lambda)^{-1}\widehat{Qu}(\xi)|^2 d\xi = \int_{S_{n-1}} \int_0^R \rho^{n-1}|(\rho^2 - r(\lambda)^2)^{-1}\hat{v}(\rho\omega)|^2 d\rho dS_\omega$$

$$= \int_{S_{n-1}} \int_0^R \left| \frac{\hat{f}(\rho, \omega)}{\rho - r(\lambda)} \right|^2 d\rho dS_\omega.$$

Since $Qu \in \mathrm{Ran}(H_0 - \lambda I)$, it follows that $\hat{f}(\rho, \omega)$ vanishes at $\rho = r(\lambda)$, and we can continue

$$\left[\int_{S_{n-1}} \int_0^R \left| \frac{\hat{f}(\rho, \omega)}{\rho - r(\lambda)} \right|^2 d\rho dS_\omega \right]^{\frac{1}{2}}$$

$$= \left[\int_{S_{n-1}} \int_0^R \left| \int_0^1 \frac{\partial \hat{f}}{\partial \rho}(t(\rho - r(\lambda)) + r(\lambda), \omega) dt \right|^2 d\rho ds_\omega \right]^{\frac{1}{2}}$$

$$\leq \int_0^1 \left[\int_{S_{n-1}} \int_0^R \left| \frac{\partial \hat{f}}{\partial \rho}(t(\rho - r(\lambda)) + r(\lambda), \omega) \right|^2 d\rho ds_\omega \right]^{\frac{1}{2}} dt$$

$$\leq 2 \left[\int_{S_{n-1}} \int_0^R \left| \frac{\partial \hat{f}}{\partial \rho}(\rho, \omega) \right|^2 d\rho ds_\omega \right]^{\frac{1}{2}}$$

$$\leq \left[\int_{S_{n-1}} \int_0^R \left| \rho^{\frac{n-3}{2}} ((n-3)\rho + (n-1)r(\lambda))(\rho + r(\lambda))^{-2} \hat{v}(\rho\omega) \right|^2 d\rho ds_\omega \right]^{\frac{1}{2}}$$

$$+ 2 \left[\int_{S_{n-1}} \int_0^R \left| \rho^{\frac{n-1}{2}} (\rho + r(\lambda))^{-1} \frac{\partial}{\partial \rho} \hat{v}(\rho\omega) \right|^2 d\rho ds_\omega \right]^{\frac{1}{2}}.$$

Taking into account that $|\frac{\partial}{\partial \rho} \hat{v}(\rho\omega)| \leq |\nabla \hat{v}|$, we get

$$\left[\int_U \left| (h_0(\xi) - \lambda)^{-1} \widehat{Qu}(\xi) \right|^2 d\xi \right]^2$$

$$\leq 2r(\lambda) \left[\int_U \left| \frac{(n-3) \mid \xi \mid + (n-1)r(\lambda)}{|\xi|(|\xi| + r(\lambda))^2} \hat{v}(\xi) \right|^2 d\xi \right]^{\frac{1}{2}} + 2 \left[\int_U \left| \frac{\nabla \hat{v}(\xi)}{|\xi| + r(\lambda)} \right|^2 d\xi \right]^{\frac{1}{2}}.$$

Since the expressions $(n-3) \mid \xi \mid + (n-1)r(\lambda), (\mid \xi \mid + r(\lambda))^{-1} \ (\lambda \in \Lambda; \xi \in U)$ and each element of the matrix-valued function $h_0(\xi) - \lambda \ (\lambda \in \Lambda; \xi \in U)$ are bounded on $\Lambda \times U$ there exist constants $c_1 > 0$ and $c_2 > 0$ such that

$$\left[\int_U |(h_0(\xi) - \lambda)^{-1} \widehat{Qu}(\xi) \mid^2 d\xi \right]^{\frac{1}{2}}$$

$$\leq c_1 \left[\int_U \left| \mid \xi \mid^{-1} \widehat{Qu}(\xi) \right|^2 d\xi \right]^{\frac{1}{2}} + c_2 \left[\int_U |\nabla \widehat{Qu}(\xi)|^2 d\xi \right]^{\frac{1}{2}}.$$

We claim that the integral operators with kernels

$$\mid \xi \mid^{-1} Q(x)e^{-i(x,\xi)}, x^l Q(x)e^{-i(x,\xi)} \ (\mid l \mid = 1; x \in \mathbb{R}^n, \xi \in U)$$

are compact operators in the space $L_2(\mathbb{R}^n; \mathbb{C}^m)$. The compactness of them can be proved by applying the criteria obtained in [4] (or, also, by applying the lemma from [8], page 45).

In addition, we note that the integral operator K_V with the kernel

$$(h_0(\xi) - \lambda)^{-1} Q(x)e^{-i(x,\xi)} \ (x \in \mathbb{R}^n; \xi \in V)$$

represents also a compact operator. To see this fact, it suffices to show that

$$\| (I - P_h)K_V \| \to 0 \ as \ h \to \infty, \tag{2.2}$$

where $(P_h u)(x) = u(x)$ for $\mid x \mid \leq h$ and $(P_h u)(x) = 0$ for $\mid x \mid > h$.

Since each element of the matrix-valued function $(h_0(\xi) - \lambda)^{-1}$ behaves as $\mid \xi \mid^{-1}$ at the infinite, it follows the evaluation

$$\| (I - P_h)K_V u \|^2 \leq c \int_{|\xi| > h} |(1 + \mid \xi \mid)^{-1} \widehat{Qu}(\xi)|^2 d\xi \leq c(1 + h)^{-2} \| u \|^2,$$

and so (2.2) is realized.

Thus, taking into account (2.1), we obtain an estimate like that from (1.6) (see Remark 1.2) and, therefore Theorem 1.1 can be applied. This completes the proof of Theorem 2.1. $\qquad\square$

As an application of Theorem 2.1 we give a result concerning the particular case of the Hamiltonian of a Dirac particle in an electromagnetic field. The Dirac operator in this case is typically written in the physics literature (see, for instance, [11], [15]) as follows

$$Hu = \sum_{j=1}^{3} \alpha_j(D_j - A_j(x))u + \alpha_4 u + q(x)u, u \in W_2^1(\mathbb{R}^3; \mathbb{C}^4), \qquad (2.3)$$

where $A(x) = (A_1(x), A_2(x), A_3(x))$ (the vector potential) and $q(x)$ (the scalar potential) are given functions on $\mathbb{R}^3$.

Theorem 2.2. *If*

$$\lim_{|x|\to\infty} |x|A_j(x) = 0 \ (j = 1, 2, 3), \ \lim_{|x|\to\infty} |x|q(x) = 0,$$

then the point spectrum of the Dirac operator defined by (2.1) is discrete having only ± 1 as accumulation points. Each eigenvalue can be only of a finite multiplicity.

References

[1] A. Bertheir, V. Georgescu, *On the point spectrum for Dirac operators*, J. Func. Anal. **71** (1987), 309–338.

[2] M.S. Birman, *On the spectrum of Schrodinger and Dirac operators*, Dokl. Acad. Nauk **129**, no.2 (1959), 239–241.

[3] M.S. Birman, *On the spectrum of singular boundary-value problems*, Mat. Sb. **55** (1961), no. 2, 125–174.

[4] M.S. Birman, G.E. Karadzhov, M.Z. Solomyak, *Boundedness conditions and spectrum estimates for the operators $b(X)a(D)$ and their analogs*, Adv. Soviet. Math., **7** (1991), 85–106.

[5] M.S. Birman, A. Laptev, *Discrete spectrum of the perturbed Dirac operator*, Ark. Mat. **32**(1) (1994), 13–32.

[6] P.A. Cojuhari, *On the discrete spectrum of a perturbed Wiener-Hopf operator*, Mat. Issled. **54** (1980), 50–55 (Russian).

[7] P.A. Cojuhari, *On the point spectrum of the Dirac operator*, Dif. Urav., Minsk, Dep. VINITI 24.06.87, N4611-B87, (1987), 1–17 (Russian).

[8] P.A. Cojuhari, *On the finiteness of the discrete spectrum of some matrix pseudodifferential operators*, Vyssh. Uchebn. Zaved. Mat., **1** (1989), 42–50 (Russian).

[9] P.A. Cojuhari, *The problem of the finiteness of the point spectrum for self-adjoint operators. Perturbations of Wiener-Hopf operators and applications to Jacobi Matrices*, Operator Theory, Adv. and Appl. **154** (2004), 35–50.

[10] W.D. Evans, B.J. Harris, *Bounds for the point spectra of separated Dirac operators*, Proc. Roy. Soc. Edinburgh, Sect. A **88** (1980), 1–15.

[11] A. Fock, *Introduction to Quantum Mechanics,* Nauka, Moscow, 1976 (Russian).

[12] D.B. Hinton, A.B. Mingarelli, T.T. Read, J.K. Shaw, *On the number of eigenvalues in the spectral gap of a Dirac system,* Proc. Edinburgh Math. Soc. **29**(2) (1986), 367–378.

[13] M. Klaus, *On the point spectrum of Dirac operators,* Helv. Phys. Acta. **53** (1980), 453–462.

[14] O.I. Kurbenin, *The discrete spectra of the Dirac and Pauli operators,* Topics in Mathematical Physics, vol. 3, Spectral Theory, (1969), 43–52.

[15] B. Thaller, *The Dirac Equation,* Springer, Berlin-Heidelberg-New York, 1992.

Petru A. Cojuhari
Department of Applied Mathematics
AGH University of Science and Technology
Al. Mickievicza 30
30-059 Cracow, Poland
e-mail: `cojuhari@uci.agh.edu.pl`

Operator Theory:
Advances and Applications, Vol. 174, 9–20

A Mathematical Study of Quantum Revivals and Quantum Fidelity

Monique Combescure

Abstract. In this paper we present some results obtained recently, partly in collaboration with Didier Robert, about "quantum revivals" and "quantum fidelity", mainly in the semiclassical framework. We also describe the exact properties of the quantum fidelity (also called Loschmidt echo) for the case of explicit quadratic plus inverse quadratic time-periodic Hamiltonians and establish that the quantum fidelity equals one for exactly the times where the classical fidelity is maximal.

Mathematics Subject Classification (2000). Primary 99Z99; Secondary 00A00.

Keywords. Class file, Birkart.

1. Introduction

The quantum return probability is the modulus of the overlap between an initial wavepacket and its time evolution governed by Schrödinger equation. When this quantity happens to equal one for some time t, then the system is said to exhibit "quantum revivals".

$$R(t) := |\langle \psi, U(t, 0)\psi \rangle|.$$

When the Hamiltonian $\hat{H}(t)$ (possibly time-dependent) is assumed to be perturbed $\hat{H}_g(t) := \hat{H}(t) + gV$, then we can compare the evolutions generated by $\hat{H}(t)$ and $\hat{H}_g(t)$ respectively, acting on the **same** initial state ψ:

$$F(t) := \langle U(t, 0)\psi, U_g(t, 0)\psi \rangle$$

which is the measure of the "quantum fidelity" in the state ψ along the evolution. Of course $F(0) = 1$ and $F(t) \equiv 1$ if $g = 0$.

Thus the decrease in t of $F(t)$ measures the lack of fidelity due to the perturbation.

Both quantum revivals and quantum fidelity have attracted much recent interest in the physics literature (see references). Notably, it has been heuristically claimed that the decrease in time of the quantum fidelity allows to distinguish between systems having regular versus chaotic classical evolution.

In our study, we consider the semiclassical regime for both quantities, using coherent states as initial wavepackets ψ.

We also perform exact calculations of the quantum fidelity in the case of the singular time-periodic harmonic oscillator, with initial wavepackets ψ being "generalized coherent states" in the sense of Perelomov, showing that they do not decrease to zero as time evolves, but present recurrences to 1 exactly at the values of times where the classical fidelity is maximal. More specifically we consider Hamiltonians:

$$\hat{H}(t) := \frac{P^2}{2} + f(t)\frac{Q^2}{2}$$

and

$$\hat{H}_g(t) := \hat{H}(t) + \frac{g^2}{Q^2}$$

where the real constant g is the size of the perturbation, and f is a T-periodic function of time, and we perform an exact calculation of:

$$F(t) := \langle U(t,0)\psi, U_g(t)\psi \rangle$$

where $U(t,0)$ (resp. $U_g(t,0)$) is the quantum evolution generated by $\hat{H}(t)$, (resp. $\hat{H}_g(t)$).

2. Semiclassical quantum revivals

We use the "coherent states" of the harmonic oscillator: let φ_0 be the ground state of the harmonic oscillator $\hat{P}^2/2 + \hat{Q}^2/2$ in dimension n, where $\hat{Q}$ (resp. $\hat{P}$) is the operator of multiplication by x (resp. the derivation operator $\hat{P} := -i\hbar\nabla$) in the Hilbert space of quantum states $\mathcal{H} = L^2(\mathbb{R}^n)$.

The Weyl-Heisenberg unitary translation operator by $\alpha := (q,p)$ is defined as:

$$\hat{T}(\alpha) := \exp\left(i\frac{p.\hat{Q} - q.\hat{P}}{\hbar}\right)$$

and the coherent state φ_α is defined as follows:

$$\varphi_\alpha := \hat{T}(\alpha)\varphi_0.$$

Now consider a classical trajectory in phase space induced by the Hamiltonian $H := \frac{p^2}{2} + V(q)$:

$$\alpha \mapsto \alpha_t := (q_t, p_t).$$

Let $S(t)$ be the classical action along this trajectory, and M_t be the Hessian matrix of H taken at point α_t:

$$(M_t)_{i,j} := \frac{\partial^2 H}{\partial\alpha_i\partial\alpha_j}(q_t, p_t).$$

It is a real symmetric $2n \times 2n$ matrix. The linearized flow (or stability matrix) is obtained by solving the differential equation:

$$\dot{F}_t = J M_t F_t$$

where J is the symplectic matrix:

$$J = \begin{pmatrix} 0 & \mathbb{1} \\ -\mathbb{1} & 0 \end{pmatrix}$$

with initial data

$$F_0 = \mathbb{1}.$$

Clearly F_t is a symplectic matrix, i.e., satisfies for any t

$$\tilde{F}_t J F_t = J.$$

Namely

$$\frac{d}{dt}(\tilde{F}_t J F_t) = -\tilde{F}_t M_t J J F_t + \tilde{F}_t J J M_t F_t = \tilde{F}_t M_t F_t - \tilde{F}_t M_t F_t = 0$$

and therefore $\tilde{F}_t J F_t$ is a constant $2n \times 2n$ matrix equal to J since $F_0 = \mathbb{1}$.

To this symplectic matrix can be associated a unitary operator $\hat{R}(F)$ in the Hilbert space $\mathcal{H}$, acting as expected:

$$\hat{R}(F)^{-1} \begin{pmatrix} \hat{Q} \\ \hat{P} \end{pmatrix} \hat{R}(F) = F \begin{pmatrix} \hat{Q} \\ \hat{P} \end{pmatrix}.$$

Now we are in a position to give the semiclassical approximation for the quantum evolution of coherent states (see [8])

Proposition 2.1. *Under suitable assumptions on V, there exists a constant $\epsilon(\hbar, t)$, small as $\hbar$ goes to zero, such that if $U(t,0)$ is the quantum evolution operator associated to the Weyl quantization of H one has:*

$$\|U(t,0)\varphi_\alpha - e^{i\delta_t/\hbar}\hat{T}(\alpha_t)\hat{R}(F_t)\varphi_0\| \leq \epsilon(\hbar, t)$$

where $\delta_t := S(t) - (p_t.q_t - p.q)/2$.

Thus, up to a controllable error, the recurrence probability $|\langle \varphi_\alpha, U(t,0)\varphi_\alpha \rangle|$ can be replaced with

$$\tilde{R}(t) := |\langle \hat{T}(\alpha)\varphi_0, \hat{T}(\alpha_t)\hat{R}(F_t)\varphi_0 \rangle| = |\langle \hat{T}(\alpha - \alpha_t)\varphi_0, \hat{R}(F_t)\varphi_0 \rangle|.$$

We shall now make use of a beautiful result by Mehlig and Wilkinson [22], that gives the Weyl covariant symbol of the metaplectic operators. For a complete mathematical proof see our paper [9].

Proposition 2.2. *Let F be a symplectic $2n \times 2n$ matrix not having 1 as eigenvalue. Then the associated metaplectic operator $\hat{R}(F)$ can be written as*

$$\hat{R}(F) = \frac{h^{-n}\gamma_F}{|\det(\mathbb{1} - F)|^{1/2}} \int_{\mathbb{R}^{2n}} dz\, \hat{T}(z) e^{iz.Az/2\hbar}$$

where γ_F is a complex number of modulus 1, and

$$A := \frac{J}{2}(F + \mathbb{1})(F - \mathbb{1})^{-1}.$$

We now set

$$z_t := \alpha - \alpha_t.$$

Then using Proposition 2.2, we have:

$$\tilde{R}(t) = h^{-n}|\det(\mathbb{1} - F_t)|^{-1/2}\left|\int dz \langle \hat{T}(z_t)\varphi_0, \hat{T}(z)\varphi_0\rangle e^{iz.Az/2\hbar}\right|$$

$$= h^{-n}|\det(\mathbb{1} - F_t)|^{-1/2}\left|\int dz \exp\left(\frac{iz.Az}{2\hbar} + \frac{iz.Jz_t}{2\hbar} - \frac{1}{4\hbar}(z - z_t)^2\right)\right|$$

$$= h^{-n}|\det(\mathbb{1} - F_t)|^{-1/2}e^{-z_t^2/4\hbar}\left|\int dz \exp\left(-\frac{1}{2\hbar}z.(\frac{1}{2} - iA)z + \frac{1}{2\hbar}z.(J + i\mathbb{1})z_t\right)\right|.$$

Now by using the calculus of Fourier transforms of Gaussians, we get:

$$\tilde{R}(t) = |\det(\mathbb{1} - F_t)\left(\frac{\mathbb{1}}{2} - iA\right)|^{-1/2}\left|\exp\left(-\frac{z_t^2}{4\hbar} - \frac{1}{4\hbar}z_t.Kz_t\right)\right|$$

where the matrix K is given by

$$K := (J - i\mathbb{1})(\mathbb{1} - 2iA)^{-1}(J + i\mathbb{1}).$$

Now we have the following remarkable result:

$$\frac{\mathbb{1}}{2} - iA = N(F - \mathbb{1})^{-1}$$

where

$$N := \frac{1}{2}(F(\mathbb{1} - iJ) - (\mathbb{1} + iJ))$$

so that

$$|\det(\mathbb{1} - F)\left(\frac{\mathbb{1}}{2} - iA\right)|^{-1/2} = |\det N|^{-1/2}.$$

But N has the important following property

Lemma 2.3.

$$|\det N| \geq 1$$

and

$|\det N| = 1 \qquad \Longleftrightarrow \quad F$ *is unitary*

This result has been established in full generality in [9], but we shall here indicate the calculus in dimension $n = 1$. The symplectic matrix F has now the simple form

$$F = \begin{pmatrix} a & b \\ c & d \end{pmatrix}$$

with $ad - bc = 1$ to ensure the symplecticity. Using the form given above for N we easily get

$$\det N = -\frac{1}{2}(a + d + i(b - c))$$

so that

$$|\det N|^2 = \frac{1}{4}((a+d)^2 + (b-c)^2) = \frac{1}{4}(4 + (a-d)^2 + (b+c)^2) \geq 1$$

with equality to 1 $\iff$ $a = d$, $b = -c$, in which case F is just unitary (rotation).

Thus we get the following result

Theorem 2.4. *Denoting by z_t the following distance $z_t := \alpha - \alpha_t$, the complete return probability has the following semiclassical estimate:*

$$\left| R(t) - |\det N|^{-1/2} \left| \exp\left(-\frac{z_t^2}{4\hbar} - \frac{1}{4\hbar} z_t.K z_t \right) \right| \right| \leq \epsilon(\hbar, t).$$

Thus if α lies on a classical periodic orbit γ with period T_γ, the exponential is just 1; furthermore the prefactor is 1 $\iff$ F_{T_γ} is unitary, in which case we have almost semiclassical recurrence.

$$R(T_\gamma) \geq 1 - \epsilon(\hbar, t).$$

Note that $\epsilon(\hbar, T_\gamma) = O(\hbar^\varepsilon)$ provided $T_\gamma \leq \lambda|\log \hbar|$ for some λ given by the classical dynamics.

3. The quantum fidelity

Let $\hat{H}$ be a quantum Hamiltonian, and $\hat{H}_g := \hat{H} + gV$ be a perturbation of it (g small). The quantum fidelity at time t in the state ψ is given as

$$F(t) := |\langle e^{-it\hat{H}/\hbar}\psi, e^{-it\hat{H}_g/\hbar}\psi \rangle|.$$

Remark 3.1. *If $\psi = \psi_j$ is an eigenstate of $\hat{H}$ (resp. $\hat{H}_g$), then the fidelity is nothing else that the "return probability" $|\langle \psi, e^{-it\hat{H}_g/\hbar}\psi \rangle|$ (resp. $|\langle \psi, e^{-it\hat{H}/\hbar}\psi \rangle|$).*

Remark 3.2. *Clearly $F(0) = 1$, and $F(t) \equiv 1$ if $g = 0$. One expects that if $g \neq 0$ then $F(t)$ will decrease as time increases. Furthermore it is believed that this decrease could be significantly different for an associated classical dynamics being regular versus chaotic. No exact result has been established up to now.*

One can semiclassically estimate this quantum fidelity along the same lines as the return probability above.

Theorem 3.3. *Let α_t (resp. $\alpha'(t)$) be the classical phase-space point reached by the trajectory governed by Hamiltonian H (resp. H_g), starting from the **same** point α at time 0. Then the following estimate holds true:*

$$|F(t) - C_t \exp\left\{ -\frac{1}{4\hbar}(F_t'(\alpha_t' - \alpha_t))^2 K(\alpha, g, t) \right\}| \leq \hbar^{1/2} L(\alpha, g, t)$$

where F_t' is the stability matrix for the dynamics generated by H_g, and C_t, $K(\alpha, g, t)$, $L(\alpha, g, t)$ are positive controllable constants.

Remark 3.4. *The proof of this statement is contained in our paper* [9]. *The important fact to notice is that the classical* **infidelity** $\alpha_t - \alpha_t'$ *is an important quantity to estimate in t and in g.*

We now come to an interesting particular case where the quantum fidelity can be computed exactly. The Hamiltonians considered are time-periodic and have the following form:

$$H_0(t) := \frac{p^2}{2} + f(t)\frac{x^2}{2}$$

$$H_g(t) := H_0(t) + \frac{g^2}{x^2}$$

where $t \mapsto f$ is a T-periodic function, and g a real constant.

We denote by $\hat{H}_0(t)$ and $\hat{H}_g(t)$ the corresponding selfadjoint operators in $\mathcal{H} = L^2(\mathbb{R})$. The initial states we shall consider are generalized coherent states in the sense of Perelomov ([23]) adapted to the underlying algebra SU(1,1). Let

$$K_0 := \frac{\hat{Q}^2 + \hat{P}^2}{4} + \frac{g^2}{2\hat{Q}^2} = \frac{1}{2}\hat{H}_g$$

$$K_{\pm} = \frac{\hat{P}^2 - \hat{Q}^2}{4} \mp i\frac{\hat{Q}.\hat{P} + \hat{P}.\hat{Q}}{4} - \frac{g^2}{2\hat{Q}^2}.$$

They satisfy:

$$[K_0, K_{\pm}] = \pm K_{\pm}, \qquad [K_-, K_+] = 2K_0, \qquad K_- = K_+^*.$$

Let ψ be the ground state of $\hat{H}_g$:

$$\hat{H}_g\psi = (\alpha + \frac{1}{2})\psi$$

with $\alpha = \frac{1}{2} + \sqrt{\frac{1}{4} + 2g^2}$. It also annihilates K_-:

$$K_-\psi = 0$$

and has the following form:

$$\psi(x) = cx^{\alpha}e^{-x^2/2}$$

where c is a normalization constant such that $\|\psi\| = 1$.
For $\beta \in \mathbb{C}$, we define the unitary operator

$$\hat{S}(\beta) := \exp(\beta K_+ - \bar{\beta}K_-)$$

and the generalized coherent states as

$$\psi_\beta := \hat{S}(\beta)\psi.$$

Let $U_0(t,0)$ and $U_g(t,0)$ be the quantum unitary evolution operators generated by $\hat{H}_0(t)$ and $\hat{H}_g(t)$. We shall study the quantum fidelity (without absolute value):

$$F(t) := \langle U_0(t,0)\psi_\beta, U_g(t,0)\psi_\beta \rangle.$$

We shall first study the particular case $g = 1$ (whence $\alpha = 2$). Then ψ is obviously a simple linear combination of the eigenstates φ_0 and φ_2 of the harmonic oscillator.

We have the following explicit result (see [7]):

Proposition 3.5. *Let $z(t)$ be a complex solution of the linear differential equation* (*Hill's equation*):

$$\ddot{z} + fz = 0.$$

We define its polar decomposition by

$$z(t) = e^{u+i\theta}$$

where $t \mapsto u$ and $t \mapsto \theta$ are real, and consider the following initial data:

$$u(0) = u_0, \qquad \dot{u}(0) = \dot{u}_0, \qquad \theta(0) = \theta_0, \qquad \dot{\theta}(0) = e^{-2(u_0-\epsilon)}.$$

Let $\hat{H}_g = \frac{\hat{P}^2+\hat{Q}^2}{2} + \frac{g^2}{\hat{Q}^2}$. Then we have:

$$U_g(t,0) = e^{i\dot{u}\hat{Q}^2/2}e^{-i(u-\epsilon)(\hat{Q}.\hat{P}+\hat{P}.\hat{Q})/2}e^{-i(\theta-\theta_0)\hat{H}_g}e^{i(u_0-\epsilon)(\hat{Q}.\hat{P}+\hat{P}.\hat{Q})/2}e^{-i\dot{u}_0\hat{Q}^2/2}.$$

The same formula holds true for $U_0(t,0)$ with $\hat{H}_g(t)$ replaced by $\hat{H}_0(t)$.

An important fact to notice is that the constants u_0, $\dot{u}_0$, θ_0, ϵ can be adjusted, given any $\beta \in \mathbb{C}$ such that

$$\hat{S}(\beta)\psi = e^{i\dot{u}_0\hat{Q}^2/2}e^{-i(u_0-\epsilon)(\hat{Q}.\hat{P}+\hat{P}.\hat{Q})/2}e^{-i\theta_0\hat{H}_g}\psi.$$

Then $U_g(t,0)\psi$ and $U_0(t,0)\psi$ have simple explicit form (see [7]), leading to a very simple form of the fidelity:

Theorem 3.6. *Let $g = 1$. Then the fidelity is just given by*

$$F(t) = \frac{2}{3} + \frac{1}{3}e^{-2i(\theta(t)-\theta(0))}.$$

Let us remark that $\theta(t)$ is just given by the formula

$$\theta(t) - \theta(0) = e^{2\epsilon}\int_0^t e^{-2u(s)}ds.$$

Study of the Hill's equation

$$\ddot{z} + fz = 0.$$

• **Stable case:**

u is T-periodic, and thus $\theta(t)$ grows from $-\infty$ to $+\infty$ when t varies from $-\infty$ to $+\infty$. Therefore there exists an infinite sequence $(t_k)_{k\in\mathbb{Z}}$ such that

$$\theta(t_k) - \theta(0) = 2k\pi$$

in which case $F(t_k) = 1$, i.e., the quantum fidelity is perfect. Moreover there exists an infinite sequence $(t'_k)_{k\in\mathbb{Z}}$ such that

$$\theta(t'_k) - \theta(0) = (2k+1)\pi$$

in which case $F(t'_k) = 1/3$. We thus have the following picture:

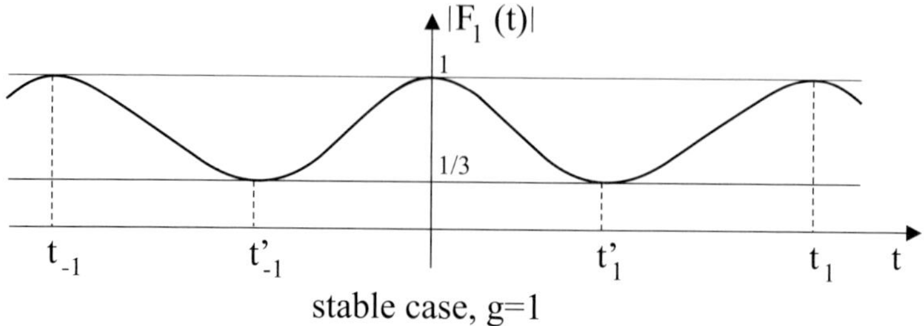

stable case, g=1

• Unstable case:

In this case there is some Lyapunov exponent $\lambda > 0$, and solutions of Hill's equation $c(s) = p(t)e^{\lambda t}$, $\quad s(t) = q(t)e^{-\lambda t}$, with $t \mapsto p, q$ real T-periodic functions such that $z(t)$ is a complex linear combination of them. Depending on the instability zone of Hill's equation, this yields for $|z(t)|^{-2}$ integrability or not at time $t \to \infty$. If $|z(t)|^{-2}$ is not integrable, then $|\theta(t)| \to \infty$ and the quantum fidelity behaves as in the picture for the stable case. If we are in the instability zone where $|z(t)|^{-2}$ is integrable, then $\theta(t) \to \theta_{\pm}$, as $t \to \pm\infty$, in which case we have the following picture:

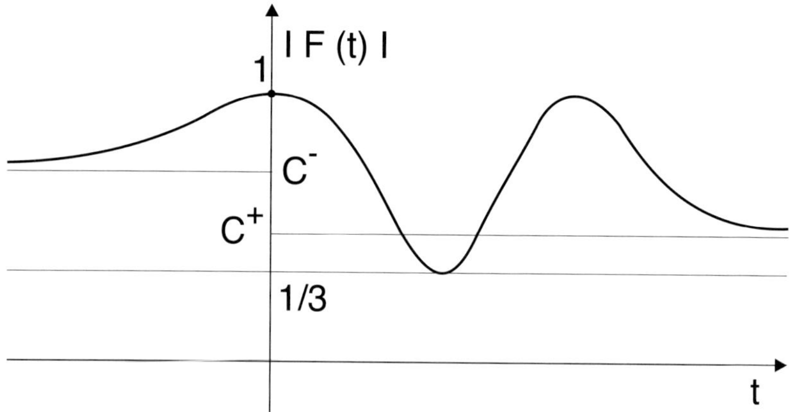

General case

We assume now that g is an arbitrary real constant. Then ψ no longer has a finite linear decomposition on the Hermite functions, but instead

$$\psi = \sum_{0}^{\infty} \lambda_n \varphi_n$$

with $\sum |\lambda_n|^2 = 1$. Then the following result holds true:

Theorem 3.7.

$$\langle U_0(t,0)\psi_\beta, U_g(t,0)\psi_\beta\rangle = e^{-i(\theta(t)-\theta(0))\alpha} \sum_0^\infty |\lambda_n|^2 e^{-in(\theta(t)-\theta(0))}.$$

Therefore $|F(t)| = 1$ *if* $\theta(t) - \theta(0) = 0 \pmod{2\pi}$.

If g is such that $\alpha := \frac{1}{2} + \sqrt{\frac{1}{4} + 2g^2} = \frac{p}{q} \in \mathbb{Q}$, *then* $F(t) = 1$ *if* $\theta(t) - \theta(0) = 0$, $\pmod{2q\pi}$.

Exact classical fidelity implies exact quantum fidelity

Let $x(t)$ and $y(t)$ be real classical solutions for the Hamiltonians $H_0(t)$ and $H_g(t)$ respectively such that $x(0) = y(0)$ and $\dot{x}(0) = \dot{y}(0)$ which means that the trajectories **merge from the same point in phase space at** $t = 0$.

t is said a time of **classical fidelity** if $x(t) = y(t)$ and $\dot{x}(t) - \dot{y}(t) = 0$, and the **classical infidelity** at time t is measured by the distance $|x(t) - y(t)|$.

Theorem 3.8. *Let $x(t)$ be a real solution of Hill's equation $\ddot{x} + fx = 0$. We write it as*

$$x(t) = e^{u(t)} \cos \tilde{\theta}(t)$$

with u and $\tilde{\theta}$ real functions and $\tilde{\theta}(t) = g\sqrt{2} \int_0^t ds\, e^{-2u(s)}$. Then $y(t) := e^{u(t)}$ is a solution of equation

$$\ddot{y} + fy - \frac{2g^2}{y^3} = 0$$

such that $x(0) = y(0)$ and $\dot{x}(0) = \dot{y}(0)$. This means that $y(t)$ is a real trajectory for Hamiltonian $H_g(t)$, merging from the same point in phase space as $x(t)$. We clearly have

$$|x(t) - y(t)| = y(t)(1 - \cos \tilde{\theta}(t))$$

which vanishes for $\tilde{\theta}(t) = 0 \pmod{2\pi}$.

Remark 3.9. *By choosing ϵ such that $e^{2\epsilon} = g\sqrt{2}$, we clearly have*

$$\tilde{\theta}(t) = \theta(t) - \theta(0).$$

Corollary 3.10. *If $\theta(t) - \theta(0) = 0 \pmod{2\pi}$, then the classical fidelity is zero, and the quantum fidelity equals 1 (at least in absolute value in the case of general g).*

The proof is very elementary. Let $z(t)$ be a complex solution of Hill's equation of the form

$$z(t) := e^{u(t) + i\tilde{\theta}(t)}$$

with u and $\tilde{\theta}$ **real** functions. Since f is real the Wronskian of z and z^* is constant, and we assume that it equals $2ig\sqrt{2}$:

$$\dot{z}z^* - \dot{z}^*z = 2i\dot{\tilde{\theta}}|z|^2 = 2ig\sqrt{2}.$$

Therefore

$$\frac{d}{dt}\tilde{\theta} = g\sqrt{2}e^{-2u}$$

and therefore Hill's equation for z implies:

$$\ddot{z} + fz = 0 = \left[\ddot{u} + \dot{u}^2 - \left(\frac{d}{dt}\tilde{\theta} \right)^2 + f + i \left(\frac{d^2}{dt^2}\tilde{\theta} + 2\dot{u}\frac{d}{dt}\tilde{\theta} \right) \right] z$$

whence the equation for u:

$$\ddot{u} + \dot{u}^2 - 2g^2 e^{-4u} = -f$$

and thus for $y := e^u$:

$$\ddot{y} + fy - \frac{2g^2}{y^3} = 0$$

which is nothing but Newton's equation for Hamiltonian $H_g(t)$. Furthermore assuming that $\tilde{\theta}(0) = 0$, $x(t) := e^{u(t)}\cos\tilde{\theta}(t)$ and $y(t)$ have the same initial data. This completes the proof, noting that

$$\tilde{\theta}(t) = \theta(t) - \theta(0) = g\sqrt{2} \int_0^t ds\, e^{-2u(s)}.$$

References

[1] G. Benenti, G. Casati, *Sensitivity of Quantum motion for Classically Chaotic Systems*, arXiv: quant-ph/0112060 (2001).

[2] G. Benenti, G. Casati, *Quantum-classical Correspondence in Perturbed Chaotic Systems*, Phys. Rev. E, **65** (2002), 066205-1.

[3] G. Benenti, G. Casati, G. Veble, *Asymptotic Decay of the Classical Loschmidt Echo in Chaotic Systems*, arXiv: nlin.CD/0208003 (2002).

[4] G. Benenti, G. Casati, G. Veble, *On the Stability of Classically Chaotic Motion under System's Perturbations*, Phys. Rev. E **67** (2003), 055202(R).

[5] N. Cerruti, S. Tomsovic, *Sensitivity of Wave Field Evolution and Manifold Stability in Chaotic Systems*, Phys. Rev. Lett. **88** (2002), 054103.

[6] N. Cerruti, S. Tomsovic, *A Uniform Approximation for the Fidelity in Chaotic Systems*, J. Phys. A: Math. Gen. **36** (2003), 3451–3465.

[7] M. Combescure, *The quantum fidelity for the time-dependent Singular Quantum Oscilator*, J. Math. Phys. **47**, 032102 1–12, (2006).

[8] M. Combescure, D. Robert, *Semiclassical Spreading of Quantum Wavepackets and Applications near Unstable Fixed Points of the Classical Flow*, Asymptotic Analysis, **14** (1997), 377–404.

[9] M. Combescure, D. Robert, *A Phase-Space Study of the Loschmidt Echo in the Semiclassical limit*, Ann. Henri Poincare, to appear (2006).

[10] F.M. Cucchietti, H.M. Pastawski, D.A. Wisniacki, *Decoherence as Decay of the Loschmidt Echo in a Lorentz Gas*, Phys. Rev. E, **65** (2002), 045206(R).

[11] F.M. Cucchietti, H.M. Pastawski, R.A. Jalabert, *Universality of the Lyapunov Regime for the Loschmidt Echo*, arXiv: cond-mat/0307752 (2003).

[12] F.M. Cucchietti, D.A. Dalvit, J.P. Paz, W.H. Zurek, *Decoherence and the Loschmidt Echo*, arXiv:quant-phys/0306142 (2003).

[13] B. Eckhardt, *Echoes in Classical Dynamical Systems*, J. Phys. A: Math. and general, **36** (2003), 371–380.

[14] J. Emerson, Y. Weinstein, S. Lloyd, D. Cory, *Fidelity Decay as an Indicator of Quantum Chaos*, Phys. Rev. Lett. **89** (2002), 284102.

[15] G.A. Fiete, E.J. Heller, *Semiclassical Theory of Coherence and Decoherence*, Phys. Rev. A **68** (2003), 022112.

[16] V. Giovannetti, S. Llyod, L. Maccone, *Quantum Limits to Dynamical Evolution*, Phys. Rev. A **67** (2003), 052109.

[17] L. Hörmander, *Symplectic Classification of Quadratic Forms and General Mehler Formulas* Math. Z. 219 (1995), 413–449.

[18] P. Jacquod, I. Adagideli, C.W. Beenakker, *Decay of the Loschmidt Echo for Quantum States with sub-Planck scale Structures*, Phys. Rev. Lett. **89** (2002), 154103.

[19] P. Jacquod, I. Adagideli, C.W. Beenakker, *Anomalous Power Law of Quantum Reversibility for Classically Regular Dynamics*, Europhys. Lett. **61** (2003), 729–735.

[20] R.A. Jalabert, H.M. Pastawski, *Environment-independent decoherence Rate in Classically Chaotic Systems*, arXiv: cond-mat/0010094 (2001).

[21] R.A. Jalabert, H.M. Pastawski, *Environment-Independent Decoherence Rate in Classically Chaotic Systems*, Phys. Rev. Lett. **86** (2001), 2490–2493.

[22] B. Mehlig, M. Wilkinson, *Semiclassical trace formulae using coherent states* Ann. Phys. (Leipzig) **10** (2001), 541.

[23] A. Perelomov, *Generalized Coherent States and their Applications*, Springer-Verlag, 1986.

[24] A. Peres, *Stability of Quantum motion in Chaotic and Regular Systems* Phys. Rev. A **30** (1984), 1610–1615.

[25] T. Prosen, *On General Relation Between Quantum Ergodicity and Fidelity of Quantum Dynamics*, arXiv:quant-ph/0106149 (2001).

[26] P. Prosen, T.H. Seligman, *Decoherence of Spin Echoes*, arXiv:nlin.CD/0201038 (2002).

[27] T. Prosen, T.H. Seligman, M. Znidaric, *Stability of Quantum Coherence and Correlation Decay*, Phys. Rev. A **67** (2003), 042112.

[28] T. Prosen, T.H. Seligman, M. Znidaric, *Theory of Quantum Loschmidt Echoes*, arXiv: quant-ph/0304104, (2003).

[29] T. Prosen, M. Znidaric, *Stability of Quantum Motion and Correlation Decay*, J. Phys. A: Math. Gen, **35** (2002), 1455–1481.

[30] R.W. Robinett, *Quantum wave packet revivals*, Phys. Rep. **392** (2004), 1–119.

[31] R. Sankaranarayanan, A. Lakshminarayan, *Recurrence of Fidelity in Near-Integrable Systems*, Phys. Rev. E **68** (2003), 036216.

[32] S. Schlunk, M.B. d'Arcy, S.A. Gardiner, D. Cassettari, R.M. Godun, G.S. Summy, *Signatures of quantum stability in a classically chaotic system*, Phys. Rev. Lett. **90** (2003), 124102.

[33] P.G. Silvestrov, J. Tworzydlo, C.W. Beenakker, *Hypersensitivity to Perturbations of Quantum-Chaotic Wavepacket Dynamics*, Phys. Rev. Lett. **67** (2003), 025204(R).

[34] J. Vanicek, D. Cohen, *Survival Probability and Local Density of States for One-dimensional Hamiltonian Systems*, J. Phys. A: Math. Gen. **36** (2003), 9591–9608.

[35] J. Vanicek, E.J. Heller, *Semiclassical Evaluation of Fidelity in the Fermi-Golden-Rule and Lyapunov Regimes*, Phys. Rev. E **68** (2003), 056208.

[36] J. Vanicek, E.J. Heller, *Uniform Semiclassical Wave Function for Coherent 2D Electron Flow*, arXiv:nlin.CD/0209001 (2002).

[37] G. Veble, T. Prosen T., *Faster than Lyapunov Decays of Classical Loschmidt Echo*, Phys. Rev. Lett. **92** (2004), 034101.

[38] W.-g. Wang, B. Li, *Crossover of Quantum Loschmidt Echo from Golden Rule Decay to Perturbation-Independent Decay*, Phys. Rev. E **66** (2002), 056208.

[39] W.-g. Wang, G. Casati, B. Li, *Stability of Quantum Motion: Beyond Fermi-golden-rule and Lyapunov Decay*, arXiv:quant-ph/0309154 (2003).

[40] Y. Weinstein, S. Lloyd, C. Tsallis, *The Edge of Quantum Chaos*, Phys. Rev. Lett. **89** (2002), 214101.

[41] Y. Weinstein, J. Emerson, S. Lloyd, D. Cory, *Fidelity Decay Saturation Level for Initial Eigenstates*, arXiv:quant-ph/0210063 (2002).

[42] Y. Weinstein, S. Lloyd, C. Tsallis, *Border between Regular and Chaotic Quantum Dynamics*, Phys. Rev. Lett. **89** (2002), 214101-1.

[43] D. Wisniacki, *Short time Decay of the Loschmidt Echo*, Phys. Rev. E **67** (2003), 016205.

[44] D. Wisniacki, D. Cohen, *Quantum Irreversibility, Perturbation Independent Decay, and the parametric Theory of the Local Density of States*, Phys. Rev. E **66** (2002), 046209.

[45] M. Znidaric, T. Prosen, *Fidelity and Purity Decay in Weakly Coupled Composite Systems*, J. Phys. A **36** (2003), 2463–2481.

Monique Combescure
Laboratoire de Physique Théorique
CNRS - UMR 8627
Université de Paris XI
Bâtiment 210
F-91405 Orsay Cedex, France

and

IPNL
Bâtiment Paul Dirac
4 rue Enrico Fermi
Université Lyon-1
F-69622 Villeurbanne Cedex, France
e-mail: `monique.combescure@ipnl.in2p3.fr`

Operator Theory:
Advances and Applications, Vol. 174, 21–34

On Relations Between Stable and Zeno Dynamics in a Leaky Graph Decay Model

Pavel Exner, Takashi Ichinose and Sylwia Kondej

Abstract. We use a caricature model of a system consisting of a quantum wire and a finite number of quantum dots, to discuss relation between the Zeno dynamics and the stable one which governs time evolution of the dot states in the absence of the wire. We analyze the weak coupling case and argue that the two time evolutions can differ significantly only at times comparable with the lifetime of the unstable system undisturbed by perpetual measurement.

Mathematics Subject Classification (2000). Primary 81V99; Secondary 47D08, 35J10.

Keywords. Zeno dynamics, Schrödinger operator, singular interactions.

1. Introduction

It is well known that the decay of an unstable system can be slowed down, or even fully stopped in the ideal case, if one checks frequently whether the system is still undecayed. The first proper statement of this fact is due to Beskow and Nilsson [2] and a rigorous mathematical proof was given by Friedman [11], but it became popular only after Misra and Sudarshan [14] invented the name "quantum Zeno effect" for it. In recent years this subject attracted a new wave of interest – a rich bibliography can be found, e.g., in [10, 15].

The motivation of this interest is twofold. On one hand the progress in experimental methods makes real the possibility to observe the effect as a phenomenon really existing in the nature, and ultimately to make use of it. On the other hand, the problem presents also interesting mathematical challenges. The most important among them is obviously the question about the *quantum Zeno dynamics:* if the perpetual measurement keeps the state of the system within the Hilbert space associated with the unstable system, what is then the time evolution of such a state? Some recent results [7, 8] give partial answers to this question, which we shall describe below, and there are counterexamples [13], see also [6, Rem. 2.4.9], which point out the borders beyond which it has no sense.

In this note we are going to address a different question. Suppose that at the beginning the interaction responsible for the decay is absent, so state vectors evolve within the mentioned space which we below call $P\mathcal{H}$. Switching the interaction with the "environment" in, we allow the system to decay which means the state vectors may partially or fully leave the space $P\mathcal{H}$. If we now perform the Zeno-style monitoring, the system is forced to stay within $P\mathcal{H}$ and to evolve there, but what is in this case the relation of its dynamics to the original "stable" one?

A general answer to this question is by no means easy and we do not strive for this ambitious goal here. Our aim is to analyze a simple example which involves a Schrödinger operator in $L^2(\mathbb{R}^2)$ with a singular interaction supported by a line and a finite family of points [9]. This model is explicitly solvable and can be regarded as a caricature description of a system consisting of a quantum wire and dots which are not connected mutually but can interact by means tunnelling. The main result of this paper given in Theorem 6.1 below is that in the model *the two dynamics do not differ significantly during time periods short at the scale given by the lifetime of the system unperturbed by the perpetual observation.*

Let us briefly summarize the contents of the paper. First we recall basic notions concerning Zeno dynamics; we will prove the needed existence result in case when the state space of the unstable system has a finite dimension. Sections 3–5 are devoted to the mentioned solvable model. We will introduce its Hamiltonian and find its resolvent. Then we will show that in the "weak-coupling" case when the points are sufficiently far from the line the model exhibit resonances, and in Section 5 we will treat the model from the decay point of view, showing how the point-interaction eigenfunctions dissipate due to the tunnelling between the points and the line; in the appendix we demonstrate that in the weak-coupling case the decay is approximately exponential. The main result is stated and proved in Section 6.

2. Quantum Zeno dynamics

Following general principles of quantum decay kinematics [6, Chap. 1] we associate with an unstable system three objects: the state Hilbert space $\mathcal{H}$ describing all of its states including the decayed ones, the full Hamiltonian H on $\mathcal{H}$ and the projection P which specifies the subspace of states of the unstable system alone. H is, of course, a self-adjoint operator, we need to assume that it is *bounded from below*.

The question about the existence of Zeno dynamics mentioned above can be then stated in this context generally as follows: does the limit

$$(Pe^{-iHt/n}P)^n \longrightarrow e^{-iH_Pt} \tag{2.1}$$

hold as $n \to \infty$, in which sense, and what is in such a case the operator H_P? Let us start from the end and consider the quadratic form $u \mapsto \|H^{1/2}Pu\|^2$ with the form domain $D(H^{1/2}P)$ which is closed but in general it may not be densely defined.

The classical results of Chernoff [3, 4] suggest that the operator associated with this form, $H_P := (H^{1/2}P)^*(H^{1/2}P)$, is a natural candidate for the generator of the Zeno dynamics, and the counterexamples mentioned in the introduction show that the limit may not exist if H_P is not *densely defined,* so we adopt this assumption.

Remark 2.1. Notice that the operator H_P is an extension of PHP, but in general a nontrivial one. This can be illustrated even in the simplest situation when $\dim P = 1$, because if H is unbounded $D(H)$ is a proper subspace of $D(H^{1/2})$. Take $\psi_0 \in D(H^{1/2}) \setminus D(H)$ such that $H^{1/2}\psi_0$ is nonzero, and let P refer to the one-dimensional subspace spanned by ψ_0. This means that PHP cannot be applied to any nonzero vector $\psi\, (= \alpha\psi_0)$ of $P\mathcal{H}$ while $H_P\psi$ is well defined and nonzero.

It is conjectured that formula (2.1) will hold under the stated assumptions in the strong operator topology. Proof of this claim remains an open question, though. The best result to the date [7] establishes the convergence in a weaker topology which includes averaging of the norm difference with respect to the time variable. While this may be sufficient from the viewpoint of physical interpretation, mathematically the situation is unsatisfactory, since other results available to the date require modifications at the left-hand side of (2.1), either by replacement of the exponential by another Kato function, or by adding a spectral projection interpreted as an additional energy measurement – see [8] for more details.

There is one case, however, when the formula can be proven, namely the situation when $\dim P < \infty$ and the density assumption simply means that $P\mathcal{H} \subset \mathcal{Q}(H)$, where $\mathcal{Q}(H)$ is the form domain of H. Since this exactly what we need for our example, let us state the result.

Theorem 2.2. *Let H be a self-adjoint operator in a separable Hilbert space $\mathcal{H}$, bounded from below, and P a finite-dimensional orthogonal projection on $\mathcal{H}$. If $P\mathcal{H} \subset \mathcal{Q}(H)$, then for any $\psi \in \mathcal{H}$ and $t \geq 0$ we have*

$$\lim_{n\to\infty} (Pe^{-iHt/n}P)^n\psi = e^{-iH_P t}\psi, \tag{2.2}$$

uniformly on any compact interval of the time variable t.

Proof. The claim can be proved in different ways, see [7] and [8]. Here we use another argument the idea of which was suggested by G.M. Graf and A. Guekos [12]. Notice first that without loss of generality we may suppose that H is strictly positive, i.e., $H \geq \delta I$ for some positive number δ. The said argument is then based on the observation that

$$\lim_{t\to 0} t^{-1} \left\| Pe^{-iHt}P - Pe^{-itH_P t}P \right\| = 0 \tag{2.3}$$

implies $\left\| (Pe^{-iHt/n}P)^n - Pe^{-iH_P t} \right\| = n\,o(t/n)$ as $n \to \infty$ by means of a natural telescopic estimate. To establish (2.3) one has first to check that

$$t^{-1}\left[(\phi, Pe^{-iHt}P\psi) - (\phi, \psi) - it(H^{1/2}P\phi, H^{1/2}P\psi) \right] \to 0$$

as $t \to 0$ for all ϕ, ψ from $D(H^{1/2}P)$ which coincides in this case with $P\mathcal{H} \oplus (I-P)\mathcal{H}$ by the closed-graph theorem. The last expression is equal to

$$\left(H^{1/2}P\phi, \left[\frac{e^{-iHt} - I}{Ht} - i \right] H^{1/2}P\psi \right)$$

and the square bracket tends to zero strongly by the functional calculus, which yields the sought conclusion. In the same way we find that

$$t^{-1} \left[(\phi, Pe^{-iH_P t}P\psi) - (\phi, \psi) - it(H_P^{1/2}\phi, H_P^{1/2}\psi) \right] \to 0$$

holds as $t \to 0$ for any vectors $\phi, \psi \in P\mathcal{H}$. Next we note that $(H_P^{1/2}\phi, H_P^{1/2}\psi) = (H^{1/2}P\phi, H^{1/2}P\psi)$ by definition, and consequently, the expression contained in (2.3) tends to zero weakly as $t \to 0$, however, in a finite dimensional $P\mathcal{H}$ the weak and operator-norm topologies are equivalent. $\square$

Remark 2.3. It is clear that the finite dimension of P is essential for the proof. The same results holds for the backward time evolution, $t \leq 0$. Moreover, the formula (2.2) has non-symmetric versions with the operator product replaced with $(Pe^{-iHt/n})^n$ and $(e^{-iHt/n}P)^n$ tending to the same limit – see [7].

3. A model of leaky line and dots

Before coming to the proper decay problem let us describe the general setting of the model. We will consider a generalized Schrödinger operator in $L^2 \equiv L^2(\mathbb{R}^2)$ with a singular interaction supported by a set consisting of two parts. One is a straight line, the other is a finite family of points situated in general outside the line, hence formally we can write our Hamiltonian as

$$-\Delta - \alpha\delta(x - \Sigma) + \sum_{i=1}^{n} \tilde{\beta}_i \delta(x - y^{(i)}), \tag{3.1}$$

where $\alpha > 0$, $\Sigma := \{(x_1, 0); x_1 \in \mathbb{R}^2\}$, and $\Pi := \{y^{(i)}\}_{i=1}^{n} \subset \mathbb{R}^2 \setminus \Sigma$. The formal coupling constants of the two-dimensional δ potentials are marked by tildes because they are not identical with the proper coupling parameters β_i which define these point interaction by means of appropriate boundary conditions.

Following the standard prescription [1] one can define the operator rigorously [9] by introducing appropriated boundary conditions on $\Sigma \cup \Pi$. Consider functions $\psi \in W_{\text{loc}}^{2,2}(\mathbb{R}^2 \setminus (\Sigma \cup \Pi)) \cap L^2$ which are continuous on Σ. For a small enough $\rho > 0$ the restriction $\psi \restriction_{\mathcal{C}_{\rho,i}}$ to the circle $\mathcal{C}_{\rho,i} \equiv \mathcal{C}_\rho(y_i) := \{q \in \mathbb{R}^2 : |q - y^{(i)}| = \rho\}$ is well defined; we will say that ψ belongs to $D(\dot{H}_{\alpha,\beta})$ iff $(\partial_{x_1}^2 + \partial_{x_2}^2)\psi$ on $\mathbb{R}^2 \setminus (\Sigma \cup \Pi)$ in the sense of distributions belongs to L^2 and the limits

$$\Xi_i(\psi) := -\lim_{\rho \to 0} \frac{1}{\ln \rho} \psi \restriction_{\mathcal{C}_{\rho,i}}, \quad \Omega_i(\psi) := \lim_{\rho \to 0}[\psi \restriction_{\mathcal{C}_{\rho,i}} + \Xi_i(\psi) \ln \rho], \quad i = 1, \ldots, n,$$

$$\Xi_\Sigma(\psi)(x_1) := \partial_{x_2}\psi(x_1, 0+) - \partial_{x_2}\psi(x_1, 0-), \quad \Omega_\Sigma(\psi)(x_1) := \psi(x_1, 0)$$

exist, they are finite, and satisfy the relations

$$2\pi\beta_i\Xi_i(\psi) = \Omega_i(\psi)\,, \quad \Xi_\Sigma(\psi)(x_1) = -\alpha\Omega_\Sigma(\psi)(x_1)\,, \tag{3.2}$$

where $\beta_i \in \mathbb{R}$ are the true coupling parameters; we put $\beta \equiv (\beta_1, \dots, \beta_n)$ in the following. On this domain we define the operator $\dot{H}_{\alpha,\beta} : D(\dot{H}_{\alpha,\beta}) \to L^2$ by

$$\dot{H}_{\alpha,\beta}\psi(x) = -\Delta\psi(x) \quad \text{for} \quad x \in \mathbb{R}^2 \setminus (\Sigma \cup \Pi)\,.$$

It is now a standard thing to check that $\dot{H}_{\alpha,\beta}$ is essentially self-adjoint [9]; we identify its closure denoted as $H_{\alpha,\beta}$ with the formal Hamiltonian (3.1).

To find the resolvent of $H_{\alpha,\beta}$ we start from $R(z) = (-\Delta-z)^{-1}$ which is for any $z \in \mathbb{C} \setminus [0,\infty)$ an integral operator with the kernel $G_z(x,x') = \frac{1}{2\pi}K_0(\sqrt{-z}|x-x'|)$, where K_0 is the Macdonald function and $z \mapsto \sqrt{z}$ has conventionally a cut at the positive half-line; we denote by $\mathbf{R}(z)$ the unitary operator with the same kernel acting from L^2 to $W^{2,2} \equiv W^{2,2}(\mathbb{R}^2)$. We introduce two auxiliary Hilbert spaces, $\mathcal{H}_0 := L^2(\mathbb{R})$ and $\mathcal{H}_1 := \mathbb{C}^n$, and the corresponding trace maps $\tau_j : W^{2,2} \to \mathcal{H}_j$ which act as

$$\tau_0\psi := \psi\!\restriction_\Sigma\,, \quad \tau_1\psi := \psi\!\restriction_\Pi = \big(\psi\!\restriction_{\{y^{(1)}\}}, \dots, \psi\!\restriction_{\{y^{(n)}\}}\big)\,,$$

respectively; they allow us to define the canonical embeddings of $\mathbf{R}(z)$ to $\mathcal{H}_i$, i.e.,

$$\mathbf{R}_{iL}(z) = \tau_i R(z) : L^2 \to \mathcal{H}_i\,, \quad \mathbf{R}_{Li}(z) = [\mathbf{R}_{iL}(z)]^* : \mathcal{H}_i \to L^2\,,$$

and $\mathbf{R}_{ji}(z) = \tau_j \mathbf{R}_{Li}(z) : \mathcal{H}_i \to \mathcal{H}_j$, all expressed naturally through the free Green function in their kernels, with the variable range corresponding to a given $\mathcal{H}_i$. The operator-valued matrix $\Gamma(z) = [\Gamma_{ij}(z)] : \mathcal{H}_0 \oplus \mathcal{H}_1 \to \mathcal{H}_0 \oplus \mathcal{H}_1$ is defined by

$$\begin{aligned}
\Gamma_{ij}(z)g &:= -\mathbf{R}_{ij}(z)g && \text{for} \quad i \neq j \quad \text{and} \quad g \in \mathcal{H}_j\,,\\
\Gamma_{00}(z)f &:= \left[\alpha^{-1} - \mathbf{R}_{00}(z)\right]f && \text{if} \quad f \in \mathcal{H}_0\,,\\
\Gamma_{11}(z)\varphi &:= \left[s_{\beta_l}(z)\delta_{kl} - G_z(y^{(k)},y^{(l)})(1-\delta_{kl})\right]_{k,l=1}^{n}\varphi && \text{for} \quad \varphi \in \mathcal{H}_1\,,
\end{aligned}$$

where $s_{\beta_l}(z) = \beta_l + s(z) := \beta_l + \frac{1}{2\pi}(\ln\frac{\sqrt{z}}{2i} - \psi(1))$ and $-\psi(1)$ is the Euler number.

For z from $\rho(H_{\alpha,\beta})$ the operator $\Gamma(z)$ is boundedly invertible. In particular, $\Gamma_{00}(z)$ is invertible and it makes sense to define $D(z) \equiv D_{11}(z) : \mathcal{H}_1 \to \mathcal{H}_1$ by

$$D(z) = \Gamma_{11}(z) - \Gamma_{10}(z)\Gamma_{00}(z)^{-1}\Gamma_{01}(z) \tag{3.3}$$

which we call the *reduced determinant* of Γ; it allows us to write the inverse of $\Gamma(z)$ as $[\Gamma(z)]^{-1} : \mathcal{H}_0 \oplus \mathcal{H}_1 \to \mathcal{H}_0 \oplus \mathcal{H}_1$ with the "block elements" defined by

$$\begin{aligned}
[\Gamma(z)]_{11}^{-1} &= D(z)^{-1}\,,\\
[\Gamma(z)]_{00}^{-1} &= \Gamma_{00}(z)^{-1} + \Gamma_{00}(z)^{-1}\Gamma_{01}(z)D(z)^{-1}\Gamma_{10}(z)\Gamma_{00}(z)^{-1}\,,\\
[\Gamma(z)]_{01}^{-1} &= -\Gamma_{00}(z)^{-1}\Gamma_{01}(z)D(z)^{-1}\,,\\
[\Gamma(z)]_{10}^{-1} &= -D(z)^{-1}\Gamma_{10}(z)\Gamma_{00}(z)^{-1}\,;
\end{aligned}$$

in the above formulae we use notation $\Gamma_{ij}(z)^{-1}$ for the inverse of $\Gamma_{ij}(z)$ and $[\Gamma(z)]_{ij}^{-1}$ for the matrix element of $[\Gamma(z)]^{-1}$.

Before using this to express $R_{\alpha,\beta}(z) \equiv (H_{\alpha,\beta} - z)^{-1}$ we introduce another notation which allow us to write $R_{\alpha,\beta}(z)$ through a perturbation of the "line only" Hamiltonian $\tilde{H}_\alpha$ the resolvent of which is the integral operator

$$R_\alpha(z) = R(z) + R_{L0}(z)\Gamma_{00}^{-1}R_{0L}(z)$$

for $z \in \mathbb{C} \setminus [-\frac{1}{4}\alpha^2, \infty)$. We define $\mathbf{R}_{\alpha;L1}(z) : \mathcal{H}_1 \to L^2$ and $\mathbf{R}_{\alpha;1L}(z) : L^2 \to \mathcal{H}_1$ by

$$\mathbf{R}_{\alpha;1L}(z)\psi := R_\alpha(z)\psi \!\restriction_\Pi \quad \text{for } \psi \in L^2$$

and $\mathbf{R}_{\alpha;L1}(z) := \mathbf{R}_{\alpha;1L}^*(z)$; the resolvent difference between $H_{\alpha,\beta}$ and $\tilde{H}_\alpha$ is given then by Krein's formula. Now we can state the result; for the proof and a more detailed discussion we refer to [9].

Theorem 3.1. *For any $z \in \rho(H_{\alpha,\beta})$ with $\mathrm{Im}\, z > 0$ we have*

$$R_{\alpha,\beta}(z) = R(z) + \sum_{i,j=0}^{1} \mathbf{R}_{Li}(z)[\Gamma(z)]_{ij}^{-1}\mathbf{R}_{jL}(z) = R_\alpha(z) + \mathbf{R}_{\alpha;L1}(z)D(z)^{-1}\mathbf{R}_{\alpha;1L}(z)\,.$$

These formulæ make it possible to analyze spectral properties of the operator $H_{\alpha,\beta}$, see again [9] for more details. In this paper we will be concerned with one aspect of this problem only, namely with perturbations of embedded eigenvalues.

4. Resonance poles

The decay in our model is due to the tunnelling between the points and the line. This interaction is "switched off" if the line is removed (formally speaking, put to an infinite distance). Consequently, the free Hamiltonian from the decay point of view is the point interaction only $\tilde{H}_\beta := H_{0,\beta}$. Depending on the configuration of the set Π and the coupling parameters β this operator has m eigenvalues, $1 \le m \le n$. We *will always assume* in the following that they satisfy the condition

$$-\frac{1}{4}\alpha^2 < \epsilon_1 < \cdots < \epsilon_m < 0 \quad \text{and} \quad m > 1\,, \tag{4.1}$$

i.e., the discrete spectrum of $\tilde{H}_\beta$ is simple, contained in (the negative part of) $\sigma(\tilde{H}_\alpha) = \sigma_{ac}(H_{\alpha,\beta}) = (-\alpha^2/4, \infty)$, and consists of more than a single point. Let us specify the interactions sites by their Cartesian coordinates, $y^{(i)} = (c_i, a_i)$. We also introduce the notations $a = (a_1, \ldots, a_n)$ and $d_{ij} = |y^{(i)} - y^{(j)}|$ for the distances between point interactions.

To find resonances in our model we will rely on a Birman-Schwinger type argument[1]. More specifically, our aim is to find poles of the resolvent through zeros of the operator-valued function (3.3). First we have to find a more explicit form of $D(\cdot)$; having in mind that resonance poles have to be looked for on the second sheet we will derive the analytical continuation of $D(\cdot)$ to a subset Ω_- of the lower half-plane across the segment $(-\alpha^2/4, 0)$ of the real axis; for the sake of definiteness we employ the notation $D(\cdot)^{(l)}$ where $l = -1, 0, 1$ refers to the

---

[1]We will follow here the idea which was precisely discussed in [9].

argument z from Ω_-, the segment $(-\alpha^2/4, 0)$, and the upper half-plane, $\operatorname{Im} z > 0$, respectively. Using the resolvent formula of the previous section we see that the first component of the operator $\Gamma_{11}(\cdot)^{(l)}$ is the $n \times n$ matrix with the elements

$$\Gamma_{11;jk}(\cdot)^{(l)} = -\frac{1}{2\pi} K_0(d_{jk}\sqrt{-\cdot}) \quad \text{for } j \neq k$$

and

$$\Gamma_{11;jj}(\cdot)^{(l)} = \beta_j + 1/2\pi(\ln\sqrt{(-\cdot)} - \psi(1))$$

for every l. To find an explicit form of the second component let us introduce

$$\mu_{ij}(z,t) := \frac{i\alpha}{2^5\pi} \frac{\left(\alpha - 2i(z-t)^{1/2}\right) e^{i(z-t)^{1/2}(|a_i|+|a_j|)}}{t^{1/2}(z-t)^{1/2}} e^{it^{1/2}(c_i-c_j)}$$

and $\mu_{ij}^0(\lambda, t) := \lim_{\eta\to 0+} \mu_{ij}(\lambda + i\eta, t)$ cf. [9]. Using this notation we can rewrite the matrix elements of $(\Gamma_{10}\Gamma_{00}^{-1}\Gamma_{01})^{(\cdot)}(\cdot)$ in the following form,

$$\theta_{ij}^{(0)}(\lambda) \;=\; \mathcal{P}\int_0^\infty \frac{\mu_{ij}^0(\lambda,t)}{t - \lambda - \alpha^2/4}\, dt + g_{\alpha,ij}(\lambda)\,, \qquad \lambda \in \left(-\frac{\alpha^2}{4}, 0\right)$$

$$\theta_{ij}^{(l)}(z) \;=\; l\int_0^\infty \frac{\mu_{ij}(z,t)}{t - z - \alpha^2/4}\, dt + (l-1)g_{\alpha,ij}(z) \quad \text{for } l = 1, -1$$

where $\mathcal{P}$ means the principal value and

$$g_{\alpha,ij}(z) := \frac{i\alpha}{(z+\alpha^2/4)^{1/2}}\, e^{-\alpha(|a_i|+|a_j|)/2}\, e^{i(z+\alpha^2/4)^{1/2}(c_i-c_j)}\,.$$

Proceeding in analogy with [9] we evaluate the determinant of $D(\cdot)^{(\cdot)}$ as

$$d(z)^{(l)} \equiv d(a,z)^{(l)} = \sum_{\pi\in\mathcal{P}_n} \operatorname{sgn}\pi \left(\sum_{j=1}^n (-1)^j (S_{p_1,\ldots,p_n}^j)^{(l)} + \Gamma_{11;1p_1}\ldots\Gamma_{11;np_n}\right)(z)\,,$$

where $\mathcal{P}_n$ denotes the permutation group of n elements, $\pi = (p_1,\ldots,p_n)$, and

$$(S_{p_1,\ldots,p_n}^j)^{(l)} = \theta_{jp_1}^{(l)} A_{p_2,\ldots,p_n}^j$$

with

$$A_{i_2,\ldots,i_n}^j := \begin{cases} \Gamma_{11;1i_2}\ldots\Gamma_{11;j-1,i_j}\Gamma_{11;j+1,i_{j+1}}\ldots\Gamma_{11;ki_k} & \text{if} \quad j > 1 \\ \Gamma_{11;2i_2}\ldots\Gamma_{11;ki_k} & \text{if} \quad j = 1\,. \end{cases}$$

After this preliminary we want to find roots of the equation $d(a,z)^{(l)}(z) = 0$. On a heuristic level the resonances are due to tunnelling between the line and the points, thus it is convenient to introduce the following reparametrization,

$$\tilde{b}(a) \equiv (b_1(a),\ldots,b_n(a)) \quad b_i(a) = e^{-|a_i|\sqrt{-\epsilon_i}}$$

and to put $\eta(\tilde{b}, z) = d^{(-1)}(a, z)$. As we have said the absence of the straight-line interaction can be regarded in a sense as putting the line to an infinite distance from the points, thus corresponding to $\tilde{b} = 0$. In this case we have

$$\eta(0, z) = \sum_{\pi\in\mathcal{P}_n} \operatorname{sgn}\pi\left(\Gamma_{11;1p_1}\ldots\Gamma_{11;np_n}\right)(z) = \det\Gamma_{11}(z)\,,$$

so the roots of the equation $\eta(0, z) = 0$ are nothing else than the eigenvalues of the point-interaction Hamiltonian $\tilde{H}_\beta$; with the condition (4.1) in mind we have

$$\eta(0, \epsilon_i) = 0\,, \quad i = 1, \ldots, m\,.$$

Now one proceeds as in [9] checking that the hypotheses of the implicit-function theorem are satisfied; then the equation $\eta(\tilde{b}, z) = 0$ has for all the b_i small enough just m zeros which admit the following weak-coupling asymptotic expansion,

$$z_i(b) = \epsilon_i + \mathcal{O}(b) + i\mathcal{O}(b) \quad \text{where} \quad b := \max_{1 \le i \le m} b_i\,. \tag{4.2}$$

Remark 4.1. If $n \ge 2$ there can be eigenvalues of $\tilde{H}_\beta$ which remain embedded under the line perturbation due to a symmetry; the simplest example is a pair of point interactions with the same coupling and mirror symmetry with respect to Σ. From the viewpoint of decay which is important in this paper they represent a trivial case which we exclude in the following. Neither shall we consider resonances which result from a slight violation of such a symmetry – cf. a discussion in [9].

5. Decay of the dot states

As usual the resonance poles discussed above can be manifested in two ways, either in scattering properties, here of a particle moving along the "wire" Σ, or through the time evolution of states associated with the "dots" Π. By assumption (4.1) there is a nontrivial discrete spectrum of $\tilde{H}_\beta$ embedded in $(-\frac{1}{4}\alpha^2, 0)$. Let us denote the corresponding normalized eigenfunctions ψ_j, $j = 1, \ldots, m$, given by

$$\psi_j(x) = \sum_{i=1}^{m} d_i^{(j)} \phi_i^{(j)}(x)\,, \quad \phi_i^{(j)}(x) := \sqrt{-\frac{\epsilon_j}{\pi}}\, K_0(\sqrt{-\epsilon_j}|x - y^{(i)}|) \tag{5.1}$$

in accordance with [1, Sec. II.3], where the vectors $d^{(j)} \in \mathbb{C}^m$ satisfy the equation

$$\Gamma_{11}(\epsilon_j) d^{(j)} = 0 \tag{5.2}$$

and a normalization condition which in view of $\|\phi_i^{(j)}\| = 1$ reads

$$|d^{(j)}|^2 + 2\mathrm{Re} \sum_{i=2}^{m} \sum_{k=1}^{i-1} \overline{d_i^{(j)}} d_k^{(j)} (\phi_i^{(j)}, \phi_k^{(j)}) = 1\,. \tag{5.3}$$

In particular, if the distances between the points of Π are large (the natural length scale is given by $(-\epsilon_j)^{-1/2}$), the cross terms are small and $|d^{(j)}|$ is close to one.

Let us now *specify the unstable system* of our model by identifying its state Hilbert space $P\mathcal{H}$ with the span of the vectors $\psi_1, \ldots, \psi_m$. Suppose that it is prepared at the initial instant $t = 0$ at a state $\psi \in P\mathcal{H}$, then the decay law describing the probability of finding the system undecayed at a subsequent measurement performed at t, without disturbing it in between [6], is

$$P_\psi(t) = \|Pe^{-iH_{\alpha,\beta}t}\psi\|^2\,. \tag{5.4}$$

We are particularly interested in the *weak-coupling situation* where the distance between Σ and Π is a large at the scale given by $(-\epsilon_m)^{-1/2}$. Since our model bears resemblance with the (multidimensional) Friedrichs model one can conjecture in analogy with [5] that the leading term in $P_\psi(t)$ will come from the appropriate semigroup evolution on $P\mathcal{H}$, in particular, for the basis states ψ_j we will have a dominantly exponential decay, $P_{\psi_j}(t) \approx \mathrm{e}^{-\Gamma_j t}$ with $\Gamma_j = 2\,\mathrm{Im}\,z_j(b)$. A precise discussion of this question is postponed to appendix – see Section 7 below.

Remark 5.1. The quantities Γ_j^{-1} provide thus a natural time scale for the decay and we will use $\max_j \Gamma_j^{-1}$ as a measure of the system lifetime. A *caveat* is needed, however, with respect to the notion of lifetime [6] which is conventionally defined as $T_\psi = \int_0^\infty P_\psi(t)\,\mathrm{d}t$. It has been shown in [9] that $P\mathcal{H}$ is not contained is the absolutely continuous subspace of $H_{\alpha,\beta}$ if $n = 1$, and the argument easily extends to any $n \in \mathbb{N}_0$. This means that a part of the original state survives as $t \to \infty$, even if it is a small one in the weak-coupling case. It is a long-time effect, of course, which has no relevance for the problem considered here.

6. Stable and Zeno dynamics in the model

Suppose now finally that we perform the Zeno time at our decaying system characterized by the operator $H_{\alpha,\beta}$ and the projection P. The latter has by assumption the dimension $1 < m < \infty$ and it is straightforward to check that $P\mathcal{H} \subset \mathcal{Q}(H_{\alpha,\beta})$. Moreover the form associated with generator H_P has in the quantum-dot state basis the following matrix representation

$$(\psi_j, H_P \psi_k) = \delta_{jk}\epsilon_j - \alpha \int_\Sigma \bar{\psi}_j(x_1,0)\psi_k(x_1,0)\,\mathrm{d}x_1\,, \tag{6.1}$$

where the first term corresponds, of course, to the "dots-only" operator $\tilde{H}_\beta$.

Theorem 6.1. *The two dynamics do not differ significantly for times satisfying*

$$t \ll C\,\mathrm{e}^{2\sqrt{-\epsilon}|\tilde{a}|}\,, \tag{6.2}$$

where C is a positive constant and $|\tilde{a}| = \min_i |a_i|$, $\epsilon = \max_i \epsilon_i$.

Proof. The difference is characterized by the operator $\mathcal{U}_t := (\mathrm{e}^{-i\tilde{H}_\beta t} - \mathrm{e}^{-iH_P t})P$. Taking into account the unitarity of its parts together with a functional calculus estimate based on $|\mathrm{e}^{iz} - 1| \leq |z|$ we find that the norm of $\mathcal{U}_t$ remains small as long as $t\|(\tilde{H}_\beta - H_P)P\| \ll 1$. Thus to check (6.2) we have to estimate norm of the operator $(\tilde{H}_\beta - H_P)P$ acting in $P\mathcal{H}$; in the basis of the vectors $\{\psi_j\}_{j=1}^m$ it is represented by $m \times m$ matrix with the elements

$$s_{ij} = \alpha(\psi_i, \psi_j)_\Sigma\,,$$

where $(\psi_i, \psi_j)_\Sigma := \int_\Sigma \bar{\psi}_i(x_1,0)\psi_j(x_1,0)\,\mathrm{d}x_1$. Using the representation (5.1) we obtain

$$s_{ij} = \alpha \sum_{(l,k)\in M\times M} \bar{d}_l^{(i)} d_k^{(j)} (\phi_l^{(i)}, \phi_k^{(j)})_\Sigma$$

where M is a shorthand for $(1, \ldots, m)$. To proceed further we use Schur-Holmgren bound by which the norm of $(\tilde{H}_\beta - H_P)P$ does not exceed mS, where $S := \max_{(i,j)\in M\times M} |s_{ij}|$, and the last named quantity can be estimated by

$$S \leq \alpha m^2 \max_{(i,j,k,l)\in M^4} |\bar{d}_l^{(i)} d_j^{(k)} (\phi_l^{(i)}, \phi_k^{(j)})_\Sigma| \, .$$

The final step is to estimate the expressions $(\phi_l^{(i)}, \phi_k^{(j)})_\Sigma$. Using the momentum representation of Macdonald function we obtain

$$(\phi_l^{(i)}, \phi_k^{(j)})_\Sigma = \frac{\sqrt{\epsilon_i \epsilon_j}}{2} \int_\mathbb{R} \frac{e^{-((p_1^2-\epsilon_i)^{1/2}|a_l| - (p_1^2-\epsilon_j)^{1/2}|a_k|)}}{(p_1^2 - \epsilon_i)^{1/2}(p_1^2 - \epsilon_j)^{1/2}} \, e^{ip_1(c_k - c_l)} \, \mathrm{d}p_1 \, ,$$

where $y^{(i)} = (c_i, a_i)$ as before. A simple estimate of the above integral yields

$$(\phi_l^{(i)}, \phi_k^{(j)})_\Sigma \leq \frac{\pi}{2} \frac{\epsilon_{\min}}{\sqrt{-\epsilon}} e^{-2\sqrt{-\epsilon}|a|}$$

where $\epsilon_{\min} = \min_i \epsilon_i$, $|\tilde{a}| = \min_i |a_i|$, and $\epsilon = \max_i \epsilon_i$. In conclusion, we get the bound

$$\|(\tilde{H}_\beta - H_P)P\| \leq Ce^{-2\sqrt{-\epsilon}|a|} \, ,$$

where $C := \frac{1}{2}\pi m^3 \alpha \, \epsilon_{\min}(-\epsilon)^{-1/2} \max_{(i,j,k,l)\in M^4} |\bar{d}_l^{(i)} d_j^{(k)}|$. $\square$

7. Appendix: pole approximation for the decaying states

Let us now return to the claim that the decay is approximately exponential when the distances of the points from the line are large. Let ψ_j be the jth eigenfunction of the point-interaction Hamiltonian $\tilde{H}_\beta$ with the eigenvalue ϵ_j; the related one-dimensional projection will be denoted P_j. Then we make the following claim.

Theorem 7.1. *Suppose that $H_{\alpha,\beta}$ has no embedded eigenvalues. Then in the limit $b \to 0$ where b is defined in (4.2) we have, pointwise in $t \in (0, \infty)$,*

$$\|P_j e^{-iH_{\alpha,\beta}t}\psi_j - e^{-iz_j t}\psi_j\| \to 0 \, .$$

To prove the theorem we need some preliminaries. For simplicity, we denote $U_t(\epsilon) := e^{-i\epsilon t}$ for a fixed $t > 0$. It was shown in [9] that the operator $H_{\alpha,\beta}$ has at least one and at most n isolated eigenvalues. We denote them by $\epsilon_{\alpha\beta,k}$, $k = 1, \ldots, l$ with $l \leq n$, and use $\psi_{\alpha\beta,k}$ as symbols for the corresponding (normalized) eigenfunctions. Then the spectral theorem gives

$$P_j \, e^{-iH_{\alpha,\beta}t}\psi_j = \sum_{k=1}^{m} U_t(\epsilon_{\alpha\beta,k})|(\psi_j, \psi_{\alpha\beta,k})|^2\psi_j + P_j \int_{-\alpha^2/4}^{\infty} U_t(\lambda)\mathrm{d}E(\lambda)\psi_j \, , \quad (7.1)$$

where $E(\cdot) \equiv E_{\alpha,\beta}(\cdot)$ is the spectral measure of $H_{\alpha,\beta}$. By assumption there are no embedded eigenvalues (cf. Remark 4.1) and by [9] also the singularly continuous component is void, hence the second term is associated solely with $\sigma_{\mathrm{ac}}(H_{\alpha,\beta})$. Let us first look at this contribution to the reduced evolution. The key observation is that one has a spectral concentration in the set $\triangle_\varepsilon \equiv \triangle_\varepsilon(b) := (\epsilon_j - \varepsilon(b), \epsilon_j + \varepsilon(b))$ with a properly chosen $\varepsilon(b)$; we denote its complement as $\bar{\triangle}_\varepsilon := \sigma_{\mathrm{ac}}(H_{\alpha,\beta}) \setminus \triangle_\varepsilon$.

Lemma 7.2. *Suppose that $\varepsilon(b) \to 0$ and $\varepsilon(b)^{-1}b \to 0$ holds as $b \to 0$, then we have*

$$\left\| P_j \int_{\bar{\triangle}_\varepsilon} U_t(\lambda) \mathrm{d}E(\lambda)\psi_j \right\| \to 0 \,.$$

Proof. Given an arbitrary Borel set $\triangle \subset \sigma_{\mathrm{ac}}(H_{\alpha,\beta})$ and a projection P we have the following simple inequality,

$$\left\| P \int_{\triangle} U_t(\lambda) \mathrm{d}E(\lambda)f \right\| \leq \|E(\triangle)f\| \,, \tag{7.2}$$

and another straightforward application of the spectral theorem gives

$$\|(H_{\alpha,\beta} - \epsilon_j)f\|^2 \geq \int_{\bar{\triangle}_\varepsilon} |\lambda - \epsilon_j|^2 (\mathrm{d}E(\lambda)f, f) \geq \varepsilon(b)^2 \|E(\bar{\triangle}_\varepsilon)f\|^2 \tag{7.3}$$

for any $f \in D(H_{\alpha,\beta})$. To make use of the last inequality we need a suitable function from the domain of $H_{\alpha,\beta}$. It is clear that one cannot use ψ_j directly because it does not satisfy the appropriate boundary conditions at the line Σ, thus we take instead its modification $f_b = \psi_j + \phi_b$, where $\phi_b \in L^2(\mathbb{R}^2)$ vanishes on $\Pi \cup \Sigma$ and satisfies the following assumptions:

(a1) $\Xi_\Sigma(\phi_b) = -\alpha\Omega_\Sigma(\psi_j)$
(a2) $\|\phi_b\| = \mathcal{O}(b)$ and $\|\Delta\phi_b\| = \mathcal{O}(b)$.

In view of (3.2) the first condition guarantees that $f_b \in D(H_{\alpha,\beta})$, while the second one expresses "smallness" of the modification. It is not difficult to construct such a family. For instance, one can take for ϕ_b a family of C^2 functions with supports in a strip neighborhood of Σ of width d_Σ assuming that ϕ_b behaves in the vicinity of Σ as $\frac{1}{2}\alpha\Omega_\Sigma(\psi_j)(x_1)|x_2|$. Since $|\Omega_\Sigma(\psi_j)| \leq Cb$, where C is positive constant we can choose $d_\Sigma = \mathcal{O}(b)$. Using (a1) and $(\tilde{H}_\beta - \epsilon_j)\psi_j = 0$ we get

$$(H_{\alpha,\beta} - \epsilon_j)f_b = -\Delta\phi_b - \epsilon_j\phi_b \,,$$

so the condition (a2) gives

$$\|(H_{\alpha,\beta} - \epsilon_j)f_b\| = \mathcal{O}(b) \,.$$

This relation together with (7.3) yields $\|E(\bar{\triangle}_\varepsilon)f_b\| = \mathcal{O}(b)\varepsilon(b)^{-1}$. Combining it further with (7.2) and using the inequality

$$\|E(\bar{\triangle}_\varepsilon)\psi_j\| \leq \|\phi_b\| + \|E(\bar{\triangle}_\varepsilon)f_b\|$$

and the condition (a2) we get the sought result. $\qquad\square$

The next step is to show that the main contribution to the reduced evolution of the unstable state comes from the interval $\triangle_\varepsilon$.

Lemma 7.3. *Under the assumptions of Lemma 7.2 we have*

$$\left\| P_j \int_{\triangle_\varepsilon} U_t(\lambda) \mathrm{d}E(\lambda)\psi_j - U_t(z_j)\psi_j \right\| \to 0$$

for any fixed $t > 0$ in the limit $b \to 0$.

Proof. Let $R^{\mathrm{II}}_{\alpha,\beta}$ stand for the second-sheet continuation of the resolvent of $H_{\alpha,\beta}$. Using the results of Section 4 we can write it for a fixed j as

$$R^{\mathrm{II}}_{\alpha,\beta}(z) = \sum_{k=1}^{m} \frac{B_b^{(k)}}{z - z_k} + A_b(z), \qquad (7.4)$$

where $B_b^{(k)}$ is a one-parameter family of rank-one operators and $A_b(\cdot)$ is a family of analytic operator-valued functions to be specified later. Mimicking now the argument of [6, Sec. 3.1] which relies on Stone's formula and Radon-Nikodým theorem we find that the spectral-measure derivative acts at the vector ψ_j as

$$\frac{\mathrm{d}E(\lambda)}{\mathrm{d}\lambda}\,\psi_j = \left[\frac{1}{2\pi i}\sum_{k=1}^{m}\left(\frac{(B_b^{(k)})^*}{\lambda - \bar{z}_k} - \frac{B_b^{(k)}}{\lambda - z_k}\right) + \frac{1}{\pi}\,\mathrm{Im}\,A_b(\lambda)\right]\psi_j. \qquad (7.5)$$

This makes it possible to estimate $P_j \int_{\triangle_\varepsilon} U_t(\lambda)\mathrm{d}E(\lambda)\psi_j$. Using the explicit form of $R^{\mathrm{II}}_{\alpha,\beta}$ derived in Section 4 one can check that $A_b(\cdot)$ can be bounded on a compact interval uniformly for b small enough, which means that the contribution to the integral from the last term in (7.5) tends to zero as $\varepsilon(b) \to 0$. The rest is dealt with by means of the residue theorem in the usual way: we can extend the integration to the whole real line and perform it by means of the integral over a closed contour consisting of a real axis segment and a semicircle in the lower half-plane, using the fact that the contribution from the latter vanishes when the semicircle radius tends to infinity. It is clear that only the m poles in (7.5) contained in the lower half-plane contribute, the kth one giving $U_t(z_k)P_j B_b^{(k)}\psi_j$; an argument similar to Lemma 7.2 shows that the integral over $\mathbb{R} \setminus \triangle_\varepsilon$ vanishes as $b \to 0$, and likewise, the integral over semicircle vanishes in the limit of infinite radius.

Furthermore, since P_j is one-dimensional we have $P_j B_b^{(k)}\psi_j = c_b^{(k)}\psi_j$ where $b \mapsto c_b^{(k)}$ are continuous complex functions, well defined for b small enough. Hence the above discussion allows us to conclude that

$$\left\|P_j \mathrm{e}^{-iH_{\alpha,\beta}t}\psi_j - \sum_{k=1}^{m} c_b^{(k)}\mathrm{e}^{-iz_k t}\psi_j\right\| \to 0 \quad \text{as} \quad b \to 0. \qquad (7.6)$$

Our next task is show that for $k \neq j$ we have $c_b^{(k)} \to 0$ as $b \to 0$ and $c_b^{(j)} \to 1$ at the same time. To this aim it suffices to check that $B_b^{(k)}$ converges to P_k for $b \to 0$. First we observe that the terms involved in the resolvent $R_{\alpha,\beta}$ derived in Theorem 3.1 satisfy the following relations

$$D(z) \to \Gamma_{11}(z), \quad \mathbf{R}_{\alpha;1L}(z) \to \mathbf{R}_{1L}(z) \quad \text{as} \quad b \to 0$$

in the operator-norm sense; the limits are uniform on any compact subset of the upper half-plane as well as for the analytical continuation of $R_{\alpha,\beta}$. Consequently, the second component of the resolvent tends $\mathbf{R}_{L1}(z)[\Gamma_{11}(z)]^{-1}\mathbf{R}_{1L}(z)$ which obviously has a singular part equal to $\sum_{k=1}^{m}(z - \epsilon_k)^{-1}P_k$; this proves the claim. $\qquad\square$

Proof of Theorem 7.1. In view of (7.1) together with Lemmata 7.2, 7.3 it remains to demonstrate that the contribution from the discrete spectrum to (7.1) vanishes as $b \to 0$, i.e., that

$$\left| \sum_{k=1}^{m} U_t(\epsilon_{\alpha\beta,k}) |(\psi_j, \psi_{\alpha\beta,k})|^2 \right| \to 0. \tag{7.7}$$

This is a direct consequence of the following relation,

$$0 = (H_{\alpha,\beta}\psi_{\alpha\beta,k}, f_b) - (\psi_{\alpha\beta,k}, H_{\alpha,\beta}f_b) = (\epsilon_{\alpha\beta,k} - \epsilon_j)(\psi_{\alpha\beta,k}, f_b) + \mathcal{O}(b),$$

where $k = 1, \ldots, l$, and f_b is the function introduced in the proof of Lemma 7.2. In combination with (4.1) we get $|(\psi_j, \psi_{\alpha\beta,k})| = \mathcal{O}(b)$ which in turn implies (7.7). $\square$

Acknowledgment

The research was supported in part by the Czech Academy of Sciences and Ministry of Education, Youth and Sports within the projects A100480501 and LC06002 and by the Polish Ministry of Scientific Research and Information Technology under the (solicited) grant no PBZ-Min-008/PO3/2003. Two of the authors (P.E. and S.K.) are grateful to the organizing committee of OTAMP2004 for supporting their participation in the conference, as well as for the hospitality at the University of Kanazawa, where a substantial part of this work was done.

References

[1] S. Albeverio, F. Gesztesy, R. Høegh-Krohn, H. Holden, *Solvable Models in Quantum Mechanics.* 2nd edition with an appendix by P. Exner, AMS Chelsea Publ., Providence, R.I., 2005.

[2] J. Beskow and J. Nilsson, *The concept of wave function and the irreducible representations of the Poincaré group, II. Unstable systems and the exponential decay law.* Arkiv Fys. **34**, 561–569 (1967).

[3] P.R. Chernoff, *Note on product formulas for operator semigroups.* J. Funct. Anal. **2** (1968), 238–242.

[4] P.R. Chernoff, *Product Formulas, Nonlinear Semigroups, and Addition of Unbounded Operators.*, Mem. Amer. Math. Soc. **140**, Providence, R.I., 1974.

[5] M. Demuth, *Pole approximation and spectral concentration.* Math. Nachr. **73** (1976), 65–72.

[6] P. Exner, *Open Quantum Systems and Feynman Integrals.* D. Reidel, Dordrecht 1985.

[7] P. Exner, T. Ichinose, *A product formula related to quantum Zeno dynamics.* Ann. H. Poincaré **6** (2005), 195–215.

[8] P. Exner, T. Ichinose, H. Neidhardt, V.A. Zagrebnov, *New product formulæ and quantum Zeno dynamics with generalized observables*, Integral Equations and Operator Theory (2007), to appear.

[9] P. Exner, S. Kondej, *Schrödinger operators with singular interactions: a model of tunneling resonances.* J. Phys. A: Math. Gen. **37** (2004), 8255–8277.

[10] P. Facchi, G. Marmo, S. Pascazio, A. Scardicchio, E.C.G. Sudarshan, *Zeno dynamics and constraints.* J. Opt. B: Quant. Semiclass. **6** (2004), S492–S501.

[11] C. Friedman, *Semigroup product formulas, compressions, and continual observations in quantum mechanics.* Indiana Math. J. **21** (1971/72), 1001–1011.

[12] G.M. Graf, A. Guekos, *private communication.*

[13] M. Matolcsi and R. Shvidkoy, *Trotter's product formula for projections.* Arch. der Math. **81** (2003), 309–317.

[14] B. Misra, E.C.G. Sudarshan, *The Zeno's paradox in quantum theory.* J. Math. Phys. **18** (1977), 756–763.

[15] A.U. Schmidt, *Mathematics of the quantum Zeno effect.* In "Mathematical Physics Research on Leading Edge" (Ch. Benton, ed.), Nova Sci, Hauppauge, N.Y., 2004; pp. 113-143.

Pavel Exner
Department of Theoretical Physics
Nuclear Physics Institute
Academy of Sciences
25068 Řež near Prague, Czechia

   and

Doppler Institute
Czech Technical University
Břehová 7
11519 Prague, Czechia
e-mail: `exner@ujf.cas.cz`

Takashi Ichinose
Department of Mathematics
Faculty of Science
Kanazawa University
Kanazawa 920-1192, Japan
e-mail: `ichinose@kappa.s.kanazawa-u.ac.jp`

Sylwia Kondej
Institute of Physics
University of Zielona Góra
ul. Szafrana 4a
65246 Zielona Góra, Poland
e-mail: `skondej@proton.if.uz.zgora.pl`

Operator Theory:
Advances and Applications, Vol. 174, 35–50
© 2007 Birkhäuser Verlag Basel/Switzerland

On the Spectrum of Partially Periodic Operators

Rupert L. Frank and Roman G. Shterenberg

Abstract. We consider Schrödinger operators $H = -\Delta + V$ in $L_2(\Omega)$ where the domain $\Omega \subset \mathbb{R}_+^{d+1}$ and the potential $V = V(x, y)$ are periodic with respect to the variable $x \in \mathbb{R}^d$. We assume that Ω is unbounded with respect to the variable $y \in \mathbb{R}$ and that V decays with respect to this variable. V may contain a singular term supported on the boundary.

We develop a scattering theory for H and present an approach to prove absence of singular continuous spectrum. Moreover, we show that certain repulsivity conditions on the potential and the boundary of Ω exclude the existence of surface states. In this case, the spectrum of H is purely absolutely continuous and the scattering is complete.

Mathematics Subject Classification (2000). Primary 35J10; Secondary 35J25, 35P05, 35P25.

Keywords. Scattering theory, periodic operator, Schrödinger operator.

Introduction

In the last decade the interest in periodic operators of mathematical physics has grown and led to a number of new results. We refer to [13] for a recent survey. The main method to investigate periodic operators is the decomposition into a direct integral. Namely, let $M = M(x, D)$, $x \in \mathbb{R}^d$, $D = -i\nabla$, be a pseudo-differential operator and assume that M is periodic with respect to a (d-dimensional) lattice Γ. Then M is unitarily equivalent to the direct integral $\int_{\Xi} \oplus M(k)\, dk$, where Ξ is an elementary cell of Γ and the operators $M(k) = M(x, D + k)$ act on the torus $\mathbb{R}^d/\Gamma$. For most of the operators of mathematical physics it turns out that, since Ξ is bounded, the spectrum of the corresponding operators $M(k)$ is discrete and depends analytically on the parameter k. This allows to use effectively the direct integral decomposition to investigate the spectrum of the operator M.

Let $\{E_j(k)\}$ be the eigenvalues of the operators $M(k)$ arranged in non-decreasing order. Then the spectrum of the operator M has band structure and consists of the union of the ranges of the band functions $E_j(\cdot)$. Sufficiently general considerations show that the singular continuous spectrum of M is empty (for an elementary proof of this fact see [7]). Herewith, if a band function E_j is non-degenerate then it contributes to the absolutely continuous spectrum of M. If $E_j(k) = \lambda_j \equiv const$ then λ_j is an eigenvalue of M of infinite multiplicity. Often one can prove absence of such degenerate bands (and thus absolute continuity of the spectrum of M) by complexification of the parameter k and estimates of the resolvent of $M(k)$ for large imaginary values of k. This method was suggested in [17] and is now known as the Thomas scheme.

The progress in the investigation of periodic operators led to the study of partially periodic operators, where the coefficients are periodic only in some variables and tend to constants in others. One can again decompose the operator M into a direct integral but now the cell Ξ is unbounded and hence the operators $M(k)$ have rich continuous spectrum. In this situation the problem of scattering for the operator M is also non-trivial. In general, there appear so called surface states, i.e., waves which propagate along the hyper-surface of periodicity of the operator.

Under not very restrictive conditions on the coefficients of M scattering theory allows to describe the structure of the spectrum of the operators $M(k)$. The following situation is typical: The spectrum of $M(k)$ consists of an absolutely continuous part, isolated eigenvalues of finite multiplicity and possibly embedded eigenvalues. The singular continuous spectrum is empty. The isolated eigenvalues depend analytically on k and can be taken into account similarly as for fully periodic operators. The main difficulties are caused by embedded eigenvalues of $M(k)$ since their dependence on k is very hard to control.

In the present paper we introduce some ways to investigate the structure of the spectrum of partially periodic operators. For this purpose we use the Schrödinger operator as an example. At the moment there exist only few papers concerning this problem and they are very disconnected (for references see our discussion in Subsection 1.5). Here, we present probably for the first time a sufficiently abstract method. Note that most of the elements of this method were already used earlier in different particular cases including the investigation of fully periodic operators. The combination of these elements with results from scattering theory allows us to advance in the study of partially periodic operators.

Let $\Omega \subset \mathbb{R}^{d+1}_+ := \{(x,y) \in \mathbb{R}^d \times \mathbb{R} : y > 0\}$ be a connected open set which is $(2\pi\mathbb{Z})^d$-periodic with respect to the variable x and unbounded with respect to the variable y. Assume also that $\mathbb{R}^d \times [a, \infty) \subset \Omega$ for some $a > 0$. We consider the Schrödinger-type operator

$$Hu = -\Delta u + Vu \qquad \text{in } \Omega$$

together with the boundary conditions

$$u = 0 \text{ on } \Gamma_D, \qquad \frac{\partial u}{\partial \nu} + \sigma u = 0 \text{ on } \Gamma_N, \qquad \partial\Omega = \Gamma_D \cup \Gamma_N. \qquad (0.1)$$

Here ν is the external unit normal on $\partial\Omega$. The functions V, σ as well as the parts $\Gamma_{D,N}$ of the boundary are assumed to be $(2\pi\mathbb{Z})^d$-periodic with respect to the variable x. The precise definition of the operator H is given in Subsection 1.2.

Our first result is the following. Assume V has bounded support with respect to the variable y. Then under not very restrictive conditions on σ and $\partial\Omega$ we prove absence of singular continuous spectrum of the operator H (see Theorem 2.1 below). Note that this result can be applied without significant changes to many other operators of mathematical physics with constant coefficients outside some bounded interval in y. The proof can be modified to deal also with the case where y is multi-dimensional.

Our second result (see Theorems 3.1 and 3.3 below) is more specific and provides sufficient conditions for complete scattering and, in particular, the absolute continuity of the spectrum of the operator H.

1. Initial results

This section is a survey. In Subsections 1.1–1.4 we introduce the basic definitions and state some initial results. Proofs are only indicated. In Subsection 1.5 we discuss the results obtained so far and give references for more detailed results.

1.1. Notation

For an open set $U \subset \mathbb{R}^n$ the index in the notation of the norm $\|.\|_{L_2(U)}$ is usually dropped.

We denote by $\mathcal{D}[a]$ the domain of a quadratic form a and by $\mathcal{N}(A), \mathcal{R}(A)$ the kernel and range, respectively, of a linear operator A.

Statements and formulae which contain double indices are understood as two independent assertions.

1.2. Definition of the operator H. Initial results about scattering

Let $\Omega \subset \mathbb{R}_+^{d+1} := \{(x,y) \in \mathbb{R}^d \times \mathbb{R} : y > 0\}$ be a connected open set which is periodic with respect to the variable x and unbounded with respect to the variable y, i.e.,

$$(x + 2\pi n, y) \in \Omega \qquad \text{whenever } (x,y) \in \Omega, n \in \mathbb{Z}^d,$$

$$\mathbb{R}^d \times [a, \infty) \subset \Omega \qquad \text{for some } a > 0.$$

We assume that $\partial\Omega$ is Lipschitz and that there is a decomposition $\partial\Omega = \Gamma_D \cup \Gamma_N$ where Γ_N is an open subset of $\partial\Omega$ (possibly empty) and $\Gamma_D = \partial\Omega \setminus \Gamma_N$. Both parts of the boundary are assumed to be periodic,

$$(x + 2\pi n, y) \in \Gamma_{N,D} \qquad \text{whenever } (x,y) \in \Gamma_{N,D}, n \in \mathbb{Z}^d.$$

Moreover, let $V : \Omega \to \mathbb{R}$ be a measurable function satisfying

$$V(x + 2\pi n, y) = V(x,y), \qquad (x,y) \in \Omega, n \in \mathbb{Z}^d,$$

$$|V(x,y)| \le C(1 + |y|)^{-\rho}, \qquad (x,y) \in \Omega, \qquad \text{for some } \rho > 1,$$

and let $\sigma : \Gamma_N \to \mathbb{R}$ be a measurable function satisfying

$$\sigma(x + 2\pi n, y) = \sigma(x, y), \qquad (x, y) \in \Gamma_N, n \in \mathbb{Z}^d,$$
$$\sigma \in L_{q,loc}(\Gamma_N) \qquad q > 1 \text{ if } d = 1, q = d \text{ if } d \geq 2.$$

(The space $L_{q,loc}(\Gamma_N)$ is of course defined with respect to the surface measure on Γ_N, which we denote by ds.) These assumptions are assumed to hold throughout.

It follows from the embedding theorems that the quadratic form

$$\mathcal{D}[h] := \{u \in H^1(\Omega) : u|_{\Gamma_D} = 0\},$$
$$h[u] := \int_\Omega (|Du|^2 + V|u|^2)\, dxdy + \int_{\Gamma_N} \sigma |u|^2\, ds \tag{1.1}$$

is lower semibounded and closed in the Hilbert space $L_2(\Omega)$. We denote the corresponding self-adjoint operator by H. Functions u in its domain satisfy the boundary conditions (0.1) in a generalized sense.

In the case $\Omega = \mathbb{R}_+^{d+1}$, $\Gamma_N = \emptyset$, $V = 0$, $\sigma = 0$ we denote the operator by H_0. This is the *Dirichlet Laplacian* in the half-space. Its spectrum coincides with $[0, +\infty)$ and is purely absolutely continuous of infinite multiplicity.

Our first goal is to compare the operator H with the "unperturbed" operator H_0 by means of scattering theory. Since these operators act in different spaces we use as identification operator $J : L_2(\mathbb{R}_+^{d+1}) \to L_2(\Omega)$ the restriction

$$Ju := u|_\Omega, \qquad u \in L_2(\mathbb{R}_+^{d+1}).$$

Recall the definition (in case of existence) of the *wave operators* (see, e.g., [18])

$$W_\pm := W_\pm(H, H_0, J) = s - \lim_{t \to \pm\infty} \exp(itH)J\exp(-itH_0). \tag{1.2}$$

The basis of all our further considerations is the following

Theorem 1.1. *The wave operators $W_\pm$ exist, are isometric and satisfy $\mathcal{R}(W_+) = \mathcal{R}(W_-)$.*

We indicate one possible proof of this result in Subsection 1.4 below.

The question of completeness of $W_\pm$ as well as spectral consequences of Theorem 1.1 will be discussed in Subsection 1.5.

1.3. Definition of the operators $H(k)$. Direct integral decomposition

Because of periodicity the operator H can be partially diagonalized. We introduce the notations

$$\Pi := \{(x, y) \in \Omega : x \in (-\pi, \pi)^d\}, \qquad \gamma_{D,N} := \Gamma_{D,N} \cap \overline{\Pi}.$$

By $\tilde{H}^1(\Pi)$ we denote the subspace of functions $u \in H^1(\Pi)$ the periodic extension of which belongs to $H^1_{\mathrm{loc}}(\Omega)$.

Let $k \in Q := [-\frac{1}{2}, \frac{1}{2}]^d$ and V, σ as above. Again by the embedding theorems the quadratic form

$$\mathcal{D}[h(k)] := \{u \in \tilde{H}^1(\Pi) : u|_{\gamma_D} = 0\},$$

$$h(k)[u] := \int_{\Pi} \left(|(D_x + k)u|^2 + |D_y u|^2 + V|u|^2\right) dx dy + \int_{\gamma_N} \sigma |u|^2 ds \tag{1.3}$$

is lower semibounded and closed in the Hilbert space $L_2(\Pi)$. We denote the corresponding self-adjoint operator by $H(k)$. In the special case $\Pi = (-\pi, \pi)^d \times \mathbb{R}_+$, $\gamma_N = \emptyset$ (i.e., $\Omega = \mathbb{R}_+^{d+1}$, $\Gamma_N = \emptyset$), $V = 0$, $\sigma = 0$ we denote the operator by $H_0(k)$. The *Gelfand transformation* is initially defined for $u \in C_0^\infty(\Omega)$ by

$$(\mathcal{U}u)(k, x, y) := \sum_{n \in \mathbb{Z}^d} e^{-i\langle k, x + 2\pi n\rangle} u(x + 2\pi n, y), \qquad k \in Q, \; (x, y) \in \Pi,$$

and extended by continuity to a *unitary* operator $\mathcal{U} : L_2(\Omega) \rightarrow \int_Q \oplus L_2(\Pi) \, dk$. Moreover, it turns out that

$$\mathcal{U} H \mathcal{U}^* = \int_Q \oplus H(k) \, dk. \tag{1.4}$$

For a careful presentation of these facts in domains with Lipschitz boundaries we refer to [2]. (The proofs there extend without problems to the case of an unbounded period cell.)

The relation (1.4) allows us to investigate the operator H by studying the fibers $H(k)$.

1.4. Results about scattering for the fiber operators

Information about the operators $H(k)$ can be obtained by developing a scattering theory for the pair $(H(k), H_0(k))$. Note that the operator $H_0(k)$ can be diagonalized explicitly. Its spectrum is purely absolutely continuous and coincides with $[|k|^2, +\infty)$. The spectral multiplicity is finite and changes at the points $|n + k|^2$, $n \in \mathbb{Z}^d$.

We use the same notation J for the restriction operator $J : L_2((-\pi, \pi)^d \times \mathbb{R}_+) \rightarrow L_2(\Pi)$,

$$Ju := u|_\Pi, \qquad u \in L_2((-\pi, \pi)^d \times \mathbb{R}_+).$$

The results about the continuous spectrum of $H(k)$ and the wave operators

$$W_\pm(k) := W_\pm(H(k), H_0(k), J) = s - \lim_{t \to \pm\infty} \exp(itH(k)) J \exp(-itH_0(k))$$

are summarized in

Proposition 1.2. *Let $k \in Q$. Then the wave operators $W_\pm(k)$ exist, are isometric and complete. In particular, $\sigma_{ac}(H(k)) = [|k|^2, +\infty)$. Moreover, $\sigma_{sc}(H(k)) = \emptyset$.*

We will only make some remarks about the proof of this proposition. It consists in a straightforward modification of the methods developed for bounded obstacle scattering. One may apply either time-dependent techniques (see, e.g., Examples

5.2 and 6.2 in [16]) or stationary techniques (see [14] where ideas of smooth scattering theory are combined with ideas of extension theory from [1]).

Finally, Theorem 1.1 can be deduced from Proposition 1.2 in the same way as in [8]. We only note that

$$W = \mathcal{U}^* \left(\int_Q \oplus W(k)\, dk \right) \mathcal{U}.$$

1.5. Discussion

Let us pause for a moment and summarize the information about the spectrum of H that we have obtained so far. We have the decomposition

$$L_2(\Omega) = \mathcal{R}(W_\pm) \oplus \mathfrak{C} \oplus \mathfrak{T},$$

where $\mathfrak{C} := \mathcal{R}(W_\pm)^\perp \cap \mathcal{R}(P_{ac})$, $\mathfrak{T} := \mathcal{R}(P_{sing})$ and P_{ac}, P_{sing} are the projections onto the absolutely continuous and the singular subspaces of H. This decomposition reduces H. Its part on $\mathcal{R}(W_\pm)$ is unitarily equivalent to H_0 and has purely absolutely continuous spectrum $[0, +\infty)$. From the results of Subsections 1.3, 1.4 it is easy to derive the time-dependent characterization

$$\mathfrak{C} \oplus \mathfrak{T} = \{u \in L_2(\Omega) : \lim_{a \to +\infty} \sup_{t \in \mathbb{R}} \int_{\{y > a\}} |\exp(-itH)u|^2 \, dx dy = 0\},$$

see [3], [15]. Hence elements in this subspace represent *surface states*. They constitute the characteristic feature of partially periodic operators. Obviously, *the spectrum of the operator H is purely absolutely continuous iff $\mathfrak{T} = \{0\}$, and the wave operators are complete iff $\mathfrak{C} = \{0\}$*. This leads to two problems.

The *problem of absolute continuity*: Prove $\mathfrak{T} = \{0\}$. This has only been achieved in the case $\Omega = \mathbb{R}^{d+1}_+$ and either $\partial\Omega = \Gamma_D$ or $\partial\Omega = \Gamma_N$. If $\sigma \equiv 0$, the spectrum of H is purely absolutely continuous provided V is super-exponentially decaying. (This follows from [5] by extending V to an even (with respect to y) function on $\mathbb{R}^{d+1}$.) The same conclusion holds if $V \equiv 0$ and σ satisfies some mild regularity conditions, see [10], [9].

The problem of absolute continuity for partially periodic operators seems to be more difficult than the corresponding problem for fully periodic operators. This is due to the fact that the period cell is unbounded and the fiber operators $H(k)$ have continuous spectrum. Since eigenvalues may be embedded in the continuous spectrum (see [10] for examples), one cannot (directly) apply the Thomas approach to prove absolute continuity. The existing proofs ([5], [6], [10], [9]) rely on a separation of the (possibly embedded) eigenvalues from the continuous spectrum. Then one applies the Thomas method and the abstract Proposition 2.2 stated below.

In Section 2 we will use this proposition to prove absence of *singular continuous* spectrum. This is only a small step towards the main goal, but still it seems to be the first result about partially periodic operators in domains with curved boundary.

The second problem is to give sufficient conditions for the *completeness* or *non-completeness* of the wave operators $W_\pm$. In [9] it is proved that the fiber operators $H(k)$ have non-empty spectrum in $(-\infty, |k|^2)$ if $\Omega = \mathbb{R}^{d+1}_+$, $\partial\Omega = \Gamma_N$ and the average of σ over a period is non-positive. The same conclusion is proved in [12] if $d = 1$, $k \neq 0$, $\partial\Omega = \Gamma_N$, $\sigma \equiv 0$ and $\partial\Omega$ is not a straight line. If the spectrum of H is purely absolutely continuous (at least in $(-\infty, 0)$ in the first case and in $[0, \epsilon)$, $\epsilon > 0$, in the second case), this gives sufficient conditions for the non-completeness of the wave operators.

The completeness of the wave operators, and simultaneously absolute continuity of the spectrum, (i.e., the absence of surface states) has been verified in [10], [9] in the case $\Omega = \mathbb{R}^{d+1}_+$, $\partial\Omega = \Gamma_N$ and $\sigma \geq 0$. In Section 3 we will generalize this result to the case where $V \not\equiv 0$ but satisfies a certain *repulsivity condition*. If we restrict ourselves to Dirichlet boundary conditions we can allow curved boundaries. The proof shows that the fiber operators $H(k)$ have no eigenvalues. It is close in spirit to the proof of Rellich's theorem on absence of positive eigenvalues for the Dirichlet Laplacian in unbounded domains and to one possible proof of the Virial Theorem (see [4]).

2. Absence of singular continuous spectrum

2.1. Statement of the result

Assume that

$$\operatorname{supp}(V) \subset \mathbb{R}^d \times [0, b] \qquad \text{for some } b > 0. \tag{2.1}$$

Theorem 2.1. *Under the assumption* (2.1) *the singular continuous spectrum of the operator H is empty.*

The proof will be given in Subsection 2.4 below after some preparations.

We note that $\sigma_{sc}(H) \cap (-\infty, 0) = \emptyset$ holds even without the assumption (2.1). (This follows from the fact that the spectrum of $H(k)$ in $(-\infty, 0)$ is discrete with piecewise analytic eigenvalues and from a "localized" version of Proposition 2.2.) The point of Theorem 2.1 is that it can treat eigenvalues of $H(k)$ embedded in the continuum $[|k|^2, +\infty)$.

Moreover, we emphasize that this theorem requires only minimal assumptions on Ω. For much more detailed results in the case $\Omega = \mathbb{R}^{d+1}_+$ see the discussion in Subsection 1.5.

2.2. An abstract result

Let $\mathfrak{H}$ be a separable Hilbert space, $O \subset \mathbb{R}^d$ a connected open set and

$$T = \int_O \oplus T(\zeta)\, d\zeta \qquad \text{in} \int_O \oplus \mathfrak{H}\, d\zeta$$

the direct integral of a measurable family of self-adjoint operators $T(\zeta)$ in $\mathfrak{H}$. We prove the following result about the spectrum of the (self-adjoint) operator T.

Proposition 2.2. *Assume that*

$$\sigma_{sc}\left(T(\zeta)\right) = \emptyset \qquad \text{for all } \zeta \in O \tag{2.2}$$

and that there exist countable families of connected open sets $U_j \subset \mathbb{R}$, $V_j \subset O$ and non-constant real-analytic functions $h_j : U_j \times V_j \to \mathbb{C}$ such that

$$\{(\lambda,\zeta) \in \mathbb{R} \times O : \lambda \in \sigma_p\left(T(\zeta)\right)\} \subset \bigcup_j \{(\lambda,\zeta) \in U_j \times V_j : h_j(\lambda,\zeta) = 0\}. \tag{2.3}$$

Then $\sigma_{sc}(T) = \emptyset$.

This result is well known if the resolvent of the operators $T(\zeta)$ is compact and depends analytically on ζ. (For an elementary proof see, e.g., [7].) In this case the assumptions (2.2), (2.3) are automatically fulfilled. However, we are interested in the case where the absolutely continuous spectrum of the operators $T(\zeta)$ is non-empty.

Proof. Put $\Lambda_0 := \bigcup_j \{\lambda \in U_j : h_j(\lambda,\zeta) = 0 \text{ for all } \zeta \in V_j\}$ and let $\Lambda \subset \mathbb{R} \setminus \Lambda_0$ with $\operatorname{meas} \Lambda = 0$. If $E(\Lambda)$, $E(\Lambda,\zeta)$ denote the spectral projections of T, $T(\zeta)$, respectively, corresponding to Λ then

$$E(\Lambda) = \int_O \oplus E(\Lambda,\zeta)\, d\zeta \tag{2.4}$$

and we claim that this operator is equal to 0.

Indeed, write $O = O_1 \cup O_2$ where

$$O_1 := \{\zeta \in O : \sigma_p\left(T(\zeta)\right) \cap \Lambda = \emptyset\}, \qquad O_2 := O \setminus O_1.$$

Since $\sigma_{sc}\left(T(\zeta)\right) = \emptyset$ we immediately obtain $E(\Lambda,\zeta) = 0$ for $\zeta \in O_1$. On the other hand,

$$O_2 \subset \bigcup_j \{\zeta \in V_j : h_j(\lambda,\zeta) = 0 \text{ for some } \lambda \in U_j \cap \Lambda\}.$$

Since h_j is real-analytic, $\lambda \notin \Lambda_0$ and $\operatorname{meas} \Lambda = 0$, it follows from Theorem A in [11] that $\operatorname{meas}\{\zeta \in V_j : h_j(\lambda,\zeta) = 0 \text{ for some } \lambda \in U_j \cap \Lambda\} = 0$ for every j. Hence $\operatorname{meas} O_2 = 0$ and the operator in (2.4) is 0.

Further, since h_j is non-constant the set $\{\lambda \in U_j : h_j(\lambda,\zeta) = 0 \text{ for all } \zeta \in V_j\}$ is countable for every j, and hence Λ_0 is so. Thus, we conclude that $\sigma_{sc}(T) = \emptyset$.　　□

2.3. Characterization of eigenvalues of $H(k)$

Our proof of Theorem 2.1 relies on Proposition 2.2. In order to construct the functions h_j we include now the eigenvalues λ of $H(k)$ into the zero-eigenvalue set of analytic auxiliary operators $A(\lambda,k)$ defined below.

We assume that (2.1) holds and write $\Pi_b := \{(x,y) \in \Pi : y < b\}$ where b is as in (2.1). Assume also that b is so large that $\mathbb{R}^d \times [b,\infty) \subset \Omega$.

Let $\lambda \in \mathbb{R}$, $k \in Q$ and define for any $n \in \mathbb{Z}^d$

$$\beta_n(\lambda,k) := \begin{cases} \sqrt{\sum_{j=1}^{d}(n_j + k_j)^2 - \lambda} & \text{if } \sum_{j=1}^{d}(n_j + k_j)^2 > \lambda, \\ -\sqrt{\lambda - \sum_{j=1}^{d}(n_j + k_j)^2} & \text{if } \sum_{j=1}^{d}(n_j + k_j)^2 \le \lambda. \end{cases} \tag{2.5}$$

In the Hilbert space $L_2(\mathbb{T}^d)$ we consider the quadratic form

$$\mathcal{D}[b(\lambda,k)] := H^{1/2}(\mathbb{T}^d),$$
$$b(\lambda,k)[f] := \sum_{n\in\mathbb{Z}^d} \beta_n(\lambda,k)\,|\hat{f}_n|^2, \tag{2.6}$$

where $\hat{f}_n := (2\pi)^{-d/2} \int_{\mathbb{T}^d} f(x)e^{-inx}\,dx$, $n \in \mathbb{Z}^d$, are the Fourier coefficients of f. We note that

$$c_0\|f\|^2_{H^{1/2}} \le b(\lambda,k)[f] + c_1\|f\|^2 \le c_2\|f\|^2_{H^{1/2}}, \qquad f \in H^{1/2}(\mathbb{T}^d), \tag{2.7}$$

for some constants $c_0, c_1, c_2 > 0$ (depending on λ, k).

Finally, we consider in the Hilbert space $L_2(\Pi_b)$ the quadratic form

$$\mathcal{D}[a(\lambda,k)] := \{u \in \tilde{H}^1(\Pi_b) : u|_{\Gamma_D} = 0\},$$
$$a(\lambda,k)[u] := \int_{\Pi_b} \left(|(D_x + k)u|^2 + |D_y u|^2 + V|u|^2 - \lambda|u|^2\right)\,dxdy + \tag{2.8}$$
$$+ \int_{\gamma_N} \sigma|u|^2\,ds + b(\lambda,k)[u|_{y=b}].$$

From our assumptions on V and σ and from (2.7) and the boundedness of the trace operator $\tilde{H}^1(\Pi_b) \ni u \mapsto u|_{y=b} \in H^{1/2}(\mathbb{T}^d)$ it follows that the forms $a(\lambda,k)$ are lower semibounded and closed, so they generate self-adjoint operators $A(\lambda,k)$.

The compactness of the embedding of $\tilde{H}^1(\Pi_b)$ in $L_2(\Pi_b)$ implies that the operators $A(\lambda,k)$ have compact resolvent.

Now we characterize the eigenvalues of the operator $H(k)$ as the values λ for which 0 is an eigenvalue of the operators $A(\lambda,k)$. More precisely, we have

Proposition 2.3. *Assume (2.1) and let $k \in Q$ and $\lambda \in \mathbb{R}$.*

1. *Let $u \in \mathcal{N}(H(k) - \lambda I)$ and define*

$$v(x,y) := u(x,y), \qquad (x,y) \in \Pi_b, \tag{2.9}$$

$$\hat{v}_n(b) := (2\pi)^{-d/2} \int_{\mathbb{T}^d} v(x,b)e^{-i\langle n,x\rangle}\,dx, \qquad n \in \mathbb{Z}^d. \tag{2.10}$$

Then $v \in \mathcal{N}(A(\lambda,k))$, $\hat{v}_n(b) = 0$ if $|n+k|^2 \le \lambda$ and, moreover,

$$u(x,y) = \frac{1}{(2\pi)^{d/2}} \sum_{|n+k|^2 > \lambda} \hat{v}_n(b)\, e^{i\langle n,x\rangle}\, e^{-\beta_n(\lambda,k)\,y}, \qquad (x,y) \in \Pi \setminus \overline{\Pi_b}. \tag{2.11}$$

2. *Let $v \in \mathcal{N}(A(\lambda,k))$ and assume that $\hat{v}_n(b) = 0$ if $|n+k|^2 \le \lambda$, where $\hat{v}_n(b)$ are given by (2.10). Define u by (2.9),(2.11).*
   Then $u \in \mathcal{N}(H(k) - \lambda I)$.

The proof of this proposition is straightforward and will be omitted.

Remark 2.4. Obviously, the statement of Proposition 2.3 does not depend on the definition of $\beta_n(\lambda, k)$ for $|n + k|^2 \le \lambda$. Our choice is useful in the following sense. Let $\Lambda = (\lambda_-, \lambda_+)$ be an open interval. A Birman-Schwinger-type argument as in Theorem 4.1 of [10] using Proposition 2.3 and the monotonicity of $A(\lambda, k)$ with respect to λ yields

$$\sharp_{cm}\{\lambda \in (\lambda_-, \lambda_+): \ \lambda \text{ is an eigenvalue of } H(k)\}$$
$$\le \sharp_{cm}\{\mu < 0: \ \mu \text{ is an eigenvalue of } A(\lambda_+, k)\} \tag{2.12}$$
$$-\sharp_{cm}\{\mu \le 0: \ \mu \text{ is an eigenvalue of } A(\lambda_-, k)\}.$$

Here $\sharp_{cm}\{\dots\}$ means that the cardinality of $\{\dots\}$ is determined according to multiplicities. The RHS of (2.12) is finite since $A(\lambda_\pm, k)$ are lower semibounded and have compact resolvent. In particular, the eigenvalues of $H(k)$ have no finite accumulation point. This result cannot be obtained by scattering theory (see Subsection 1.4) and improves the corresponding results in [15].

2.4. Proof of Theorem 2.1

We will apply Proposition 2.2. In order to cover "threshold eigenvalues" we include the functions h_n, $n \in \mathbb{Z}^d$, defined by $h_n(\lambda, k) := \sum_{j=1}^d (n_j + k_j)^2 - \lambda$ in our collection. By a covering argument it suffices to prove the following statement: For all $k^0 \in Q$, $\lambda_0 \in \sigma_p\left(H(k^0)\right) \setminus \{|n + k^0|^2 : n \in \mathbb{Z}^d\}$ there exist neighborhoods $U \subset \mathbb{R}$, $V \subset \mathbb{R}^d$ of λ_0, k^0 and a non-constant real-analytic function $h : U \times V \to \mathbb{C}$ such that

$$h(\lambda, k) = 0 \qquad \text{if } (\lambda, k) \in U \times V, \ \lambda \in \sigma_p\left(H(k)\right).$$

For this we construct an analytic extension of the operators $A(\lambda, k)$ near $(\lambda, k) = (\lambda_0, k^0)$. Since $|n + k^0|^2 \ne \lambda_0$ for all $n \in \mathbb{Z}^d$, there exist neighborhoods $\tilde{U} \subset \mathbb{C}$, $\tilde{V} \subset \mathbb{C}^d$ of λ_0, k^0 such that the functions β_n, $n \in \mathbb{Z}^d$, admit an analytic continuation to $\tilde{U} \times \tilde{V}$. Then we can define sectorial and closed forms $a(z, \kappa)$ for $z \in \tilde{U}$, $\kappa \in \tilde{V}$ by (2.8) and obtain corresponding m-sectorial operators $A(z, \kappa)$. These operators have compact resolvent and it is well known that (after possibly decreasing $\tilde{U}$ and $\tilde{V}$) there is an analytic function $h : \tilde{U} \times \tilde{V} \to \mathbb{C}$ such that $h(z, \kappa) = 0$ iff $0 \in \sigma\left(A(z, \kappa)\right)$. It is easy to see (cf. Lemma 4.3 in [10]) that h is non-constant. Now it suffices to note that $0 \in \sigma\left(A(\lambda, k)\right)$ whenever $\lambda \in \sigma_p\left(H(k)\right)$ according to Proposition 2.3. This completes the proof of Theorem 2.1.

3. Absolute continuity in the repulsive case

In this section we give sufficient conditions on V, σ and Ω that exclude surface states. This implies purely absolutely continuous spectrum of H and complete scattering. Our conditions say basically that the surface $\partial\Omega$ is *repulsive*.

3.1. Statement of the result

Our first result concerns smooth curved boundaries with Dirichlet conditions.

We assume that

$$\partial\Omega = \Gamma_D \qquad \text{and} \qquad \Gamma_D \in C^2, \tag{3.1}$$

and that, with the exterior unit normal $\nu(x,y)$ at $(x,y) \in \Gamma_D$ and the unit vector $e_y = (0,\ldots,0,1)$ in y-direction,

$$\langle \nu(x,y), e_y \rangle \leq 0, \qquad (x,y) \in \Gamma_D. \tag{3.2}$$

Note that these assumptions are satisfied in the special case $\Omega = \{(x,y) : y > h(x)\}$ with a periodic C^2-function h.

Introduce a modified distance to $\partial\Omega$,

$$d(x,y) := \min\{|(x,y') - (x,y)| : (x,y') \in \partial\Omega\}, \qquad (x,y) \in \Omega.$$

Concerning the electric potential V we will assume the *repulsivity condition*

$$\frac{\partial V}{\partial y}(x,y) \leq \frac{1}{2y\,d(x,y)^2}, \qquad (x,y) \in \Omega. \tag{3.3}$$

This relation is understood in the sense of distributions, i.e.,

$$\int_\Omega V \frac{\partial\varphi}{\partial y}\,dxdy \geq -\frac{1}{2}\int_\Omega \frac{\varphi}{y\,d^2}\,dxdy$$

for all $\varphi \in C_0^\infty(\Omega)$, $\varphi \geq 0$. Our result is

Theorem 3.1. *Assume* (3.1), (3.2), (3.3). *Then the wave operators $W_\pm$ are unitary and satisfy*

$$H = W_\pm H_0 W_\pm^*.$$

In particular, the spectrum of H is purely absolutely continuous and there exist no surface states.

Remark 3.2. The conclusion is not true if $\partial\Omega = \Gamma_N$, $\sigma \equiv 0$ and Γ_N is not a hyperplane. This is an easy generalization of [12] to the case $d \geq 2$.

A result for general boundary conditions can be obtained when $\partial\Omega$ is straight, say,

$$\Omega = \mathbb{R}_+^{d+1}. \tag{3.4}$$

We write $\sigma(x)$ instead of $\sigma(x,0)$ and assume that

$$\sigma \geq 0 \qquad \text{a.e. on } \Gamma_N. \tag{3.5}$$

In addition we assume that V satisfies the following repulsivity condition: There exists a measurable periodic function $\omega : \Gamma_N \to \mathbb{R}_+$ such that

$$\frac{\partial V}{\partial y}(x,y) \leq \begin{cases} \frac{\omega(x)\sigma(x)^2}{2\omega(x)+\pi/2} \frac{1}{y(\omega(x)^2+\sigma(x)^2 y^2)} & \text{if } (x,0) \in \Gamma_N, \\ \frac{1}{2y^3} & \text{if } (x,0) \in \Gamma_D. \end{cases} \tag{3.6}$$

Similarly to (3.3), this relation is understood in the sense of distributions. Note that the allowed oscillations of $V(x,y)$ are larger the larger $\sigma(x)$ is. If first $\sigma \to \infty$ and then $\omega \to \infty$, we obtain $\frac{\omega(x)\sigma(x)^2}{2\omega(x)+\pi/2} \frac{1}{y(\omega(x)^2+\sigma(x)^2 y^2)} \to \frac{1}{2y^3}$ and we recover the required decay in the case of Dirichlet boundary conditions.

Theorem 3.3. *Assume* (3.4), (3.5), (3.6). *Then the conclusions of Theorem* 3.1 *hold.*

Remark 3.4. This answers a question posed by Prof. P. Exner. Theorem 3.3 generalizes the results of [10], [9], where only the case $V \equiv 0$ and $\Gamma_D = \emptyset$ was considered.

3.2. A virial inequality

The first part of the proofs of Theorems 3.1, 3.3 can be given simultaneously. We assume in this subsection that $\partial\Omega$ is of class C^2 (at least in a neighbourhood of Γ_D) and that $a : \overline{\Omega} \to \mathbb{R}^{d+1}$ is a periodic (with respect to x) C^1-vector field such that $\mathrm{supp}(a)$ is bounded with respect to the variable y and such that

$$a = 0 \qquad \text{in a neighborhood of } \Gamma_N. \tag{3.7}$$

We denote the Jacobian of a by J_a.

In order to simplify the notation we write in this and the next subsection $D + k$ instead of $(D_x + k, D_y)$ when $k \in Q$. Moreover, by $\tilde{C}^\infty(\overline{\Pi})$ we denote the class of functions in $C^\infty(\overline{\Pi})$ with periodic extension (with respect to the variables x) in $C^\infty(\overline{\Omega})$. The following is sometimes called *Rellich's identity* (at least when $k = 0$).

Lemma 3.5. *Let* $\lambda \in \mathbb{R}$, $k \in Q$. *Then for any* $v \in \tilde{C}^\infty(\overline{\Pi})$ *with bounded support and* $v|_{\gamma_D} = 0$ *one has*

$$2\,\mathrm{Im}\int_{\Pi} ((D + k)^2 - \lambda)v \overline{\langle (D + k)v, a \rangle}\, dxdy$$

$$= \int_{\Pi} \left(2\,\mathrm{Re}\langle J_a(D + k)v, (D + k)v \rangle - \mathrm{div}\, a(|(D + k)v|^2 - \lambda|v|^2) \right) dxdy$$

$$- \int_{\gamma_D} \langle a, \nu \rangle \left| \frac{\partial v}{\partial \nu} \right|^2 ds.$$

Proof. This is a straightforward calculation using that $\nabla v|_{\gamma_D} = \frac{\partial v}{\partial \nu}\nu$. $\qquad\square$

Now let $W \in L_{1,loc}(\Omega)$ be periodic with respect to x and assume that

$$\langle \nabla V, a \rangle \leq W \tag{3.8}$$

as distribution in Ω, i.e.,

$$\int_{\Omega} V\,\mathrm{div}(a\varphi)\, dxdy \geq -\int_{\Omega} W\varphi\, dxdy \tag{3.9}$$

for all $\varphi \in C_0^\infty(\Omega)$, $\varphi \geq 0$. Moreover, assume that

$$\left| \int_{\Pi} W|v|^2\, dxdy \right| \leq C\|v\|_{H^1}^2 \tag{3.10}$$

for all $v \in \tilde{C}^\infty(\overline{\Pi})$ with bounded support and $v|_{\gamma_D} = 0$. Then we obtain the following *virial inequality.*

Proposition 3.6. *Let $\lambda \in \mathbb{R}$, $k \in Q$ and assume (3.8), (3.10). Then for all $u \in \mathcal{N}(H(k) - \lambda I)$ one has*

$$\int_{\Pi} \left(2\operatorname{Re}\langle J_a(D+k)u, (D+k)u \rangle - W|u|^2 \right) dxdy$$

$$\leq \int_{\Pi} \operatorname{div} a \left(|(D+k)u|^2 + V|u|^2 - \lambda|u|^2 \right) dxdy + \int_{\gamma_D} \langle a, \nu \rangle \left| \frac{\partial u}{\partial \nu} \right|^2 ds.$$

Proof. Let $v \in \tilde{C}^\infty(\overline{\Pi})$ with bounded support and $v|_{\gamma_D} = 0$. Because of (3.10) we can replace φ in (3.9) by $|v|^2$ and we obtain by Lemma 3.5

$$2\operatorname{Im} \int_{\Pi} ((D+k)^2 + V - \lambda) v \overline{\langle (D+k)v, a \rangle} \, dxdy$$

$$\geq \int_{\Pi} \left(2\operatorname{Re}\langle J_a(D+k)v, (D+k)v \rangle - W|v|^2 \right) dxdy$$

$$- \int_{\Pi} \operatorname{div} a \left(|(D+k)v|^2 + V|v|^2 - \lambda|v|^2 \right) dxdy - \int_{\gamma_D} \langle a, \nu \rangle \left| \frac{\partial v}{\partial \nu} \right|^2 ds.$$

Now the boundedness of V, $\Gamma_D \in C^2$ and (3.7) imply that any $u \in \mathcal{N}(H(k) - \lambda I)$ is of class H^2 on $\operatorname{supp}(a)$. Approximating u by $v \in \tilde{C}^\infty(\overline{\Pi})$ with bounded support and $v|_{\gamma_D} = 0$ simultaneously in the H^1-norm on Π and in the H^2-norm on $\operatorname{supp}(a)$ we obtain the result. $\qquad\square$

Remark 3.7. Recall that one form of the Virial Theorem (see [4]) says

$$\int_{\Pi} \left(2\left| \frac{\partial u}{\partial y} \right|^2 - y\frac{\partial V}{\partial y}|u|^2 \right) dxdy = \int_{\gamma_D} y\langle \nu, e_y \rangle \left| \frac{\partial u}{\partial \nu} \right|^2 ds, \quad u \in \mathcal{N}(H(k) - \lambda I),$$

provided V is smooth and $\gamma_N = \emptyset$. This is equality in Proposition 3.6 with $a = ye_y$, $W = y\frac{\partial V}{\partial y}$ and serves as intuition behind our proof of Theorems 3.1 and 3.3. Note that our conditions are weaker and more effective than those in [4].

3.3. Proof of Theorems 3.1 and 3.3

We begin with the

Proof of Theorem 3.1. Taking Proposition 1.2 into account we see that it suffices to prove $\sigma_p(H(k)) = \emptyset$ for all $k \in Q$.

Let $\eta \in C^\infty(\overline{\mathbb{R}_+})$ such that $\eta \geq 0$ and $\eta(t) = 1$ if $t \leq 1$, $\eta(t) = 0$ if $t \geq 2$. We apply the results of Subsection 3.2 with the vector field $a_\epsilon(x, y) := \eta(\epsilon y)ye_y$, where $e_y = (0, \ldots, 0, 1)$ as above. Note that we have by (3.3)

$$\langle \nabla V, a_\epsilon \rangle \leq W_\epsilon, \qquad W_\epsilon(x, y) := \frac{\eta(\epsilon y)}{2d(x, y)^2},$$

and (3.10) holds because of Hardy's inequality

$$\frac{1}{4} \int_{\Pi} \frac{|v|^2}{d^2} \, dxdy \leq \int_{\Pi} |D_y v|^2 \, dxdy. \tag{3.11}$$

Hence if $u \in \mathcal{N}(H(k) - \lambda I)$, then by Proposition 3.6

$$\int_\Pi \left(2\chi_\epsilon |D_y u|^2 - W_\epsilon |u|^2 \right) \, dxdy \tag{3.12}$$

$$\leq \int_\Pi \chi_\epsilon \left(|(D+k)u|^2 + V|u|^2 - \lambda |u|^2 \right) \, dxdy + \int_{\gamma_D} \eta(\epsilon y) y \langle \nu, e_y \rangle \left| \frac{\partial u}{\partial \nu} \right|^2 \, ds$$

where $\chi_\epsilon(y) := \frac{\partial}{\partial y}(\eta(\epsilon y)y)$. Now we let ϵ tend to zero. Because of (3.2) the integral over γ_D is non-positive. Moreover, $\chi_\epsilon(y) = \epsilon y \eta'(\epsilon y) + \eta(\epsilon y)$ is bounded and tends to 1 pointwise. Hence by dominated convergence the integral over Π on the RHS of (3.12) tends to

$$\int_\Pi \left(|(D+k)u|^2 + V|u|^2 - \lambda|u|^2 \right) \, dxdy = 0.$$

Similarly by dominated convergence (recall (3.11)) the LHS of (3.12) tends to $2 \int_\Pi \left(|D_y u|^2 - \frac{|u|^2}{4d^2} \right) \, dxdy$. Hence (3.12) implies that

$$2 \int_\Pi \left(|D_y u|^2 - \frac{|u|^2}{4d^2} \right) \, dxdy \leq 0,$$

i.e., we have equality in Hardy's inequality. We conclude that $u = 0$ and hence that $\sigma_p(H(k)) = \emptyset$. $\qquad\square$

Proof of Theorem 3.3. The proof is similar to the previous one, so we only sketch the differences. We define $a_\epsilon(x,y) := \eta(\epsilon y)(1 - \eta(\epsilon^{-1}y))y e_y$ with η and e_y as before. Again we apply Proposition 3.6 and obtain as $\epsilon \to 0$

$$2 \int_\Pi \left(|D_y u|^2 - W|u|^2 \right) \, dxdy \leq - \int_{\gamma_N} \sigma |u|^2 \, dx,$$

where W denotes $y/2$ times the RHS of (3.6). The proof is completed by Lemma 3.8 below. $\qquad\square$

The following Hardy-type inequality was used in the previous proof.

Lemma 3.8. *Let $\sigma, \omega \in \mathbb{R}_+$. Then*

$$\frac{\omega \sigma^2}{2\omega + \pi/2} \int_0^\infty \frac{|u|^2}{\omega^2 + \sigma^2 y^2} \, dy \leq 2 \int_0^\infty |u'|^2 \, dy + \sigma |u(0)|^2, \qquad u \in C_0^\infty(\overline{\mathbb{R}_+}).$$

Proof. By scaling we may assume $\sigma = 1$. For $v := u - u(0)$ the "ordinary" Hardy inequality holds, so

$$\int_0^\infty |v'|^2 \, dy \geq \frac{1}{4} \int_0^\infty \frac{|v|^2}{y^2} \, dy \geq \frac{1}{4} \int_0^\infty \frac{|v|^2}{\omega^2 + y^2} \, dy.$$

It follows that for any $\epsilon > 0$

$$2 \int_0^\infty |u'|^2 \, dy \geq \frac{1}{2}(1-\epsilon) \int_0^\infty \frac{|u|^2}{\omega^2 + y^2} \, dy - \frac{1}{2}(\epsilon^{-1} - 1)|u(0)|^2 \int_0^\infty \frac{1}{\omega^2 + y^2} \, dy,$$

and the assertion follows with $\epsilon = (4\omega/\pi + 1)^{-1}$. $\qquad\square$

Acknowledgments

The authors would like to thank Prof. M.Sh. Birman and Prof. N. Filonov for consultations. The second author is grateful to A. Laptev and KTH Stockholm for hospitality.

References

[1] M.Sh. Birman, *Perturbations of the continuous spectrum of a singular elliptic operator by varying the boundary and the boundary conditions*, (Russian) Vestnik Leningrad. Univ. **17**, no. 1 (1962), 22–55.

[2] M.Sh. Birman, M.Z. Solomyak, *On the negative discrete spectrum of a periodic elliptic operator in a waveguide-type domain, perturbed by a decaying potential*, J. Anal. Math. **83** (2001), 337–391.

[3] E.B. Davies, B. Simon, *Scattering Theory for Systems with Different Spatial Asymptotics on the Left and Right*, Commun. Math. Phys. **63** (1978), 277-301.

[4] M. Eastham, H. Kalf, *Schrödinger-type operators with continuous spectra*, Research Notes in Mathematics, 65. Pitman (Advanced Publishing Program), Boston, Mass.-London, 1982.

[5] N. Filonov, F. Klopp, *Absolute continuity of the spectrum of a Schrödinger operator with a potential which is periodic in some directions and decays in others*, Documenta Math. **9** (2004), 107-121; Erratum: ibd., 135–136.

[6] N. Filonov, F. Klopp, *Absolutely continuous spectrum for the isotropic Maxwell operator with coefficients that are periodic in some directions and decay in others*, Comm. Math. Phys., to appear.

[7] N. Filonov, A.V. Sobolev, *Absence of the singular continuous component in the spectrum of analytic direct integrals*, Zap. Nauchn. Sem. S.-Petersburg. Otdel. Mat. Inst. Steklov. (POMI) **318** (2004), 298–307.

[8] R.L. Frank, *On the scattering theory of the Laplacian with a periodic boundary condition. I. Existence of wave operators*, Documenta Math. **8** (2003), 547–565.

[9] R.L. Frank, *On the Laplacian in the halfspace with a periodic boundary condition*, preprint, mp-arc 04-407.

[10] R.L. Frank, R.G. Shterenberg, *On the scattering theory of the Laplacian with a periodic boundary condition. II. Additional channels of scattering*, Documenta Math. **9** (2004), 57–77.

[11] R. Hempel, I. Herbst, *Bands and gaps for periodic magnetic Hamiltonians*, Partial differential operators and mathematical physics (Holzhau, 1994), 175–184, Oper. Theory Adv. Appl., 78, Birkhäuser, Basel, 1995.

[12] I.V. Kamotskiĭ, S.A. Nazarov, *Exponentially decreasing solutions of the problem of diffraction by a rigid periodic boundary*, Math. Notes **73**, no. 1-2 (2003), 129–131.

[13] P. Kuchment, *On some spectral problems of mathematical physics*, Partial differential equations and inverse problems, 241–276, Contemp. Math., 362, Amer. Math. Soc., Providence, RI, 2004.

[14] S.T. Kuroda, *Scattering theory for differential operators. III. Exterior problems*, Spectral theory and differential equations (Proc. Sympos., Dundee, 1974), 227–241, Lecture Notes in Math., 448, Springer, Berlin, 1975.

[15] A.W. Saenz, *Quantum-mechanical scattering by impenetrable periodic surfaces*, J. Math. Phys. **22**, no. 12 (1981), 2872–2884.

[16] B. Simon, *Phase space analysis of simple scattering systems: extensions of some work of Enss*. Duke Math. J. **46**, no. 1 (1979), 119–168.

[17] L. Thomas, *Time dependent approach to scattering from impurities in a crystal.* Comm. Math. Phys. **33** (1973), 335–343.

[18] D.R. Yafaev, *Mathematical Scattering Theory*, Amer. Math. Soc., 1992.

Rupert L. Frank
Royal Institute of Technology
Department of Mathematics
Lindstedtsvägen 25
S-10044 Stockholm, Sweden
e-mail: `rupert@math.kth.se`

Roman G. Shterenberg
Department of Mathematics
University of Wisconsin-Madison
480 Lincoln Drive Madison
WI 53706-1388, USA
e-mail: `shterenb@math.wisc.edu`

Operator Theory:
Advances and Applications, Vol. 174, 51–67
© 2007 Birkhäuser Verlag Basel/Switzerland

Functional Model for Singular Perturbations of Non-self-adjoint Operators

Alexander V. Kiselev

Abstract. We discuss the definition of a rank one singular perturbation of a non-self-adjoint operator L in Hilbert space H. Provided that the operator L is a non-self-adjoint perturbation of a self-adjoint operator A and that the spectrum of the operator L is absolutely continuous we are able to establish a concise resolvent formula for the singular perturbations of the class considered and to establish a model representation of it in the dilation space associated with the operator L.

Mathematics Subject Classification (2000). Primary 47A10; Secondary 47A55.

Keywords. Singular perturbations, functional model, non-self-adjoint operators.

1. Introduction

Let A be a self-adjoint operator acting in Hilbert space H. Let φ be a linear bounded functional on the domain of the operator, endowed with the graph norm. Then the rank one perturbation of the operator A is the operator formally given by

$$A_\alpha = A + \alpha \langle \cdot, \varphi \rangle \, \varphi, \tag{1.1}$$

where α is a real parameter. The formal expression (1.1) corresponds to a well-defined self-adjoint operator acting in Hilbert space H when φ belongs to at least the space H_{-2} (the operator is determined uniquely if φ belongs to at least the space H_{-1}, see Section 2 below) of bounded linear functionals over the domain of the operator A, see [1, 2, 3]. The theory of singular perturbations of self-adjoint operators in Hilbert spaces relies heavily on the classical extension theory for symmetric operators. In fact (see Section 2 for details) every operator (1.1) is a certain extension of a symmetric restriction of the operator A to a manifold dense in H. Thus there is no obvious way to define singular perturbations of a

The author gratefully appreciates financial support of IRCSET (Post-Doctoral Fellowships Programme) and RFBR (grant No. 03-01-00090).

non-self-adjoint operator within the framework of the extension theory. It should be mentioned however that the spectral theory of non-self-adjoint extensions of symmetric operators was developed in terms of the spaces of boundary values by Ryzhov in [16], see also [13, 19, 14]. This included construction of the functional model for operators of this class.

In the present paper we are going to extend the formalism (1.1) to the case of possibly non-self-adjoint operators, to explicitly construct the corresponding scale of spaces and to provide a direct analogue of the Krein's formula for the resolvent of the perturbed operator in both the model representation of the non-self-adjoint operator and the original Hilbert space H. To this end, we use the functional model of a non-self-adjoint operator constructed in [20, 12] for the dissipative case and then extended in [9, 10, 11] to the case of a wide class of arbitrary non-self-adjoint operators.

The functional model for an operator of the form $L = A + iV$ ($A = A^*$, $V = V^*$; see also Section 3 below for complete set of assumptions and definitions) with absolutely continuous spectrum makes two different approaches to consideration of rank one perturbations of L possible. Following the first approach, one might consider the operator A_α, corresponding to the formal expression (1.1), and then the class of operators $L_\alpha = A_\alpha + iV$ for all admissible perturbations V. This limits one to consideration of real α only and leads to different functional models for the operators L and L_α. We propose a different approach to the non-self-adjoint operator L_α, namely, we begin with the non-self-adjoint operator $L = A + iV$ which action in its model representation, provided that spectrum of L is absolutely continuous, closely resembles just the componentwise multiplication by the independent variable.[1] We then show, that all admissible singular perturbations of the operator L lead to rank one singular perturbations of the multiplication operator in the dilated Hilbert space $\mathcal{H}$ and, using the Krein's resolvent formula, "return back" to the original Hilbert space H, obtaining an explicit description of the operator L_α via its resolvent. It turns out that this approach leads to the same results as the "standard" one in at least the case when the perturbation V is bounded in H.

The present paper is organized as follows. In the second section we briefly recall relevant results of the theory of self-adjoint singular perturbations of self-adjoint operators, devised in [1, 2, 3]. We are mainly interested in the construction proposed in [2] for dealing with the case of non-semibounded self-adjoint operators.

In the third section we pass over to the case of a non-self-adjoint operator in Hilbert space H, restricting consideration to operators possessing absolutely continuous spectrum. We then provide a construction of a functional model for singular perturbations of such operators, following ideas outlined above, and prove that results obtained are consistent with ones obtained in the self-adjoint case.

---

[1] We refrain from rigorous discussion of this statement here since at this point our primary concern is to give an overall outlook of the problem. We refer the reader to Section 3 for details.

2. Singular perturbations of self-adjoint operators

In the present section we briefly describe (following [1, 2, 3]) a construction leading to the definition of a rank one self-adjoint singular perturbation of a self-adjoint operator A, acting in Hilbert space H, i.e., the self-adjoint operator corresponding to the formal expression

$$A_\alpha = A + \alpha \langle \cdot, \varphi \rangle \, \varphi, \tag{2.1}$$

where the vector φ, determining the perturbation, does not belong to Hilbert space H.

Suppose that φ is a bounded linear functional on the domain of the operator A (treated as a Hilbert space with the norm chosen to be equal to the graph norm of the operator A), i.e., $\varphi \in (D(A))^* \setminus H$. Then [2] the condition $\langle \psi, \varphi \rangle = 0$ is well defined for all $\psi \in D(A)$ and the restriction A_0 of the operator A to the domain $D_\varphi = \{\psi \in D(A) : \langle \psi, \varphi \rangle = 0\}$ is a densely defined symmetric operator in H with deficiency indices $(1, 1)$.

We choose a normalization for the vector φ in a way such that $\|(A-i)^{-1}\varphi\| = 1$. The element $(A - \lambda)^{-1}\varphi$ for all non-real values of λ has to be understood in the generalized sense, i.e., $(A - \lambda)^{-1}\varphi$ is a bounded linear functional over H which acts on every $\psi \in H$ according to the formula

$$\langle \psi, (A - \lambda)^{-1}\varphi \rangle = \langle (A - \bar{\lambda})^{-1}\psi, \varphi \rangle .$$

Henceforth we identify the functional $(A - \lambda)^{-1}\varphi$ with an element of Hilbert space H.

The domain of the adjoint operator A_0^* can be described as follows:

$$D(A_0^*) = D_\varphi \dotplus \bigvee \{(A - i)^{-1}\varphi, (A + i)^{-1}\varphi\},$$

so that for every $\psi \in D(A_0^*)$ we have the following representation:

$$\psi = \hat{\psi} + a_+(\psi)(A - i)^{-1}\varphi + a_-(\psi)(A + i)^{-1}\varphi, \tag{2.2}$$

where $\hat{\psi} \in D_\varphi$, $a_\pm(\psi) \in \mathbb{C}$. On every $\psi \in D(A_0^*)$ the operator A_0^* acts as follows:

$$A_0^*(\hat{\psi} + a_+(\psi)(A - i)^{-1}\varphi + a_-(\varphi)(A + i)^{-1}\varphi)x$$
$$= A\hat{\psi} + a_+(\psi)i(A - i)^{-1}\varphi + a_-(\psi)(-i)(A + i)^{-1}\varphi. \tag{2.3}$$

Every self-adjoint extension of the operator A_0 is a restriction of the adjoint operator A_0^* to the domain $D_v = \{\psi \in D(A_0^*) : a_-(\psi) = va_+(\psi)\}$ with a unimodular parameter $v \in \mathbb{C}$. The extension corresponding to $v = -1$ coincides with the original operator A.

Consider the standard scale of Hilbert spaces associated with the nonnegative operator $|A|$ [1]:

$$H_2(|A|) \subset H_1(|A|) \subset H \subset H_{-1}(|A|) \subset H_{-2}(|A|), \tag{2.4}$$

where $H_2(|A|) = D(A)$ and $H_1(|A|)$ coincides with the domain of the operator $|A|^{1/2}$; the spaces $H_{-1}(|A|)$ and $H_{-2}(|A|)$ are dual to $H_{1,2}(|A|)$, respectively, with respect to the inner product in H.

A vector φ is said to define a form-bounded perturbation of the operator A if it belongs to Hilbert space $H_{-1}(|A|)$. Vectors $\varphi \in H_{-2}(|A|) \setminus H_{-1}(|A|)$ are said to define form-unbounded perturbations.

If φ defines a form-bounded perturbation of the self-adjoint operator A, $\varphi \in H_{-1} \setminus H$, then the self-adjoint operator A_α corresponding to the formal expression (2.1) can be shown to be a self-adjoint extension of A_0; its domain is described by the following formula:

$$D(A_\alpha) = \{\psi \in D(A_0^*) : a_+(\psi) = -\frac{1 + \alpha \left\langle (A+i)^{-1}\varphi, \varphi \right\rangle}{1 + \alpha \left\langle (A-i)^{-1}\varphi, \varphi \right\rangle} a_-(\psi)\}. \tag{2.5}$$

If $\alpha = 0$ then the corresponding operator corresponds to the original operator A. Considering different $\alpha \in \mathbb{R} \cup \{\infty\}$ all self-adjoint extensions of the operator A_0 can be obtained.

In the form-unbounded case, $\varphi \in H_{-2}(|A|) \setminus H_{-1}(|A|)$, the formal expression (2.1) is defined on the domain of the adjoint operator A_0^* if the functional φ is defined on this domain. The domain of the operator A_0^* can be described as a one-dimensional extension of the domain $D(A)$: any element $\psi \in D(A_0^*)$ possesses the following representation:

$$\psi = \tilde{\psi} + \frac{b(\psi)}{2}((A-i)^{-1}\varphi + (A+i)^{-1}\varphi), \tag{2.6}$$

where $b(\psi) = a_+(\psi) + a_-(\psi)$ and $\tilde{\psi} = \hat{\psi} + (a_+(\psi) - a_-(\psi))i(A-i)^{-1}(A+i)^{-1}\varphi$. Therefore, since φ is a bounded functional over the domain $D(A)$, it needs to be extended to the elements $(A \pm i)^{-1}\varphi$ as a bounded functional.

All such extensions are parameterized by a single parameter $c \in \mathbb{C}$ and the extended functional φ_c acts as follows on any $\psi \in D(A_0^*)$, see (2.6):

$$\langle \psi, \varphi_c \rangle = \left\langle \tilde{\psi}, \varphi \right\rangle + cb(\psi). \tag{2.7}$$

This extension defines a real quadratic form $\langle \psi, [(A-i)^{-1} + (A+i)^{-1}]\psi \rangle$ with domain $H \dot{+} \{\varphi\}$ if and only if the parameter c is real.

Henceforth in the case of $\varphi \in H_{-2}(|A|) \setminus H_{-1}(|A|)$ we define φ_c to be the linear bounded extension of the functional φ to the domain $D(A_0^*)$, submitted to the condition

$$\left\langle [(A-i)^{-1} + (A+i)^{-1}]\varphi, \varphi_c \right\rangle = 2c, \quad c \in \mathbb{R}. \tag{2.8}$$

Then it can be shown, that the self-adjoint operator A_α corresponding to the formal expression $A_\alpha = A + \alpha \langle \cdot, \varphi_c \rangle \varphi$ is a self-adjoint extension of the operator A_0 (with $A_\alpha = A$ for $\alpha = 0$) with its domain described by the following formula:

$$D(A_\alpha) = \{\psi \in D(A_0^*) : a_-(\psi) = -\frac{1 + \alpha(c+i)}{1 + \alpha(c-i)} a_+(\psi)\}. \tag{2.9}$$

We note, that in the case of form-unbounded perturbations the definition of a singular perturbation of a self-adjoint operator is not unique. This non-uniqueness is due to the fact, that the choice of a bounded extension of the functional φ to the domain of the operator A_0^* is also not unique.

The Krein's resolvent formula holds in both cases of form-bounded and form-unbounded perturbations, that is,

$$(A_\alpha - \lambda)^{-1} = (A - \lambda)^{-1} - \frac{\alpha}{1 + \alpha F(\lambda)} \left\langle \cdot, (A - \overline{\lambda})^{-1}\varphi \right\rangle (A - \lambda)^{-1}\varphi, \qquad (2.10)$$

where

$$F(\lambda) = \left\langle (A - \lambda)^{-1}\varphi, \varphi_c \right\rangle.$$

In the case of form-bounded perturbation $\varphi_c = \varphi$, whereas in the case of form-unbounded perturbation the following representation can be obtained from the definition of the extended functional φ_c:

$$F(\lambda) = c + \left\langle (1 + \lambda A)(A - \lambda)^{-1}(A - i)^{-1}\varphi, (A - i)^{-1}\varphi \right\rangle.$$

In both cases, $F(\lambda)$ is a Nevanlinna function (a holomorphic function in the upper half-plane with positive imaginary part there).

3. The non-self-adjoint case

3.1. The functional model

In the present section, we use the functional model of a non-self-adjoint operator constructed in [20, 12] for the dissipative case and then extended in [9, 10, 11, 16] to the case of a wide class of arbitrary non-self-adjoint operators.

Let us briefly describe the corresponding results here. Consider a class of non-self-adjoint operators of the form [11]

$$L = A + iV,$$

where A is a self-adjoint operator in H defined on the domain $D(A)$ and the perturbation V admits the factorization $V = \frac{\gamma J \gamma}{2}$, where γ is a nonnegative self-adjoint operator in H and $J = J^* \equiv \text{sign}\, V$ is a unitary operator in the auxiliary Hilbert space E, defined as the closure of the range of the operator γ: $E \equiv \overline{R(\gamma)}$. This factorization corresponds to the polar decomposition of the operator V. It can also be easily generalized (when A and V are bounded) to the "node" case [21], where J acts in an auxiliary Hilbert space $\mathfrak{H}$ and $V = \gamma^* J \gamma / 2$, γ being an operator acting from H to $\mathfrak{H}$. In order that the expression $A + iV$ be meaningful, we impose the condition that V be (A)-bounded with the relative bound less than 1, i.e., $D(A) \subset D(V)$ and for some a and b $(a < 1)$ the condition

$$\|Vu\| \le a\|Au\| + b\|u\|, \quad u \in D(A)$$

is satisfied, see [8]. Then the operator L is well defined on the domain $D(L) = D(A)$.

Alongside with the operator L we are going to consider the maximal dissipative operator $L^{\|} = A + i\frac{\gamma^2}{2}$ and the one adjoint to it, $L^{-\|} \equiv L^{\|*} = A - i\frac{\gamma^2}{2}$. Since the functional model for the dissipative operator $L^{\|}$ will be used below, we require that $L^{\|}$ be completely non-self-adjoint, i.e., that it has no reducing self-adjoint parts. This requirement is not restrictive in our case due to Proposition 1 in [11].

Now we are going to briefly describe construction of the self-adjoint dilation of the completely non-self-adjoint dissipative operator $L^{\parallel}$, following [20, 12], see also [11].

The characteristic function $S(\lambda)$ of the operator $L^{\parallel}$ is a contractive, analytic operator-valued function acting on Hilbert space E, defined for $\operatorname{Im}\lambda > 0$ by

$$S(\lambda) = I + i\gamma(L^{-\parallel} - \lambda)^{-1}\gamma. \tag{3.1}$$

In the case of an unbounded γ the characteristic function is first defined by the latter expression on the manifold $E \cap D(\gamma)$ and then extended by continuity to the whole space E.

Formula (3.1) makes it possible to consider $S(\lambda)$ for $\operatorname{Im}\lambda < 0$ with $S(\bar{\lambda}) = (S^*(\lambda))^{-1}$. Finally, $S(\lambda)$ possesses boundary values almost everywhere on the real axis in the strong operator topology: $S(k) \equiv S(k + i0)$, $k \in \mathbb{R}$ (see [20]).

Consider the model space $\mathcal{H} = L_2(\begin{smallmatrix} I & S^* \\ S & I \end{smallmatrix})$, which is defined in [12] as a Hilbert space of two-component vector-functions $(\tilde{g}, g)$ on the axis $(\tilde{g}(k), g(k) \in E, k \in \mathbb{R})$ with metric

$$\left(\begin{pmatrix} \tilde{g} \\ g \end{pmatrix}, \begin{pmatrix} \tilde{g} \\ g \end{pmatrix} \right) = \int_{-\infty}^{\infty} \left(\begin{pmatrix} I & S^*(k) \\ S(k) & I \end{pmatrix} \begin{pmatrix} \tilde{g}(k) \\ g(k) \end{pmatrix}, \begin{pmatrix} \tilde{g}(k) \\ g(k) \end{pmatrix} \right)_{E \oplus E} dk.$$

It is assumed here that the set of two-component functions has been factored by the set of elements with the norm equal to zero and then closed w.r.t. the above metric. Although we consider $(\tilde{g}, g)$ as a symbol only, the formal expressions $g_- := (\tilde{g} + S^*g)$ and $g_+ := (S\tilde{g} + g)$ can be shown to represent some true $L_2(E)$-functions on the real line. In what follows we plan to deal mostly with these functions.

Define the following orthogonal subspaces in $\mathcal{H}$:

$$D_- \equiv \begin{pmatrix} 0 \\ H_2^-(E) \end{pmatrix}, \quad D_+ \equiv \begin{pmatrix} H_2^+(E) \\ 0 \end{pmatrix}, \quad K \equiv \mathcal{H} \ominus (D_- \oplus D_+),$$

where $H_2^{+(-)}(E)$ denotes the Hardy class [7] of analytic functions f in the upper (lower) half-plane taking values in the Hilbert space E.

The subspace K can be described as $K = \{(\tilde{g}, g) \in \mathcal{H} : \tilde{g} + S^*g \in H_2^-(E), S\tilde{g} + g \in H_2^+(E)\}$. Let P_K be the orthogonal projection of $\mathcal{H}$ onto K:

$$P_K \begin{pmatrix} \tilde{g} \\ g \end{pmatrix} = \begin{pmatrix} \tilde{g} - P_+(\tilde{g} + S^*g) \\ g - P_-(S\tilde{g} + g) \end{pmatrix},$$

where $P_\pm$ are orthogonal projections of $L_2(E)$ onto $H_2^\pm(E)$.

The following theorem holds [20, 12]:

Theorem 3.1. *The operator $(L^{\parallel} - \lambda_0)^{-1}$ is unitarily equivalent to the operator $P_K(k - \lambda_0)^{-1}|_K$ for all $\lambda_0, \operatorname{Im}\lambda_0 < 0$.*

This means, that the operator of multiplication by k serves as a self-adjoint dilation [20] of the operator $L^{\parallel}$. Moreover, this dilation also has the property of minimality ($\operatorname{clos}_{\operatorname{Im}\lambda \neq 0}(k - \lambda)^{-1}K = \mathcal{H}$).

Provided that the non-real spectrum of the operator L is countable, the characteristic function of the operator L is defined by the following expression:

$$\Theta(\lambda) \equiv I + iJ\gamma(L^* - \lambda)^{-1}\gamma, \qquad \operatorname{Im}\lambda \neq 0,$$

and under an additional assumption that V is a relatively compact perturbation[2] can be shown to be a meromorphic, J-contractive ($\Theta^*(\lambda)J\Theta(\lambda) \leq J$, $\operatorname{Im}\lambda > 0$) operator-function [6]. The characteristic function $\Theta(\lambda)$ admits [4, 9] a factorization (also called Ginzburg-Potapov factorization of a J-contractive function [5]) in the form of a ratio of two bounded analytic operator-functions (in the corresponding half-planes $\operatorname{Im}\lambda < 0$, $\operatorname{Im}\lambda > 0$) triangular with respect to the decomposition of the space E into the orthogonal sum

$$E = (\mathcal{X}_+ E) \oplus (\mathcal{X}_- E), \quad \mathcal{X}_\pm \equiv \frac{I \pm J}{2} :$$

$$\Theta(\lambda) = \Theta_1'^*(\overline{\lambda})(\Theta_2'^*)^{-1}(\overline{\lambda}), \qquad \operatorname{Im}\lambda > 0$$

$$\Theta(\lambda) = \Theta_2^*(\overline{\lambda})(\Theta_1^*)^{-1}(\overline{\lambda}), \qquad \operatorname{Im}\lambda < 0,$$

where the following designations have been adopted [10]:

$$\Theta_1(\lambda) = \mathcal{X}_- + S(\lambda)\mathcal{X}_+, \qquad \Theta_2(\lambda) = \mathcal{X}_+ + S(\lambda)\mathcal{X}_-;$$

$$\Theta_1'(\lambda) = \mathcal{X}_- + S^*(\overline{\lambda})\mathcal{X}_+, \qquad \Theta_2'(\lambda) = \mathcal{X}_+ + S^*(\overline{\lambda})\mathcal{X}_-,$$

and $S(\lambda)$ is defined by (3.1).

Following [10], we define the subspaces $N_\pm$ in $\mathcal{H}$ as follows:

$$\hat{N}_\pm \equiv \left\{ \begin{pmatrix} \tilde{g} \\ g \end{pmatrix} : \ \begin{pmatrix} \tilde{g} \\ g \end{pmatrix} \in \mathcal{H}, \quad P_\pm \left(\Theta_1'^*\tilde{g} + \Theta_2^* g\right) = P_\pm(\mathcal{X}_+ g_+ + \mathcal{X}_- g_-) = 0 \right\}$$

and introduce the following designation:

$$N_\pm = \operatorname{clos} P_K \hat{N}_\pm.$$

Then, as it is shown in [11], one gets for $\operatorname{Im}\lambda < 0$ ($\operatorname{Im}\lambda > 0$) and $(\tilde{g}, g) \in \hat{N}_{-(+)}$, respectively:

$$(L - \lambda)^{-1} P_K \begin{pmatrix} \tilde{g} \\ g \end{pmatrix} = P_K \frac{1}{k - \lambda} \begin{pmatrix} \tilde{g} \\ g \end{pmatrix}. \tag{3.2}$$

Conversely, the property (3.2) for $\operatorname{Im}\lambda < 0$ ($\operatorname{Im}\lambda > 0$) guarantees that the vector $(\tilde{g}, g)$ belongs to the set $\hat{N}_{-(+)}$.

The absolutely continuous and singular subspaces of the non-self-adjoint operator L were defined in [9]: let $N \equiv \hat{N}_+ \cap \hat{N}_-$, $\tilde{N}_\pm \equiv P_K \hat{N}_\pm$, $\tilde{N}_e \equiv \tilde{N}_+ \cap \tilde{N}_-$, then[3]

$$N_e \equiv \operatorname{clos}\left(\tilde{N}_+ \cap \tilde{N}_-\right) = \operatorname{clos} P_K N \equiv \operatorname{clos} \tilde{N}_e$$

$$N_i \equiv K \ominus N_e(L^*), \tag{3.3}$$

where $N_e(L^*)$ denotes the absolutely continuous subspace of the operator L^*.

---

[2]This assumption guarantees that the non-real spectrum of L is discrete.

[3]The linear set $\tilde{N}_e$ is called the set of "smooth" vectors of the operator L (see [11]).

This definition in the case of maximal dissipative operators leads to the same subspace as the classical definition by L.A. Sahnovich [18] (the latter definition introduces the absolutely continuous subspace as the maximal invariant subspace reducing the operator L to an operator with purely outer characteristic function) and was later developed by V.A. Ryzhov (in the case of more general non-dissipative operators) [16, 17] and A.S. Tikhonov (the so-called weak definition of the absolutely continuous subspace). Recently it turned out that (at least in the dissipative case) the weak definition coincides with the strong one (3.3) (see [15]).

We call an operator L an *operator with absolutely continuous spectrum* if $N_e = H$, i.e., $P_K N$ is dense in K.

3.2. Singular perturbations of non-self-adjoint operators

In this section we assume, that the non-self-adjoint operator $L = A + iV$ acting in Hilbert space H has real spectrum. We further assume (without the loss of generality, see above) that it is completely non-self-adjoint. Under these assumptions we develop an approach to the theory of singular perturbations based on the functional model.

We begin with the following Lemma, which provides an equivalent description of the scale (2.4) in terms of the non-self-adjoint operator L and its resolvent.

Proposition 3.2. *Suppose that $L = A + iV$, V being relatively bounded with respect to A with the relative bound less than 1. Let the spectrum of the operator L be real. Then*

(i) *The space $H_{-2}(|A|)$ is the set of bounded linear functionals φ over $D(A) \equiv D(L)$ such that the element $(L - \lambda)^{-1}\varphi$, $\operatorname{Im}\lambda \neq 0$ is itself a bounded linear functional over H.*

(ii) *The space $H_{-1}(|A|)$ is the set of bounded linear functionals φ over $D(A) \equiv D(L)$ such that the element $(L - \lambda)^{-1}\varphi$, $\operatorname{Im}\lambda \neq 0$ is itself a bounded linear functional over $H_1(|A|)$ (with respect to the $H_1(|A|)$-inner product).*

Proof. (i) Let $\varphi \in H_{-2}(|A|)$. For $u \in H$ consider

$$|\langle u, (A - \lambda)^{-1}\varphi\rangle|^2 = |\langle (A - \overline{\lambda})^{-1}u, \varphi\rangle|$$
$$\leq \|\varphi\|^2_{H_{-2}}\|(A - \overline{\lambda})^{-1}u\|^2_{H_2} \leq C(\lambda)\|\varphi\|^2_{H_{-2}}\|u\|^2,$$

and therefore $(A - \lambda)^{-1}\varphi$ is itself a bounded linear functional over H.

Conversely, let $(A - \lambda)^{-1}\varphi$ be a bounded linear functional over H and suppose, that $\varphi \notin H_{-2}(|A|)$. Then there exists a sequence of vectors $\{u_n\} \in D(A)$ such that $\|u_n\|_{H_2} \equiv 1$ for all n and $|\langle u_n, \varphi\rangle| \uparrow +\infty$ as $n \to \infty$.

Denote $v_n := (A - \overline{\lambda})u_n$, then $\{v_n\}$ is a uniformly bounded sequence in H. Finally,

$$|\langle (A - \overline{\lambda})^{-1}v_n, \varphi\rangle| = |\langle v_n, (A - \lambda)^{-1}\varphi\rangle| \leq C(\lambda, \varphi)\|v_n\| \leq C$$

uniformly with respect to n. On the other hand, the left-hand side tends to infinity by our assumption. Therefore, φ has to be an element of $H_{-2}(|A|)$.

We pass over to the operator L using Hilbert identities

$$(L - \lambda)^{-1} = (A - \lambda)^{-1} + (L - \lambda)^{-1}(-iV)(A - \lambda)^{-1}$$
$$(A - \lambda)^{-1} = (L - \lambda)^{-1} + (A - \lambda)^{-1}(iV)(L - \lambda)^{-1}.$$

Taking into account that the conditions imposed on the operator L lead to both operators $(L - \lambda)^{-1}V$ and $(A - \lambda)^{-1}V$ being bounded in H for all non-real values of λ (see [8]), these identities immediately imply that $(A - \lambda)^{-1}\varphi$ is a bounded functional over H (and therefore can be identified with an element from H) if and only if the functional $(L - \lambda)^{-1}\varphi$ also possesses this property.

(ii) Let $\varphi \in H_{-1}(|A|)$. Then obviously for every $u \in H_1(|A|)$

$$\langle u, \varphi \rangle = \langle u, (|A| + 1)^{-1}\varphi \rangle_{H_1},$$

where $(|A|+1)^{-1}\varphi$ is a bounded functional over $H_1(|A|)$ and can therefore itself be identified with an element from $H_1(|A|)$. It is easy to see now that $(A - \lambda)^{-1}\varphi = (|A| + 1)(A - \lambda)^{-1}(|A| + 1)^{-1}\varphi$ also belongs to Hilbert space $H_1(|A|)$.

Conversely, let $(A - \lambda)^{-1}\varphi \in H_1$ and suppose that $\varphi \notin H_{-1}(|A|)$. Then there exists a sequence of elements $\{u_n\} \in H_1(|A|)$ such that $\|u_n\|_{H_1} \equiv 1$ for all n, but $|\langle u_n, \varphi \rangle| \uparrow +\infty$ as $n \to \infty$.

Denote $v_n := (A - \bar{\lambda})(|A| + 1)^{-1}u_n$. It is easy to see that the sequence $\{v_n\} \in H_1(|A|)$ is itself uniformly bounded with respect to n. Then

$$|\langle u_n, \varphi \rangle| = |\langle v_n, (A - \lambda)^{-1}\varphi \rangle_{H_1}| \le C$$

uniformly with respect to n. On the other hand, the left-hand side tends to infinity by the assumption and therefore in fact $\varphi \in H_{-1}(|A|)$. Passing from $(A - \lambda)^{-1}$ to $(L - \lambda)^{-1}$ on the basis of Hilbert identities as in the proof of (i) above, we complete the proof. $\qquad\square$

Remark 3.3. The result obtained makes it sensible to consider the natural scale of Hilbert spaces, associated with the self-adjoint operator A, as the natural choice of the corresponding scale, associated with the non-self-adjoint operator L, provided that the assumptions of the last Proposition hold for the operator L.

Consider (see [12]) another representation of the dilation of the dissipative operator $L^{\|}$, i.e., the space $\mathfrak{H}$: $\mathfrak{H} = L_2(\mathbb{R}_-; E) \oplus H \oplus L_2(\mathbb{R}_+; E)$. There exists [11] a unitary operator $\mathcal{F}$ that maps $\mathfrak{H}$ onto $\mathcal{H}$. This mapping is defined by the following formulae:

$$\tilde{g} + S^*g = -\frac{1}{\sqrt{2\pi}}\gamma(L^{\|} - k + i0)^{-1}u + S^*(k)\hat{v}_-(k) + \hat{v}_+(k)$$
$$S\tilde{g} + g = -\frac{1}{\sqrt{2\pi}}\gamma(L^{-\|} - k - i0)^{-1}u + \hat{v}_-(k) + S(k)\hat{v}_+(k)$$

$$\tag{3.4}$$

Here $\hat{v}_\pm = (1/\sqrt{2\pi})\int_{\mathbb{R}} \exp(ik\xi)v_\pm(\xi)d\xi$ is the Fourier transform of the function $v_\pm(\xi)$ extended by zero to the complementary semiaxis, where $v_\pm(\xi)$ are two non-central elements of an element of $\mathfrak{H}$, $v_\pm \in L_2(\mathbb{R}_\pm; E)$.

Formulae (3.4) do indeed define a mapping of a vector $(v_-, u, v_+) \in \mathfrak{H}$ onto a vector $(\tilde{g}, g) \in \mathcal{H}$ due to the fact (see Theorem 2 in [11]) that every vector $(\tilde{g}, g) \in \mathcal{H}$ is uniquely determined by the corresponding true $L_2(E)$ functions $g_- = \tilde{g} + S^* g$ and $g_+ = S\tilde{g} + g$. The latter statement is an immediate consequence of the definition of the norm in $\mathcal{H}$, from which it is easy to see that

$$\|(\tilde{g}, g)\|_{\mathcal{H}} \geq \max\{\|g_-\|_{L_2(E)}, \|g_+\|_{L_2(E)}\}. \tag{3.5}$$

In the space $\mathfrak{H}$ the self-adjoint dilation $\mathfrak{L}$ of the operator $L^{\|}$ is defined on the domain

$$D(\mathfrak{L}) = \{(v_-, u, v_+) :$$
$$u \in D(A), v_- \in W_2^1(\mathbb{R}_-; E), v_+ \in W_2^1(\mathbb{R}_+; E), v_+(0) - v_-(0) = i\gamma u\}$$

and acts on it as follows:

$$\mathfrak{L}\begin{pmatrix} v_- \\ u \\ v_+ \end{pmatrix} = \begin{pmatrix} -\frac{1}{i}\frac{d}{d\xi}v_-(\xi) \\ Au + (\gamma/2)[v_+(0) + v_-(0)] \\ -\frac{1}{i}\frac{d}{d\xi}v_+(\xi) \end{pmatrix}$$

(see [11] for details).

We introduce a natural scaling procedure for the Hilbert space $\mathfrak{H}$. Namely, let $\mathfrak{H}_1 := L_2(\mathbb{R}_-; E) \oplus H_1(|A|) \oplus L_2(\mathbb{R}_+; E)$ and $\mathfrak{H}_2 := L_2(\mathbb{R}_-; E) \oplus H_2(|A|) \oplus L_2(\mathbb{R}_+; E)$, respectively; the subspaces $\mathfrak{H}_{-1}$ and $\mathfrak{H}_{-2}$ being dual to $\mathfrak{H}_1$ and $\mathfrak{H}_2$, respectively, in the usual sense.

Alongside with this scaling in $\mathfrak{H}$ consider the one, associated with the self-adjoint operator $\mathfrak{L}$ itself,

$$\mathfrak{H}_2(|\mathfrak{L}|) \subset \mathfrak{H}_1(|\mathfrak{L}|) \subset \mathfrak{H} \subset \mathfrak{H}_{-1}(|\mathfrak{L}|) \subset \mathfrak{H}_{-2}(|\mathfrak{L}|),$$

where $\mathfrak{H}_2(|\mathfrak{L}|) = D(\mathfrak{L})$ is a Hilbert space, equipped with the graph norm of the operator $\mathfrak{L}$; $\mathfrak{H}_1(|\mathfrak{L}|) = D(|\mathfrak{L}|^{1/2})$ and the dual spaces are defined accordingly.

The following lemma clarifies the relative geometry of the two scales introduced in $\mathfrak{H}$.

Lemma 3.4. *Suppose that $L = A + iV$, V being relatively bounded with respect to A with the relative bound less than 1. Let the spectrum of the operator L be real. Then the following identities hold:*

$$\mathfrak{H}_{-1} = L_2(\mathbb{R}_-; E) \oplus H_{-1} \oplus L_2(\mathbb{R}_+; E)$$
$$\mathfrak{H}_{-2} = L_2(\mathbb{R}_-; E) \oplus H_{-2} \oplus L_2(\mathbb{R}_+; E).$$

The unitary transformation $\mathcal{F}$ admits unitary continuation $\hat{\mathcal{F}}$ from $\mathfrak{H}$ to both $\mathfrak{H}_{-1}(|\mathfrak{L}|)$ and $\mathfrak{H}_{-2}(|\mathfrak{L}|)$, and

$$\hat{\mathcal{F}}\mathfrak{H}_{-1}(|\mathfrak{L}|) = L_2((\begin{smallmatrix} I & S^* \\ S & I \end{smallmatrix}); (1 + |k|)^{-1}) =: \mathcal{H}_{-1}(|k|)$$
$$\hat{\mathcal{F}}\mathfrak{H}_{-2}(|\mathfrak{L}|) = L_2((\begin{smallmatrix} I & S^* \\ S & I \end{smallmatrix}); (1 + |k|)^{-2}) =: \mathcal{H}_{-2}(|k|),$$

where $L_2((\begin{smallmatrix} I & S^* \\ S & I \end{smallmatrix}); \frac{1}{1+|k|})$ is a Hilbert space of two-component vector-functions $(\tilde{g}, g)$ on the axis $(\tilde{g}(k), g(k) \in E, k \in \mathbb{R})$ with metric

$$\left(\begin{pmatrix} \tilde{g} \\ g \end{pmatrix}, \begin{pmatrix} \tilde{g} \\ g \end{pmatrix} \right) = \int_{-\infty}^{\infty} \left(\begin{pmatrix} I & S^*(k) \\ S(k) & I \end{pmatrix} \begin{pmatrix} \tilde{g}(k) \\ g(k) \end{pmatrix}, \begin{pmatrix} \tilde{g}(k) \\ g(k) \end{pmatrix} \right)_{E \oplus E} \frac{1}{1+|k|} dk$$

and the space $L_2((\begin{smallmatrix} I & S^* \\ S & I \end{smallmatrix}); \frac{1}{(1+|k|)^2})$ is defined analogously.

Finally, the following inclusions hold:

$$\mathfrak{H}_{-1} \subset \mathfrak{H}_{-1}(|\mathcal{L}|); \quad \hat{\mathcal{F}}\mathfrak{H}_{-1} \subset \mathcal{H}_{-1}(|k|)$$

$$\mathfrak{H}_{-2} \subset \mathfrak{H}_{-2}(|\mathcal{L}|); \quad \hat{\mathcal{F}}\mathfrak{H}_{-2} \subset \mathcal{H}_{-2}(|k|).$$

Proof. The first claim of Lemma follows immediately from the definition of $\mathfrak{H}$.

The second one can be easily obtained based on the fact, that the self-adjoint dilation $\mathcal{L}$ of the dissipative operator $L^{\parallel}$ is unitarily equivalent under the unitary transformation $\mathcal{F}$ to the operator of component-wise multiplication by the independent variable in $\mathcal{H}$. Therefore, $\mathcal{F}\mathfrak{H}_2(|\mathcal{L}|)$ is equal to the domain of the multiplication operator in $\mathcal{H}$ equipped with the graph norm of the latter. It follows, that the operator $(1 + |k|^2)\mathcal{F}(\mathcal{L} - i)^{-1}(\mathcal{L} + i)^{-1}$ is the unitary continuation of $\mathcal{F}$ to the (unitary) operator, intertwining $\mathfrak{H}_{-2}(|\mathcal{L}|)$ and $\mathcal{H}_{-2}(|k|)$ (here we have used the fact that the operator $(\mathcal{L} - i)^{-1}(\mathcal{L} + i)^{-1}$ is a unitary operator from $\mathfrak{H}_{-2}(|\mathcal{L}|)$ to $\mathfrak{H}_2(|\mathcal{L}|)$ and the operator of component-wise multiplication by $(1 + |k|^2)^{-1}$ is a unitary operator from $\mathcal{H}_{-2}(|k|)$ to $\mathcal{H}_2(|k|)$, see [1]). The space $\mathfrak{H}_{-1}(|\mathcal{L}|)$ can be treated analogously.

Finally, the last part of the statement follows immediately from the obvious inclusions $\mathfrak{H}_2(|\mathcal{L}|) \subset \mathfrak{H}_2$ and $\mathfrak{H}_1(|\mathcal{L}|) \subset \mathfrak{H}_1$. $\square$

The last lemma makes it possible to consider natural imbeddings $H_{-1} \subset \mathfrak{H}_{-1} \equiv L_2(\mathbb{R}_-; E) \oplus H_{-1} \oplus L_2(\mathbb{R}_+; E)$ and $H_{-2} \subset \mathfrak{H}_{-2} \equiv L_2(\mathbb{R}_-; E) \oplus H_{-2} \oplus L_2(\mathbb{R}_+; E)$. Moreover, the unitary extension of transformation $\mathcal{F}$ justifies the procedure of "lifting" the vector $\varphi \in H_{-1}(|A|)$ ($\varphi \in H_{-2}(|A|)$) into the space $\hat{\mathcal{F}}\mathfrak{H}_{-1}$ ($\hat{\mathcal{F}}\mathfrak{H}_{-2}$, respectively). The following lemma provides an explicit and transparent description of this procedure.

Lemma 3.5. *Suppose that $L = A + iV$, V being relatively bounded with respect to A with the relative bound less than 1. Let the spectrum of the operator L be real. Then the model image $(\tilde{g}, g) := \hat{\mathcal{F}}\varphi$ of any vector $\varphi \in H_{-2}(|A|)$ ($\varphi \in H_{-1}(|A|)$, respectively) can be obtained as the limit of the transformation*

$$\tilde{g}_n + S^* g_n = -\frac{1}{\sqrt{2\pi}} \gamma (L^{\parallel} - k + i0)^{-1} \varphi_n$$

$$S\tilde{g}_n + g_n = -\frac{1}{\sqrt{2\pi}} \gamma (L^{-\parallel} - k - i0)^{-1} \varphi_n$$

$$(3.6)$$

as $H \ni \varphi_n \to \varphi$ in $H_{-2}(|A|)$ ($H_{-1}(|A|)$, respectively) norm, where the limit as $n \to \infty$ of the first of expressions in (3.6) exists in $H_2^-(E; (1+|k|)^{-2})$ ($H_2^-(E; (1+|k|)^{-1})$, respectively) and of the second – in $H_2^+(E; (1+|k|)^{-2})$ ($H_2^+(E; (1+|k|)^{-1})$, *respectively). Here $H_2^{\pm}(E; (1+|k|)^{-2})$ is a weighted Hardy class, i.e., the class of analytic in the upper (lower) half-plane E-valued functions f such that*

$$\sup_{\varepsilon>0(\varepsilon<0)} \int_{\mathbb{R}} \|f(k+i\varepsilon)\|^2 \frac{1}{(1+|k|)^2} dk < \infty.$$

Proof. Consider the case $\varphi \in H_{-2}(|A|)$ (the case $\varphi \in H_{-1}(|A|)$ is dealt with analogously). As shown in the proof of Lemma 3.4 above, the unitary transformation $\mathcal{F} : \mathfrak{H} \to \mathcal{H}$ is extended by continuity to the unitary transformation $\hat{\mathcal{F}} : \mathfrak{H}_{-2}(|\mathfrak{L}|) \to \mathcal{H}_{-2}(|k|)$ by the formula $\hat{\mathcal{F}} = (1+|k|^2)\mathcal{F}(\mathfrak{L}-i)^{-1}(\mathfrak{L}+i)^{-1}$. Moreover, it is easy to see that $\hat{\mathcal{F}}|_H = \mathcal{F}$. Let $\{\varphi_n\}$ be a sequence of elements from Hilbert space H such that $\varphi_n \to \varphi$ in $H_{-2}(|A|)$. Then on each φ_n the transformation (3.4) is well defined and coincides with the transformation (3.6). From [11] it also follows that for each n $\tilde{g}_n + S^*g_n \in H_2^-(E)$, $S\tilde{g}_n + g_n \in H_2^+(E)$. Furthermore, by Lemma 3.4 $\hat{\mathcal{F}}H_{-2}(|A|) \subset \mathcal{H}_{-2}(|k|)$, hence passing to the limit as $n \to \infty$ we arrive at the fact that there exists a unique vector $(\tilde{g}, g) \in \mathcal{H}_{-2}(|k|)$ such that $(\tilde{g}_n, g_n) \to (\tilde{g}, g)$ in $H_{-2}(|k|)$ as $n \to \infty$. It is easy to see that due to the definition of the norm in $\mathcal{H}_{-2}(|k|)$ the following generalization of (3.5) holds:

$$\|(\tilde{g}, g)\|_{\mathcal{H}_{-2}(|k|)} \geq \max\{\|g_-\|_{L_2(E;(1+|k|)^{-2})}, \|g_+\|_{L_2(E;(1+|k|)^{-2}))}\}, \tag{3.7}$$

and an analogous inequality holds for $\mathcal{H}_{-1}(|k|)$ and $L_2(E; (1+|k|)^{-1}))$, respectively. Then by the same argument as in [11] one immediately obtains on the basis of the definition of the norm in $\mathcal{H}_{-2}(|k|)$ ($\mathcal{H}_{-1}(|k|)$, respectively) that the vector $(\tilde{g}, g)$ of any of these Hilbert spaces is uniquely determined by the corresponding pair of true $L_2(E; (1+|k|)^{-2})$ ($L_2(E; (1+|k|)^{-1})$, respectively) functions $g_- \equiv \tilde{g} + S^*g$ and $g_+ \equiv S\tilde{g} + g$.

By (3.7), $\tilde{g}_n + S^*g_n \to g_-$ and $S\tilde{g}_n + g_n \to g_+$ in $L_2(E; (1+|k|)^{-2})$, and therefore also in $H_2^-(E; (1+|k|)^{-2})$ and $H_2^+(E; (1+|k|)^{-2})$, respectively. This completes the proof. $\qquad\square$

Remark 3.6. The statement of Lemma 3.5 clearly makes it possible to identify the model image $(\tilde{g}, g) = \hat{\mathcal{F}}\varphi$ of every element φ from $H_{-2}(|A|)$ ($H_{-1}(|A|)$, respectively) with a certain element of Hilbert space $H_2^-(E; (1+|k|)^{-2}) \oplus H_2^+(E; (1+|k|)^{-2})$ ($H_2^-(E; (1+|k|)^{-1}) \oplus H_2^+(E; (1+|k|)^{-1})$, respectively). This identification is via the corresponding functions $g_- \equiv \tilde{g} + S^*g$ and $g_+ \equiv S\tilde{g} + g$ belonging to the named weighted Hardy spaces, which uniquely determine the vector $\hat{\mathcal{F}}\varphi$.

Let $\mathcal{L}$ be the self-adjoint operator of component-wise multiplication by the independent variable in $\mathcal{H}$:

$$(\mathcal{L}\mathfrak{g})(k) = k\mathfrak{g}(k), \quad \mathfrak{g} \equiv \begin{pmatrix} \tilde{g} \\ g \end{pmatrix} \in D(\mathcal{L}),$$

where $D(\mathcal{L}) = \{\mathfrak{g} : k\mathfrak{g}(k) \in \mathcal{H}\}$ and consider the class of rank one perturbations for the operator $\mathcal{L}$, $\mathcal{L}_\alpha = \mathcal{L} + \alpha \langle \cdot, \varphi \rangle \varphi$, where φ is an arbitrary bounded linear functional on the domain of the operator $\mathcal{L}$. The natural scale of Hilbert spaces, associated with the operator $\mathcal{L}$, is $\mathcal{H}_2(|k|) \subset \mathcal{H}_1(|k|) \subset \mathcal{H} \subset \mathcal{H}_{-1}(|k|) \subset \mathcal{H}_{-2}(|k|)$, where by Lemmas 3.4 and 3.5 the subspaces $H_{-1}(|A|)$ and $H_{-2}(|A|)$ are naturally imbedded into $\mathcal{H}_{-1}(|k|)$ and $\mathcal{H}_{-2}(|k|)$, respectively.

Fix an element $\varphi \in \hat{\mathcal{F}} H_{-1} \backslash \mathcal{H}$ ($\varphi \in \hat{\mathcal{F}} H_{-2} \backslash \hat{\mathcal{F}} H_{-1}$). Then the Krein's formula for the action of the resolvent of the perturbed operator $\mathcal{L}_\alpha$ on any vector $\mathfrak{g} \in \mathcal{H}$ holds:

$$(\mathcal{L}_\alpha - z)^{-1} \mathfrak{g} = (\mathcal{L} - z)^{-1} \mathfrak{g} - \frac{\alpha}{1 + \alpha \langle (\mathcal{L} - z)^{-1} \varphi, \varphi_c \rangle} \langle (\mathcal{L} - z)^{-1} \mathfrak{g}, \varphi \rangle (\mathcal{L} - z)^{-1} \varphi,$$

$$(3.8)$$

where in the case of $\varphi \in \mathcal{F} H_{-1} \setminus \mathcal{H}$ one has $\varphi_c = \varphi$, whereas in the case $\varphi \in \mathcal{F} H_{-2} \setminus \mathcal{F} H_{-1}$ the identity

$$\langle (\mathcal{L} - z)^{-1} \varphi, \varphi_c \rangle = c + \langle (1 + z\mathcal{L})(\mathcal{L} - \lambda)^{-1}(\mathcal{L} + i)^{-1}(\mathcal{L} - i)^{-1} \varphi, \varphi \rangle$$

holds, see [1, 2].

Lemma 3.4 allows to extend the definition of the orthogonal projection $P_H :$ $\mathfrak{H} \to H$ to the space $\mathfrak{H}_{-2}(|\mathcal{L}|)$ by the same formula $P_H(v_-, u, v_+) = (0, u, 0)$. Then the orthogonal projection P_K admits the corresponding extension to the operator acting from $\hat{\mathcal{F}} \mathfrak{H}_{-2}(|\mathcal{L}|) = \mathcal{H}_{-2}(|k|)$ to $\hat{\mathcal{F}} H_{-2}(|A|)$. We preserve the notation P_K for this extension and expect that this will not lead to any confusion.

Consider the operator-valued analytic function $P_K(\mathcal{L}_\alpha - z)^{-1}|_K$. The following Lemma establishes its connection to the resolvent of the non-self-adjoint operator L in the case of absolute continuity of its spectrum.

Lemma 3.7. *Suppose that $L = A + iV$, V being relatively bounded with respect to A with the relative bound less than 1. Assume also that the operator L is an operator with absolutely continuous spectrum, i.e., $N_e(L) = H$. Let $\varphi \in H_{-2}(|A|) \backslash H_{-1}(|A|)$ ($\varphi \in H_{-1}(|A|) \setminus H$) and $\boldsymbol{\varphi} = \hat{\mathcal{F}} \varphi$. Then for every $u \in \tilde{N}_e$*

$$P_K(\mathcal{L}_\alpha - \lambda)^{-1} u = (L - \lambda)^{-1} u - \frac{\alpha}{D(\lambda)} \langle u, (L^* - \bar{\lambda})^{-1} \varphi \rangle (L - \lambda)^{-1} \varphi, \qquad (3.9)$$

where $D(\lambda) = 1 + \alpha \langle (L - \lambda)^{-1} \varphi, \varphi_c \rangle$ is a scalar analytic function in $\mathbb{C}_\pm$;

$$\langle (L - \lambda)^{-1} \varphi, \varphi_c \rangle = \langle (L - \lambda)^{-1} \varphi, \varphi \rangle$$

in the case $\varphi \in H_{-1}(|A|) \setminus H$ and

$$\langle (L - \lambda)^{-1} \varphi, \varphi_c \rangle = c + \langle (1 + \lambda L)(L - \lambda)^{-1}(L + i)^{-1}(L - i)^{-1} \varphi, \varphi \rangle$$

in the case $\varphi \in H_{-2}(|A|) \setminus H_{-1}(|A|)$.

Proof. Let $\{\varphi_n\} \in H$ be a sequence of elements such that $\varphi_n \to \varphi$ in $H_{-2}(|A|)$ as $n \to \infty$. On each φ_n the identity (3.9) clearly holds. Passing to the limit as $n \to \infty$ and using Proposition 3.2 and Lemmas 3.4, 3.5 one obtains (3.9) for an arbitrary $\varphi \in H_{-2}(|A|)$ ($\varphi \in H_{-1}(|A|)$). $\qquad \square$

It is therefore reasonable to expect, that the analytic operator-valued function $\Phi(\lambda)$, defined for all λ such that $\operatorname{Im} \lambda \neq 0$ by the following expression:

$$\Phi(\lambda) = (L - \lambda)^{-1} - \frac{\alpha}{D(\lambda)} \langle \cdot, (L^* - \bar{\lambda})^{-1} \varphi \rangle (L - \lambda)^{-1} \varphi, \qquad (3.10)$$

is the resolvent of some non-self-adjoint linear operator in H. Indeed, the following Theorem holds.

Theorem 3.8. *For every $\varphi \in H_{-1}(|A|) \setminus H$ ($\varphi \in H_{-2}(|A|) \setminus H_{-1}(|A|)$, respectively) the expression (3.10) is the resolvent of a densely defined non-self-adjoint operator L_α. Moreover, if the perturbation V is a bounded operator in H and α is a real number, the corresponding operator L_α coincides with the non-self-adjoint operator $A_\alpha + iV$, where $A_\alpha = A + \alpha \langle \cdot, \varphi_c \rangle \varphi$.*

Proof. Without any loss of generality, let $\alpha = 1$ (if this is not the case, scale the functional φ accordingly). Let further $\varphi \in H_{-1}(|A|) \setminus H$ (the case of form-unbounded perturbation φ can be dealt with in an analogous fashion). Let $u \in D(A_\alpha)$ and consider the expression (denoting $A_\varphi \equiv A_\alpha$) $\Phi(\lambda)(A_\varphi + iV - \lambda)u$. It is clear that since for every $u_0 \in D(A_0)$ we have $\langle u_0, \varphi \rangle = 0$, $\Phi(\lambda)(A_\varphi + iV - \lambda)u_0 = u_0$. Let $u_\varphi := (A + i)^{-1}\varphi + \beta(A - i)^{-1}\varphi$, where $\beta = -\frac{1 + \langle (A+i)^{-1}\varphi, \varphi \rangle}{1 + \langle (A-i)^{-1}\varphi, \varphi \rangle}$. From (2.2) and (2.5) it follows then, that $D(A_\varphi) = D(A_0) \dot{+} \{u_\varphi\}$.

We are going to prove that $\Phi(\lambda)(A_\varphi + iV - \lambda)u_\varphi = u_\varphi$. Indeed, consider

$$(L - \lambda)^{-1}iVu_\varphi = (L - \lambda)^{-1}iV(A + i)^{-1}\varphi + \beta(L - \lambda)^{-1}iV(A - i)^{-1}\varphi$$

$$= (i + \lambda)(L - \lambda)^{-1}(A + i)^{-1}\varphi - (L - \lambda)^{-1}\varphi$$

$$+ \beta(-i + \lambda)(L - \lambda)^{-1}(A - i)^{-1}\varphi - \beta(L - \lambda)^{-1}\varphi + u_\varphi.$$

Then

$$(L - \lambda)^{-1}(A_\varphi - \lambda + iV)\varphi$$

$$= (L - \lambda)^{-1}[(-i - \lambda)(A + i)^{-1}\varphi + \beta(i - \lambda)(A - i)^{-1}\varphi] + (L - \lambda)^{-1}iVu_\varphi$$

$$= u_\varphi - (1 + \beta)(L - \lambda)^{-1}\varphi. \tag{3.11}$$

Therefore,

$$\Phi(\lambda)(A_\varphi + iV - \lambda)u_\varphi = u_\varphi - (1 + \beta)(L - \lambda)^{-1}\varphi$$

$$- \frac{1}{1 + \langle (L - \lambda)^{-1}\varphi, \varphi \rangle} \langle u_\varphi - (1 + \beta)(L - \lambda)^{-1}\varphi, \varphi \rangle (L - \lambda)^{-1}\varphi.$$

On the other hand,

$$-(1 + \beta) - \frac{1}{1 + \langle (L - \lambda)^{-1}\varphi, \varphi \rangle} \langle u_\varphi - (1 + \beta)(L - \lambda)^{-1}\varphi, \varphi \rangle (L - \lambda)^{-1}\varphi$$

$$= \frac{-(1 + \beta) - \langle u_\varphi, \varphi \rangle}{1 + \langle (L - \lambda)^{-1}\varphi, \varphi \rangle} = 0,$$

since

$$\langle u_\varphi, \varphi \rangle = \langle (A + i)^{-1}\varphi, \varphi \rangle + \beta \langle (A - i)^{-1}\varphi, \varphi \rangle = -\beta - 1.$$

Therefore, we have established the fact that for all $u \in D(A_\varphi) \equiv D(A_\varphi + iV)$

$$\Phi(\lambda)(A_\varphi + iV - \lambda)u = u. \tag{3.12}$$

Since for λ with sufficiently large imaginary part the range of the operator $A_\varphi + iV - \lambda$ coincides with Hilbert space H, for these values of λ at least the operator $\Phi(\lambda)$ is invertible and therefore

$$(A_\varphi + iV - \lambda)u = \Phi(\lambda)^{-1}u, \quad u \in D(A_\varphi).$$

Moreover, the domain of the operator $\Phi(\lambda)^{-1}$ coincides with $D(A_\varphi)$. Indeed, the inclusion $D(A_\varphi) \subset Ran(\Phi(\lambda))$ follows from (3.12). On the other hand, from (3.11) it follows, that for all $u \in H$

$$\Phi(\lambda)u = (L - \lambda)^{-1}u$$
$$- \frac{1}{D(\lambda)(1 + \beta)} \left\langle (L - \lambda)^{-1}u, \varphi \right\rangle [u_\varphi - (L - \lambda)^{-1}(A_\varphi + iV - \lambda)u_\varphi]$$
$$\equiv u_0 + c(\lambda)u_\varphi,$$

where

$$u_0 = (L - \lambda)^{-1}u + \frac{1}{D(\lambda)(1 + \beta)} \left\langle (L - \lambda)^{-1}u, \varphi \right\rangle [(L - \lambda)^{-1}(A_\varphi + iV - \lambda)u_\varphi]$$

is a vector from the domain of the operator A. Moreover, it is easy to see that $\langle u_0, \varphi \rangle = 0$ and thus $u_0 \in D_\varphi$. This in turn implies (see (2.2)) that $\Phi(\lambda)u \in D(A_\varphi)$ for all $u \in H$.

Therefore, we have established the fact that the operator $\Phi(\lambda)^{-1}$ coincides with the operator $A_\varphi + iV - \lambda$.

In the case of arbitrary (possibly, unbounded) perturbation V and non-real values of α it is still true that the operator-function $\Phi(\lambda)$ is analytic; moreover, its kernel for all values of λ is trivial and the range is dense in the Hilbert space H. Indeed, let $u \in D(L)$ be such that $\langle u, \varphi \rangle = 0$. Then

$$\Phi(\lambda)(L - \lambda)u = u$$

by (3.10). On the other hand, such elements u are clearly dense in H. Finally, the function $\Phi(\lambda)$ clearly satisfies the Hilbert identity for the resolvents.

It follows then (see [8]), that $\Phi(\lambda)$ is the resolvent of a densely defined (in general, non-self-adjoint) linear operator L_α in Hilbert space H, which completes the proof of the Theorem. $\qquad \square$

Remark 3.9. Note, that in the last Theorem the assumption of V being a bounded operator in H cannot be dropped. Indeed, the classes of relatively bounded perturbations for the operators A and A_α do not coincide due to the fact, that the domains of these operators are different. Therefore, the classes of admissible perturbations V are necessarily different as well.

We are now able to formulate the following result, which allows us to derive a functional model representation of the singular perturbations of non-self-adjoint operators with absolutely continuous spectrum.

Theorem 3.10. *Let the spectrum of the operator $L = A + iV$ be absolutely continuous. Let the vector φ belong to the space $H_{-1}(|A|)$ ($H_{-2}(|A|)$, respectively). Then in the space $\mathcal{H}$ of the dilation associated with the dissipative operator $L^{\|}$, the resolvent $(L_\alpha - \lambda)^{-1}$ of the the singular perturbation of the operator L, formally given by the expression $L_\alpha = L + \alpha \langle \cdot, \varphi \rangle \varphi$, on all $u \in \tilde{N}_e$ (the set of smooth vectors of the operator L) acts as follows:*

$$(L_\alpha - \lambda)^{-1}u \tag{3.13}$$

$$= P_K \left[(\mathcal{L} - z)^{-1}u - \frac{\alpha}{1 + \alpha \langle (\mathcal{L} - z)^{-1}\boldsymbol{\varphi}, \boldsymbol{\varphi}_c \rangle} \langle u, (\mathcal{L} - \bar{z})^{-1}\boldsymbol{\varphi} \rangle (\mathcal{L} - z)^{-1}\boldsymbol{\varphi} \right],$$

where $\mathcal{L}$ is the operator of component-wise multiplication by the independent variable in Hilbert space $\mathcal{H}$, $\boldsymbol{\varphi}$ is the model image of the vector φ and $\boldsymbol{\varphi}_c$ is the corresponding one-dimensional extension of the latter (in the case of $\varphi \in H_{-1}$ we have $\boldsymbol{\varphi}_c \equiv \boldsymbol{\varphi}$).

The proof of this theorem follows immediately from a combination of Theorem 3.8 and Lemmas 3.5, 3.7.

Remark 3.11. The approach developed by us in the present paper allows for consideration of more general singular perturbations of non-self-adjoint operators with absolutely continuous spectrum, i.e., of the perturbations corresponding to the formal expression

$$A_\alpha = A + \alpha \langle \cdot, \varphi \rangle \psi.$$

This analysis will be carried out in a forthcoming publication on this subject.

Acknowledgements

The author is grateful to Dr. D.J. Gilbert for fruitful discussions and to Prof. S.N. Naboko and Dr. M.M. Faddeev for their constant attention to his work. The author would also like to express his deep gratitude to Dr. P.B. Kurasov for introducing him to the theory of singular perturbations of self-adjoint operators and exciting discussions that followed. Last but not the least, the author is grateful to the referees for a number of extremely helpful remarks that hopefully have helped to greatly improve the quality of the paper.

References

[1] S. Albeverio, P. Kurasov, *Singular perturbations of differential operators*, Cambridge Univ. Press, 2000.

[2] S. Albeverio, P. Kurasov, *Rank one perturbations of not semibounded operators*, Integr. equ. oper. theory **27** (1997), 379–400.

[3] S. Albeverio, P. Kurasov, *Rank one perturbations, approximations, and selfadjoint extensions*, J. Func. Anal. **148** (1997), 152–169.

[4] D.Z. Arov, *Scattering theory with dissipation of energy*, English transl. in Dokl. Akad. Nauk SSSR **216** (1974), 713–716.

[5] T. Azizov, I. Iokhvidov, *Linear operators in spaces with an indefinite metric*, Wiley-Interscience Publ., John Wiley & Sons Ltd., Chichester, 1989.

[6] M.S. Brodskij, *Triangular and Jordan representations of linear operators*, English transl. in Amer. Math. Soc., Providence, R.I., 1971.

[7] K. Hoffman, *Banach spaces of analytic functions*, Prentice-Hall, Englewood Cliffs, N.J., 1962.

[8] T. Kato, *Perturbation theory for linear operators*, Springer-Verlag, 1966.

[9] S.N. Naboko, *Absolutely continuous spectrum of a nondissipative operator and functional model. I.*, Zapiski Nauchnykh Seminarov LOMI **65** (1976), 90–102 (Russian); English translation in J. Sov. Math.

[10] S.N. Naboko, *Absolutely continuous spectrum of a nondissipative operator and functional model. II.*, Zapiski Nauchnykh Seminarov LOMI, **73** (1977), 118–135 (Russian); English translation in J. Sov. Math.

[11] S.N. Naboko, *Functional model of perturbation theory and its applications to scattering theory*, Proc. Steklov Inst. Math. **147** (1981), 85–116.

[12] B.S. Pavlov, *Conditions for separation of the spectral components of a dissipative operator*, Math. USSR-Izv. **39** (1975), 123–148.

[13] B.S. Pavlov, *Dilation theory and spectral analysis of nonselfadjoint differential operators*, AMS Transl. **115** (1980), no. 2.

[14] A.M. Petrov, *Spectral projectors of dissipative operators*, J. Soviet Math., **57** (1991), 3440–3443.

[15] R.V. Romanov, *A remark on equivalence of weak and strong definitions of absolutely continuous subspace for non-self-adjoint operators*, Operator Theory: Advances and Applications (Proceedings of the OTAMP'02 Conference), **154** (2004), 179–184.

[16] V.A. Ryzhov, *Absolutely continuous subspace of a non-self-adjoint operator and the scattering theory*, PhD Thesis, St. Petersburg, 1994.

[17] V.A. Ryzhov, *Absolutely continuous and singular subspaces of a nonselfadjoint operator*, J. Math. Sci., **87** (1997), 3886–3911.

[18] L.A. Sahnovič, *Nonunitary operators with absolutely continuous spectrum*, Izv. Akad. Nauk SSSR Ser. Mat. **33** (1969), 52–64 (Russian); English translation in Math. USSR-Izv.

[19] B.S. Solomyak, *A functional model for dissipative operators. A coordinate-free approach*, J. Soviet Math., **61** (1992), 1981–2002.

[20] B. Sz.-Nagy and C. Foiaş, *Analyse harmonique des opérateurs de l'espace de Hilbert*, Masson, Paris and Akad. Kiadó, Budapest, 1967.

[21] V.F. Veselov, *Spectral decompositions of non-self-adjoint operators with singular spectrum*, PhD Thesis, Leningrad, 1986.

Alexander V. Kiselev
School of Mathematical Sciences
DIT Kevin Street
Dublin 8, Ireland
e-mail: `akiselev@mail.ru`

Operator Theory:
Advances and Applications, Vol. 174, 69–76

Trace Formulas for Jacobi Operators in Connection with Scattering Theory for Quasi-Periodic Background

Johanna Michor and Gerald Teschl

Abstract. We investigate trace formulas for Jacobi operators which are trace class perturbations of quasi-periodic finite-gap operators using Krein's spectral shift theory. In particular we establish the conserved quantities for the solutions of the Toda hierarchy in this class.

Mathematics Subject Classification (2000). Primary 47B36, 37K15; Secondary 81U40, 39A11.

Keywords. Scattering, Toda hierarchy, Trace formulas.

1. Introduction

Scattering theory for Jacobi operators H with periodic (respectively more general) background has attracted considerable interest recently. In [14] Volberg and Yuditskii have exhaustively treated the case where H has a homogeneous spectrum and is of Szegö class. In [2] Egorova and the authors have established direct and inverse scattering theory for Jacobi operators which are short range perturbations of quasi-periodic finite-gap operators. For further information and references we refer to these articles and [12].

In the case of constant background it is well known that the transmission coefficient is the perturbation determinant in the sense of Krein [8], see, e.g., [11] or [12]. The purpose of the present paper is to establish this result for the case of quasi-periodic finite-gap background, thereby establishing the connection with Krein's spectral shift theory. For related results see also [7], [10].

Moreover, scattering theory for Jacobi operators is not only interesting in its own right, it also constitutes the main ingredient of the inverse scattering transform for the Toda hierarchy (see, e.g., [5], [4], [12], or [13]). Since the transmission

---

Work supported by the Austrian Science Fund (FWF) under Grant No. P17762.

coefficient is invariant when our Jacobi operator evolves in time with respect to some equation of the Toda hierarchy, the corresponding trace formulas provide the conserved quantities for the Toda hierarchy in this setting.

2. Notation

We assume that the reader is familiar with quasi-periodic Jacobi operators. Hence we only briefly recall some notation and refer to [2] and [12] for further information.

Let

$$H_q f(n) = a_q(n)f(n+1) + a_q(n-1)f(n-1) + b_q(n)f(n) \qquad (2.1)$$

be a quasi-periodic Jacobi operator in $\ell^2(\mathbb{Z})$ associated with the Riemann surface of the function

$$R_{2g+2}^{1/2}(z), \qquad R_{2g+2}(z) = \prod_{j=0}^{2g+1}(z - E_j), \qquad E_0 < E_1 < \cdots < E_{2g+1}, \qquad (2.2)$$

$g \in \mathbb{N}$. The spectrum of H_q is purely absolutely continuous and consists of $g+1$ bands

$$\sigma(H_q) = \bigcup_{j=0}^{g}[E_{2j}, E_{2j+1}]. \qquad (2.3)$$

For every $z \in \mathbb{C}$ the Baker-Akhiezer functions $\psi_{q,\pm}(z,n)$ are two (weak) solutions of $H_q\psi = z\psi$, which are linearly independent away from the band-edges $\{E_j\}_{j=0}^{2g+1}$, since their Wronskian is given by

$$W_q(\psi_{q,-}(z), \psi_{q,+}(z)) = \frac{R_{2g+2}^{1/2}(z)}{\prod_{j=1}^{g}(z - \mu_j)}. \qquad (2.4)$$

Here μ_j are the Dirichlet eigenvalues at base point $n_0 = 0$. We recall that $\psi_{q,\pm}(z,n)$ have the form

$$\psi_{q,\pm}(z,n) = \theta_{q,\pm}(z,n)w(z)^{\pm n},$$

where $\theta_{q,\pm}(z,n)$ is quasi-periodic with respect to n and $w(z)$ is the quasi-momentum. In particular, $|w(z)| < 1$ for $z \in \mathbb{C}\backslash\sigma(H_q)$ and $|w(z)| = 1$ for $z \in \sigma(H_q)$.

3. Asymptotics of Jost solutions

After we have these preparations out of our way, we come to the study of short-range perturbations H of H_q associated with sequences a, b satisfying $a(n) \to a_q(n)$ and $b(n) \to b_q(n)$ as $|n| \to \infty$. More precisely, we will make the following assumption throughout this paper:

Let H be a perturbation of H_q such that

$$\sum_{n\in\mathbb{Z}}\Big(|a(n) - a_q(n)| + |b(n) - b_q(n)|\Big) < \infty, \qquad (3.1)$$

that is, $H - H_q$ is trace class.

We first establish existence of Jost solutions, that is, solutions of the perturbed operator which asymptotically look like the Baker-Akhiezer solutions.

Theorem 3.1. *Assume* (3.1). *For every* $z \in \mathbb{C}\backslash\{E_j\}_{j=0}^{2g+1}$ *there exist (weak) solutions* $\psi_\pm(z,.)$ *of* $H\psi = z\psi$ *satisfying*

$$\lim_{n\to\pm\infty} w(z)^{\mp n}\left(\psi_\pm(z,n) - \psi_{q,\pm}(z,n)\right) = 0, \tag{3.2}$$

where $\psi_{q,\pm}(z,.)$ *are the Baker-Akhiezer functions. Moreover,* $\psi_\pm(z,.)$ *are continuous (resp. holomorphic) with respect to* z *whenever* $\psi_{q,\pm}(z,.)$ *are, and have the following asymptotic behavior*

$$\psi_\pm(z,n) = \frac{z^{\mp n}}{A_\pm(n)}\Big(\prod_{j=0}^{n-1}{}^* a_q(j)\Big)^{\pm 1}\Big(1 + \Big(B_\pm(n) \pm \sum_{j=1}^{n}{}^* b_q(j - {}_{1}^{0})\Big)\frac{1}{z} + O(\frac{1}{z^2})\Big), \tag{3.3}$$

where

$$A_+(n) = \prod_{j=n}^{\infty} \frac{a(j)}{a_q(j)}, \qquad B_+(n) = \sum_{m=n+1}^{\infty} (b_q(m) - b(m)),$$

$$A_-(n) = \prod_{j=-\infty}^{n-1} \frac{a(j)}{a_q(j)}, \qquad B_-(n) = \sum_{m=-\infty}^{n-1} (b_q(m) - b(m)). \tag{3.4}$$

Note that since $a_q(n)$ *are bounded away from zero,* $A_\pm(n)$ *are well defined. Here the star indicates that* $\displaystyle\sum_{j=1}^{n}{}^* = -\sum_{j=n+1}^{0}$ *for* $n < 0$ *and similarly for the product.*

Proof. The proof can be done as in the periodic case (see, e.g., [2], [6], [9] or [12], Section 7.5). There a stronger decay assumption (i.e., first moments summable) is made, which is however only needed at the band edges $\{E_j\}_{j=0}^{2g+1}$. $\qquad\square$

For later use we note the following immediate consequence

Corollary 3.2. *Under the assumptions of the previous theorem we have*

$$\lim_{n\to\pm\infty} w(z)^{\mp n}\left(\psi'_\pm(z,n) \mp n\frac{w'(z)}{w(z)}\psi_\pm(z,n) - \psi'_{q,\pm}(z,n) \pm n\frac{w'(z)}{w(z)}\psi_{q,\pm}(z,n)\right) = 0,$$

where the prime denotes differentiation with respect to z.

Proof. Just differentiate (3.2) with respect to z, which is permissible by uniform convergence on compact subsets of $\mathbb{C}\backslash\{E_j\}_{j=0}^{2g+1}$. $\qquad\square$

We remark that if we require our perturbation to satisfy the usual short range assumption as in [2] (i.e., the first moments are summable), then we even have $w(z)^{\mp n}(\psi'_\pm(z,n) - \psi'_{q,\pm}(z,n)) \to 0$.

From Theorem 3.1 we obtain a complete characterization of the spectrum of H.

Theorem 3.3. *Assume (3.1). Then we have $\sigma_{ess}(H) = \sigma(H_q)$, the point spectrum of H is confined to $\overline{\mathbb{R}\backslash\sigma(H_q)}$. Furthermore, the essential spectrum of H is purely absolutely continuous except for possible eigenvalues at the band edges.*

Proof. This is an immediate consequence of the fact that $H - H_q$ is trace class and boundedness of the Jost solutions inside the essential spectrum. $\square$

Our next result concerns the asymptotics of the Jost solutions at the *other side*.

Lemma 3.4. *Assume (3.1). Then the Jost solutions $\psi_\pm(z,.)$, $z \in \mathbb{C}\backslash\sigma(H)$, satisfy*

$$\lim_{n\to\mp\infty} |w(z)^{\mp n}(\psi_\pm(z,n) - \alpha(z)\psi_{q,\pm}(z,n))| = 0, \tag{3.5}$$

where

$$\alpha(z) = \frac{W(\psi_-(z),\psi_+(z))}{W_q(\psi_{q,-}(z),\psi_{q,+}(z))} = \frac{\prod_{j=1}^g (z-\mu_j)}{R_{2g+2}^{1/2}(z)} W(\psi_-(z),\psi_+(z)). \tag{3.6}$$

Proof. Since $H - H_q$ is trace class, we have for the difference of the Green's functions

$$\lim_{n\to\pm\infty} G(z,n,n) - G_q(z,n,n) = \lim_{n\to\pm\infty} \langle \delta_n, ((H-z)^{-1} - (H_q-z)^{-1})\delta_n \rangle = 0$$

and using

$$G_q(z,n,n) = \frac{\psi_{q,-}(z,n)\psi_{q,+}(z,n)}{W_q(\psi_{q,-}(z),\psi_{q,+}(z))}, \qquad G(z,n,n) = \frac{\psi_-(z,n)\psi_+(z,n)}{W(\psi_-(z),\psi_+(z))}$$

we obtain

$$\lim_{n\to-\infty} \psi_{q,-}(z,n)(\psi_+(z,n) - \alpha(z)\psi_{q,+}(z,n)) = 0,$$

which is the claimed result. $\square$

Note that $\alpha(z)$ is just the inverse of the transmission coefficient (see, e.g., [2] or [12], Section 7.5). It is holomorphic in $\mathbb{C}\backslash\sigma(H_q)$ with simple zeros at the discrete eigenvalues of H and has the following asymptotic behavior

$$\alpha(z) = \frac{1}{A}\left(1 + \frac{B}{z} + O(z^{-2})\right), \quad A = A_-(0)A_+(0), \quad B = B_-(1) + B_+(0), \tag{3.7}$$

with $A_\pm(n)$, $B_\pm(n)$ from (3.4).

4. Connections with Krein's spectral shift theory and trace formulas

To establish the connection with Krein's spectral shift theory we next show:

Lemma 4.1. *We have*

$$\frac{d}{dz}\alpha(z) = -\alpha(z)\sum_{n\in\mathbb{Z}}\left(G(z,n,n) - G_q(z,n,n)\right), \qquad z \in \mathbb{C}\backslash\sigma(H), \tag{4.1}$$

where $G(z,m,n)$ and $G_q(z,m,n)$ are the Green's functions of H and H_q, respectively.

Proof. Green's formula ([12], eq. (2.29)) implies

$$W_n(\psi_+(z), \psi'_-(z)) - W_{m-1}(\psi_+(z), \psi'_-(z)) = \sum_{j=m}^{n} \psi_+(z,j)\psi_-(z,j), \qquad (4.2)$$

hence the derivative of the Wronskian can be written as

$$\frac{d}{dz}W(\psi_-(z), \psi_+(z)) = W_n(\psi'_-(z), \psi_+(z)) + W_n(\psi_-(z), \psi'_+(z))$$

$$= W_m(\psi'_-(z), \psi_+(z)) + W_n(\psi_-(z), \psi'_+(z)) - \sum_{j=m+1}^{n} \psi_+(z,j)\psi_-(z,j).$$

Using Corollary 3.2 and Lemma 3.4 we have

$$
\begin{aligned}
W_m(\psi'_-(z), \psi_+(z)) \;=\;& W_m(\psi'_- + m\frac{w'}{w}\psi_-, \psi_+) \\
& -\frac{w'}{w}\big(m\,W(\psi_-, \psi_+) - a(m)\psi_-(m+1)\psi_+(m)\big) \\
\to\;& \alpha W_{q,m}(\psi'_{q,-} + m\frac{w'}{w}\psi_{q,-}, \psi_{q,+}) \\
& -\alpha\frac{w'}{w}\big(m\,W_q(\psi_{q,-}, \psi_{q,+}) - a_q(m)\psi_{q,-}(m+1)\psi_{q,+}(m)\big) \\
=\;& \alpha(z)W_m(\psi'_{q,-}(z), \psi_{q,+}(z))
\end{aligned}
$$

as $m \to -\infty$. Similarly we obtain

$$W_n(\psi_-(z), \psi'_+(z)) \to \alpha(z)W_n(\psi_{q,-}(z), \psi'_{q,+}(z))$$

as $n \to \infty$ and again using (4.2) we have

$$W_m(\psi'_{q,-}(z), \psi_{q,+}(z)) = W_n(\psi'_{q,-}(z), \psi_{q,+}(z)) + \sum_{j=m+1}^{n} \psi_{q,+}(z,j)\psi_{q,-}(z,j).$$

Collecting terms we arrive at

$$W'(\psi_-(z), \psi_+(z)) = -\sum_{j\in\mathbb{Z}} \Big(\psi_+(z,j)\psi_-(z,j) - \alpha(z)\psi_{q,+}(z,j)\psi_{q,-}(z,j)\Big)$$

$$+ \alpha(z)W'_q(\psi_{q,-}(z)\psi_{q,+}(z)).$$

Now we compute

$$
\begin{aligned}
\frac{d}{dz}\alpha(z) &= \frac{d}{dz}\Big(\frac{W}{W_q}\Big) = \Big(\frac{1}{W_q}\Big)'W + \frac{1}{W_q}W' \\
&= -\frac{W'_q}{W_q^2}W + \frac{1}{W_q}\Big(-\sum_{j\in\mathbb{Z}}\big(\psi_+\psi_- - \alpha\psi_{q,+}\psi_{q,-}\big) + \alpha W'_q\Big) \\
&= -\frac{1}{W_q}\sum_{j\in\mathbb{Z}}\big(\psi_+\psi_- - \alpha\psi_{q,+}\psi_{q,-}\big),
\end{aligned}
$$

which finishes the proof. $\qquad\square$

As an immediate consequence, we can identify $\alpha(z)$ as Krein's perturbation determinant ([8]) of the pair H, H_q.

Theorem 4.2. *The function $A\alpha(z)$ is Krein's perturbation determinant:*

$$\alpha(z) = \frac{1}{A} \det \left(\mathbb{1} + (H(t) - H_q(t))(H_q(t) - z)^{-1} \right),$$
$$A = \prod_{j \in \mathbb{Z}} \frac{a(j)}{a_q(j)}. \tag{4.3}$$

By [8], Theorem 1, $\alpha(z)$ has the following representation

$$\alpha(z) = \frac{1}{A} \exp \left(\int_{\mathbb{R}} \frac{\xi_\alpha(\lambda)d\lambda}{\lambda - z} \right), \tag{4.4}$$

where

$$\xi_\alpha(\lambda) = \frac{1}{\pi} \lim_{\epsilon \downarrow 0} \arg \alpha(\lambda + i\epsilon) \tag{4.5}$$

is the spectral shift function.

Hence

$$\tau_j = \mathrm{tr}(H^j - (H_q)^j) = j \int_{\mathbb{R}} \lambda^{j-1} \xi_\alpha(\lambda)d\lambda, \tag{4.6}$$

where τ_j/j are the expansion coefficients of $\ln \alpha(z)$ around $z = \infty$:

$$\ln \alpha(z) = -\ln A - \sum_{j=1}^{\infty} \frac{\tau_j}{j} \frac{1}{z^j}.$$

They are related to the expansion α_j coefficients of

$$\alpha(z) = \frac{1}{A} \sum_{j=0}^{\infty} \frac{\alpha_j}{z^j}, \qquad \alpha_0 = 1,$$

via

$$\tau_1 = -\alpha_1, \qquad \tau_j = -j\alpha_j - \sum_{k=1}^{j-1} \alpha_{j-k}\tau_k. \tag{4.7}$$

5. Conserved quantities of the Toda hierarchy

Finally we turn to solutions of the Toda hierarchy TL_r (see, e.g., [1], [4], [12], or [13]). Let $(a_q(t), b_q(t))$ be a quasi-periodic finite-gap solution of some equation in the Toda hierarchy, $\mathrm{TL}_r(a_q(t), b_q(t)) = 0$, and let $(a(t), b(t))$ be another solution, $\mathrm{TL}_r(a(t), b(t)) = 0$, such that (3.1) holds for one (hence any) t.

Since the transmission coefficient $T(z, t) = T(z, 0) \equiv T(z)$ is conserved (see [3] – formally this follows from unitary invariance of the determinant), so is $\alpha(z) = T(z)^{-1}$.

Theorem 5.1. *The quantities*

$$A = \prod_{j=-\infty}^{\infty} \frac{a(j,t)}{a_q(j,t)} \tag{5.1}$$

and $\tau_j = \mathrm{tr}(H^j(t) - H_q(t)^j)$, *that is,*

$$\tau_1 = \sum_{n \in \mathbb{Z}} b(n,t) - b_q(n,t)$$

$$\tau_2 = \sum_{n \in \mathbb{Z}} 2(a(n,t)^2 - a_q(n,t)^2) + (b(n,t)^2 - b_q(n,t)^2)$$

$$\vdots$$

are conserved quantities for the Toda hierarchy.

Acknowledgments

We thank the referee for several valuable suggestions for improving the presentation.

References

[1] W. Bulla, F. Gesztesy, H. Holden, and G. Teschl, *Algebro-Geometric Quasi-Periodic Finite-Gap Solutions of the Toda and Kac-van Moerbeke Hierarchies*, Memoirs of the Amer. Math. Soc. **135/641**, (1998).

[2] I. Egorova, J. Michor, and G. Teschl, *Scattering theory for Jacobi operators with quasi-periodic background*, Comm. Math. Phys. **264-3**, 811–842 (2006).

[3] I. Egorova, J. Michor, and G. Teschl, *Inverse scattering transform for the Toda hierarchy with quasi-periodic background*, Proc. Amer. Math. Soc. (to appear).

[4] L. Faddeev and L. Takhtajan, *Hamiltonian Methods in the Theory of Solitons*, Springer, Berlin, 1987.

[5] H. Flaschka, *On the Toda lattice. II*, Progr. Theoret. Phys. **51**, 703–716 (1974).

[6] J.S. Geronimo and W. Van Assche, *Orthogonal polynomials with asymptotically periodic recurrence coefficients*, J. App. Th. **46**, 251–283 (1986).

[7] F. Gesztesy and H. Holden, *Trace formulas and conservation laws for nonlinear evolution equations*, Rev. Math. Phys. **6**, 51–95 (1994).

[8] M.G. Krein, *Perturbation determinants and a formula for the traces of unitary and self-adjoint operators*, Soviet. Math. Dokl. **3**, 707–710 (1962).

[9] G. Teschl, *Oscillation theory and renormalized oscillation theory for Jacobi operators*, J. Diff. Eqs. **129**, 532–558 (1996).

[10] G. Teschl, *Trace Formulas and Inverse Spectral Theory for Jacobi Operators*, Comm. Math. Phys. **196**, 175–202 (1998).

[11] G. Teschl, *Inverse scattering transform for the Toda hierarchy*, Math. Nach. **202**, 163–171 (1999).

[12] G. Teschl, *Jacobi Operators and Completely Integrable Nonlinear Lattices*, Math. Surv. and Mon. **72**, Amer. Math. Soc., Rhode Island, 2000.

[13] M. Toda, *Theory of Nonlinear Lattices*, 2nd enl. ed., Springer, Berlin, 1989.

[14] A. Volberg and P. Yuditskii, *On the inverse scattering problem for Jacobi matrices with the spectrum on an interval, a finite systems of intervals or a Cantor set of positive length*, Commun. Math. Phys. **226**, 567–605 (2002).

Johanna Michor and Gerald Teschl
Faculty of Mathematics
Nordbergstrasse 15
A-1090 Wien, Austria

and

International Erwin Schrödinger Institute
   for Mathematical Physics
Boltzmanngasse 9
A-1090 Wien, Austria

e-mail: `Johanna.Michor@esi.ac.at`
URL: `http://www.mat.univie.ac.at/~jmichor/`

e-mail: `Gerald.Teschl@univie.ac.at`
URL: `http://www.mat.univie.ac.at/~gerald/`

Operator Theory:
Advances and Applications, Vol. 174, 77–103

Dirichlet-to-Neumann Techniques for the Plasma-waves in a Slot-diod

Anna B. Mikhailova, Boris Pavlov and Victor I. Ryzhii

Abstract. Plasma waves in a slot-diod with governing electrodes are described by the linearized hydrodynamic equations. Separation of variables in the corresponding scattering problem is generally impossible. Under natural physical assumption we reduce the problem to the second order differential equation on the slot with an operator weight, defined by the Dirichlet-to-Neumann map of the three-dimensional Laplacian on the complement of the electrodes and the slot. The reduction is based on a formula for the Poisson map for the exterior Laplace Dirichlet problem on the complement of a few standard bodies in terms of the Poisson maps on the complement of each standard body.

Mathematics Subject Classification (2000). 82D10, 47A40.

Keywords. Dirichlet-to-Neumann map, Scattered waves.

1. Introduction: basic equations

Mathematical design of optical sensors based on slot-diodes gives rise to an interesting class of mathematical problems for non-linear and linearized hydrodynamical equations describing plasma waves in a slot between basic electrodes. In this paper we consider some of these problems. Beginning from analysis of the simplest slot-diod with flat geometry, we develop a mathematical construction based on Dirichlet-to-Neumann map which allows to derive convenient equations for description of the plasma current and plasma waves in presence of governing electrodes. We compare results of direct calculation of the amplitudes of the cross-section eigenfunction based on Dirichlet-to-Neumann map machinery with calculations based on the simplest physical model with the slot substituted by the layer between two parallel flat electrodes. The comparison shows that the substitution is viable in case when governing electrodes are absent, see Appendix B, but fails when the governing electrodes are present. We suggest also a solvable model for scattering of plasma waves in a slot in presence of the governing electrodes, see the plan of our paper at the end of the section.

The hydrodynamical analogy was suggested for plasma waves in [1, 2] and was intensely used for description of the plasma-current in a two-dimensional slot of a simple configuration, with no governing electrodes, see for instance recent papers [3, 4]. In [4] the hydrodynamic electron transport model is used for description of plasma oscillations in gated 2D channel in high electron mobility transistor (HMET). Analysis of the spectrum of plasma oscillations based on hydro-dynamic analogy is applicable also to other HMET-based teraherz devices, see [5, 6].

For the simplest device constructed of two flat basic electrodes $\Gamma^{\pm}$ in the horizontal plane $S = \{z = 0\}$, separated by the slot $\Gamma : \{-l < y < l\}$ the problem is reduced, see for instance [3, 4, 5, 6], to the self-consistent calculation of the electric potential $\varphi(x, z, t) = \varphi_0(x, y, z) + \int \varphi_\omega(x, y, z)e^{i\omega t}\, d\omega$ from the system of three basic equations (1.1, 1.2, 1.3) below. The three-dimensional Poisson equation with the dielectric constant κ:

$$\triangle_3 \varphi = \frac{4\pi e}{\kappa}\, \delta(z) \,\Xi\, \mathcal{X}_\Gamma \tag{1.1}$$

connects the potential with the non-zero concentration $\delta(z)\Xi(x, y, 0, t)$ localized on a two-dimensional slot Γ situated on the horizontal plane $S : \{z = 0\}$ between the electrodes $\Gamma_\pm \in S$, $\Xi(x, y, 0, t) = \Xi_0(x, y) + \int \Xi_\omega(x, y)e^{-i\omega t}\, d\omega$. The function $\mathcal{X}_\Gamma$ is the indicator of the slot: $\mathcal{X}_\Gamma(x, y) = 1$, if $(x, y) \in \Gamma$, otherwise $\mathcal{X}_\Gamma(x, y) = 0$. The variables Ξ, φ fulfil the continuity equation:

$$\frac{\partial \Xi}{\partial t} + \mathrm{div}_2 \Xi\, u = 0, \ \ (x, y) \in \Gamma, \tag{1.2}$$

and the Euler equation on the slot, taking into account the exponential decay parameter ν, see [3]:

$$\frac{\partial u}{\partial t} + \langle u, \nabla_2 \rangle u = \frac{e}{m} \nabla_2 \varphi - \nu u, \ \ (x, y) \in \Gamma. \tag{1.3}$$

These equations describe plasma waves on the slot. They connect the electron's concentration with the electron's velocity in tangent direction, $u(x, y) \in T_\Gamma(x, y)$, on the slot. Eventually we aim on the problem with several governing electrodes γ_s, $s = 1, 2, \dots$. The system of basic electrodes $\Gamma^{\pm}$ and governing electrodes constitute the device $\{\Gamma_- \cup \Gamma \cup \Gamma_+ \cup \gamma_1 \cup \gamma_2 \cdots\} := \Omega_d$. The complement $R^3 \backslash \Omega_d := \Omega$ of the device plays the role of a basic domain where the electric potential is defined. We assume that the geometry and the physical parameters of the device are chosen in such a way that the plasma current is observed only on the two-dimensional slot $\Gamma \subset S$ between the basic electrodes $\Gamma_\pm$. In simplest case considered previously in [3, 4], when the governing electrodes are absent, the slot is a straight channel $-l < y < l$, $-\infty < x < \infty$ and the plasma waves are running in the lateral direction x with amplitudes defined by the cross-section eigenfunctions $f_l(x)$ of some spectral problem on the slot. Though the technique of Dirichlet-to-Neumann map also permits to derive convenient equations for the plasma current in general case when the surface $\Gamma^+ \cup \Gamma \cup \Gamma^- = S$ is not flat, we analyze here only devices with flat surface S. We assume that (x, y) are the coordinates on the slot $\Gamma \subset S$ and z is the normal coordinate. Each small open neighborhood of the slot $\Omega_\varepsilon \supset \Gamma : -\varepsilon < z < \varepsilon$ is cut by the surface S into two parts: the "upper" part $\Omega_\varepsilon^+ = \{\Omega_\varepsilon \cap (z > 0)\}$ and the "lower" part $\Omega_\varepsilon^- = \{\Omega_\varepsilon \cap (z < 0)\}$. For the functions defined on Ω_ε^+ we can consider the upper and lower limits as $\lim_{z \to 0^\pm} f(x, y, z) = f_\pm(x, y, 0)$.

Assuming that the speed u of the plasma flow only slightly deviates from the stationary speed $u_0(x, y)$, which is assumed bounded,

$$u(x, y, t) = u_0(x, y) + \int u_\omega(x, y) e^{-i\omega t} \, d\omega,$$

$\int |u_\omega(x, y)| d\omega \ll |u_0(x, y)|$, $u_\omega(x, y) \to 0$ when $|x| \to \infty$, and imposing similar conditions on the deviations of the potential and the concentration from the equilibrium on the slot, one can derive from the above basic equations (1.1, 1.2, 1.3) stationary equations for equilibrium values of the parameters Ξ_0, φ_0, u_0:

$$\nabla_2 \Xi_0 u_0 = 0, \quad \langle u_0, \nabla_2 \rangle u_0 = \frac{e}{m} \nabla_2 \varphi_0 - \nu u_0, \quad \Delta_3 \varphi_0 = \frac{4\pi e}{\kappa} \Xi_0 \delta(z) \mathcal{X}_\Gamma. \tag{1.4}$$

Values of the equilibrium concentration $\Xi_0(x, y)$, velocity $u_0(x, y)$ and the equilibrium potential $\varphi_0(x, y.z)$ on the slot $z = 0$ are uniformly bounded. We do not discuss here solutions of the equations (1.4), but following [3, 4, 5, 6] we assume that this non-linear system of partial differential equations, with appropriate boundary conditions on the electrodes $\Gamma_\pm, \gamma_s$

$$\varphi_0(x, y, 0) \Big|_{(x,y)\in\Gamma_\pm} = V_\pm, \quad \varphi_0(x, y, 0) \Big|_{(x,y)\in\gamma_s} = V_s, \, s = 1, 2, \ldots$$

is already solved. We consider the linear system for the amplitudes $\Xi_\omega, u_\omega, \varphi_\omega$ of the first order correcting terms with zero boundary conditions on the electrodes. Neglecting terms of higher order we may connect directly the amplitude $\Xi_\omega(x, y)$ of the electron's concentration, with the amplitude u_ω, φ_ω of the velocity and the amplitude of the potential:

$$-i\omega \Xi_\omega(x) + \nabla_2 \left[\Xi_0(x, y) u_\omega + u_0(x, y) \Xi a_\omega \right] = 0,$$

$$\Delta_3 \varphi_\omega = \frac{4\pi e}{\kappa} \Xi_\omega \, \delta(z) \, \mathcal{X}_\Gamma$$

$$(\nu - i\omega) u_\omega = \frac{e}{m} \nabla_2 \varphi_\omega - \langle u_\omega, \nabla_2 \rangle u_0 - \langle u_0, \nabla_2 \rangle u_\omega. \tag{1.5}$$

In this paper we assume that the first order correcting term of velocity is orthogonal to the plane of the device $u_\omega = u_\omega e_z$. Then we have: $\langle u_\omega, \nabla_2 \rangle u_0 = 0$. Another simplification appears when assuming that the first of equations (1.4) defines the connection between the stationary speed and the stationary concentration as $u_0 = \Xi_0^{-1} \vec{a}$ where $\vec{a}$ is a solenoidal vector field, which we assume constant hereafter.

First of the equations (1.5) can be interpreted based on the physical meaning of the concentration: one should take into account that the concentration of electrons on the slot is originated by the supply of electrons from Γ_- and is spread on the slot due to the drift defined by the stationary speed u_0. Hence the corresponding first order differential equation should be supplied with boundary data for Σ_ω on the boundary of the electrode Γ_-, where the stationary speed u_0 looks into the outgoing direction, toward Γ_+. In that case the concentration is obtained inside the slot via integration on characteristics of the first equation. Note that this algorithm of calculation of the concentration is in agreement with the algorithm of the construction of solution of partial differential equations of second order with a small coefficient in front of the higher derivatives. This algorithm was suggested in the mathematical paper [7]. In our case it also can be verified, when considering full Navier-Stokes equation with the small viscosity $\varepsilon \to 0$, instead of the Euler equation. It makes sense to set the boundary conditions for the amplitude as $\Xi_\omega\big|_{\Gamma_-} = 0$. Then the system (1.5) has a unique solution for given ν, ω, if the corresponding homogeneous system with zero boundary conditions has only trivial solution $u_\omega = \Xi_\omega = \varphi_\omega = 0$. Thus the question on solvability of the system is reduced to the corresponding spectral problem for the system (1.5) with zero boundary conditions on the electrodes. Non-trivial solutions of the system (1.5) are eigenfunctions of the spectral problem.

In this paper we explore a simplified version of the above equations (1.5). For the convenience of the reader we provide below the derivation of this basic equation.

Assuming that only component of u_ω orthogonal to the slot is non-zero, we may solve the last equation (1.5). It is convenient to represent this equation via differentiation along trajectories $x(t)$ of the stationary velocity, $dx/dt = u_0(x)$, introducing the time $t(x)$ needed to reach the point x on the trajectory, starting at $t = 0$ from the initial point on Γ_-. The integrating factor of the equation is $\exp(\nu - i\omega)\,t(x)$, and the solution is presented as an integral along the trajectory $x(t)$

$$u_\omega(x) = \int_{\Gamma_-}^{x} e^{-(\nu - i\omega)(t(x)-t(s))}\, \frac{e}{m}\, \nabla_2\, \varphi_\omega(s)\,dt(s).$$

If ω is large comparing with time needed for trajectory to pass the slot, $[t(\Gamma_+) - t(\Gamma_-)]\omega >> 1$, and φ_ω is smooth, we replace the integral in the above representation of $u_\omega(x)$ by the asymptotic

$$u_\omega(x) \approx \frac{e}{m(\nu - i\omega)}\, \varphi_\omega(x). \tag{1.6}$$

Substituting the obtained formula for u_ω we obtain from the first equation (1.5) an equation for Ξ_ω:

$$\nabla_2 \frac{\Xi_0 e}{m(\nu - i\omega)} \nabla_2 \varphi_\omega + \nabla_2 u_0 \Xi_\omega - i\omega \Xi_\omega = 0. \tag{1.7}$$

This equation can be also interpreted in terms of differentiation of the ratio Ξ_ω/Ξ_0 along the trajectory of the stationary speed. Then, introducing the time $t(x)$ along the trajectory, as before, we obtain:

$$\Xi_\omega(x) = -\Xi_0(x) \int_{\Gamma_-}^{x} e^{i(t(x)-s)} \Xi_0^{-1}(x(s))\, \nabla_2 \frac{\Xi_0 e}{m(\nu - i\omega)} \nabla_2 \varphi_\omega(x(s))\,ds. \tag{1.8}$$

Again, we replace the obtained expression for $\Xi_\omega(x,y)$ by the asymptotic for large ω, assuming additionally that Ξ_0 is slowly changing along the trajectories of the stationary speed:

$$\Xi_\omega(x) = \nabla_2 \frac{\Xi_0 e}{i\omega m(\nu - i\omega)} \nabla_2 \varphi_\omega(x). \tag{1.9}$$

Now substituting the result into the second equation we obtain:

$$\Delta_3 \varphi_\omega = \frac{4\pi e^2}{\kappa i\omega m(\nu - i\omega)} \nabla_2 \Xi_0 \nabla_2 \varphi_\omega \, \delta(z) \, \mathcal{X}_\Gamma \tag{1.10}$$

Three-dimensional Laplacian stays in the left side of the middle equation of (1.5). This equation can be transformed with use of Green formula, or just formally via integrating on the short "vertical" interval $-\delta < z < \delta$ and then taking the limit $\delta \to 0$. Then we obtain in the left side the jump of the normal derivative of the potential φ_ω on the slot, and in the right side

$$\frac{4\pi e^2}{\kappa i\omega m(\nu - i\omega)} \nabla_2 \Xi_0 \nabla_2 \varphi_\omega := \frac{4\pi e}{\kappa} \mathcal{L} \varphi_\omega,$$

$$\left[-\Lambda_+ - \Lambda_-\right] \varphi_\omega = \frac{4\pi e}{\kappa} \mathcal{L} \varphi_\omega. \tag{1.11}$$

Note that in the left side of the equation stays the construction defined by the values $\Lambda_\pm$ of the Dirichlet-to-Neumann map (DN-map) of the Laplacian, see [16, 17] on the upper and lower sides of the slot. Basic properties of the Dirichlet-to-Neumann map (DN-map) are reviewed in Appendix A. If the DN map is known, the non- trivial solutions of the above homogeneous equation (1.11) define spectral values of the parameter $\dfrac{4\pi e^2}{i\omega m \kappa (\nu - i\omega)} := \dfrac{2}{q}$. Hereafter we represent the equation (1.11) as an equation on the slot Γ:

$$\left[\Lambda_+ + \Lambda_-\right] \varphi_\omega = -\frac{2}{q} \mathrm{div}_2 \Sigma_0(x,y) \nabla_2 \varphi_\omega. \tag{1.12}$$

Subject to the above assumptions, the derived equation (1.12) is equivalent to the original problem on the plasma waves for slowly changing electron's velocity not only in case of the flat slot Γ and electrodes $\Gamma_\pm$, but also in general case when the slot and electrodes have arbitrary geometry. Nevertheless hereafter we explore the most important case of the flat geometry when S is a horizontal plane $z = 0$, but eventually consider the case when the governing electrons are present. The physical conditions of absence of the plasma current between the governing and basic electrodes will be discussed in Section 3.

Remark. Note that in [22] the problem concerning plasma current on the flat two-dimensional slot is considered based on the equation for the potential:

$$\Delta_3 \varphi_\omega = \left(\frac{4\pi e^2 \Xi_0}{m\kappa\omega(\omega - i\nu)}\right) \left(\frac{\partial^2 \varphi_\omega}{\partial x^2} + \frac{\partial^2 \varphi_\omega}{\partial y^2}\right) \delta(z) = \frac{2}{q} \Xi_0 \Delta_2 \, \varphi_\omega \, \delta(z), \tag{1.13}$$

with $\Xi_0 = \text{const}$, which is only slightly different from (1.5). But the approach to solving it differs essentially from our suggestion above: instead of using the DN-map in the left side of the equation, the authors of [22] use an integral operator in the right side:

$$\varphi_\omega(x,y) = \frac{2}{q} \int_\Gamma G(x,y,0;\xi,\eta,0) \left(\frac{\partial \varphi_\omega^2}{\partial \xi^2} + \frac{\partial \varphi_\omega^2}{\partial \eta^2} \right) d\xi\, d\eta, \quad (x,y) \in \Gamma,$$

assuming that $G(x,y,0;\xi,\eta,0)$ is the Green function of the Laplacian restricted onto the horizontal plane $z = 0$. In fact this suggestion gives a correct answer in case of flat geometry with no governing electrodes, because the restriction of the doubled free Green function onto the slot coincides with the restriction onto the slot of the Green function of the Neumann problem in Ω, which is just a kernel of the integral operator to the Dirichlet-to-Neumann map, see for instance [16, 17] and Appendix A. In [22], due to the symmetry, we have $\Lambda_+ + \Lambda_- = 2\Lambda_+$ on the slot. Then

$$2 \left[\Lambda_+ + \Lambda_- \right]^{-1} * = \Lambda_+^{-1} * = \int_{\Gamma_\pm} G^N(x,y|\xi,\eta) * d\xi\, d\eta = 2 \int_\Gamma G(x,y|\xi,\eta) * d\xi\, d\eta$$

where G is the free Green-function and the integral on $\Gamma_\pm$ is calculated on the bilateral slot. Though in that case the substitution of the Neumann Green function by the free Green function is possible, but if the governing electrodes are present and/or Γ, $\Gamma_\pm$ are non-flat, either DN-map or the corresponding inverse operator must be used.

For the slot-diod, with additional electrodes included, the Green-function $G^D(x,y,z;\xi,\eta,\zeta)$ of the homogeneous Dirichlet problem is the main tool for solution of the problem on plasma current, because all important maps used in course of solution of the equations can be obtained from it. In particular, the kernel of the Poisson map $\mathcal{P}$ is obtained via differentiation of the Green function of the Dirichlet problem in outward direction on the boundary of Ω. In our case on the upper side Γ_+ of the slot we have:

$$\mathcal{P}(x,y,z;\xi,\eta,0) = -\frac{\partial G^D}{\partial n_{\xi,\eta,0}}(x,y,z;\xi,\eta,0) = \frac{\partial G^D}{\partial \zeta}(x,y,z;\xi,\eta,0)\bigg|_{\zeta=0},$$

where $(x,y,z) \in \Omega^+$, with $z > 0$ and $(\xi,\eta,0) \in \Gamma_+$. DN-map can be presented as a formal integral operator on $\Gamma^\pm$ with the generalized kernel:

$$\Lambda(x,y,0;\xi,\eta,0) = -\frac{\partial^2 G^D}{\partial n_{x,y,0}\, \partial n_{\xi,\eta,0}}(x,y,0;\xi,\eta,0) = -\frac{\partial^2 G^D}{\partial z\, \partial \zeta}(x,y,0;\xi,\eta,0)$$

where the outward normals on Γ_+ with respect to the upper or lower neighborhoods $\Omega_\varepsilon^\pm$ of the slot Γ are used respectively for $\Lambda_\pm$. On the other hand, construction of the Green function in a domain with few standard exclusions like $\Gamma_\pm$, γ_1, γ_2, $\ldots$ may be obtained via simple iteration process, see [10]. In particular the Dirichlet Green function may be constructed for the device based on a flat horizontal slot between basic horizontal electrodes and few governing electrodes in form of straight cylindrical rods suspended parallel to the horizontal plane, see Section 3 below.

Here is the plan of our paper. In second section we review the spectral properties of the simplest problem with an infinite straight slot and no governing electrodes. In the third section we consider the modified problem with few governing electrodes. Due to a simple re-normalization of the boundary data, the DN-map of the problem with the governing electrodes is obtained from the DN-map of the flat problem via an explicit transformation derived from the Poisson maps on the complements of the additional electrodes. In the forth section, assuming that the equilibrium concentration has "bumps" at some cross-sections of the slot (caused by the governing electrodes γ_s), we reveal the role of the resonance phenomena in scattering of lateral waves. This observation permits, in principle, to manipulate the transmission coefficients of the waves in the slot. In Appendix A basic features of the Dirichlet-to-Neumann map are reviewed. In Appendix B numerical data on few cross-section eigenfunctions in the slot are supplied.

2. Flat slot-diod: basic spectral problem via DN-map

In this section we study the plasma waves in the simplest flat slot-diod. The derived formulae serve as a base for the studies of the waves in the plasma-current on the slot in the device with governing electrodes to be carried out in Section 3.

Consider the device constructed of flat basic electrodes $\Gamma_\pm$ and a slot Γ : $-l < y < l$, $-\infty < x < \infty$, all situated on the horizontal plane $z = 0$. The corresponding spectral problem is reduced to the differential equation (1.13) with zero boundary conditions on the electrodes and Meixner boundary condition on the edges of the slot. We look for the bounded solutions, which correspond to the continuous spectrum of the problem. To re-write (1.13) in form (1.12), we need the DN-map Λ_+ of the upper half-space $R_3^+ : z > 0$. It is a generalized integral operator, see Appendix A, with the distribution kernel:

$$\Lambda_+ (x, y; \xi, \eta) = -\frac{\partial}{\partial z}\mathcal{P}_+ = \frac{1}{4\pi^2} \int_\infty^\infty \int_\infty^\infty e^{ip(x-\xi)} e^{iq(y-\eta)} \sqrt{p^2 + q^2}\, dp\, dq. \quad (2.1)$$

The Laplacian on the slot with zero boundary conditions and the Meixner condition, $\int |\nabla u|^2 dx < \infty$, at the edges of the electrodes $\Gamma_\pm$, has continuous spectrum with step-wise growing multiplicity $2m$ on the spectral bands separated by spectral thresholds $\left[\frac{\pi^2 m^2}{4l^2}\right]$, $m = 1, 2, 3, \ldots$ and eigenfunctions $\psi_{m,p}(y, x) = \frac{1}{\sqrt{2l}} e^{ipx} \sin \frac{\pi m (y+l)}{2l} := \frac{1}{\sqrt{2l}} e^{ipx} \varphi_m$, $m = 1, 2, \ldots$, which correspond to the values of the spectral parameter $\lambda = \frac{\pi^2 m^2}{4l^2} + p^2$,

$$-\triangle_\Gamma = \sum_{m=1}^\infty \int_{-\infty}^\infty dp \left[\frac{\pi^2 m^2}{4l^2} + p^2\right] \psi_{m\,p}(y, x)\rangle \langle\psi_{m\,p}(y, x),$$

where $e\rangle\langle e$ is the orthogonal projection onto the one-dimensional subspace spanned by the vector e. We rewrite the equation (1.13) in form (1.12) as an infinite linear system $\mathcal{K}\varphi = 2\Xi_0 q^{-1}\varphi$ with the constant concentration Ξ_0 and the generalized matrix kernel:

$$\int_{-\infty}^{\infty} dp \int_{-l}^{l} dy \int_{-\infty}^{\infty} dq \int_{-l}^{l} d\eta \frac{\sin\frac{m\pi(y+l)}{2l}}{\sqrt{\frac{\pi^2 m^2}{4l^2}+p^2}} e^{i(px+qy)} \frac{\sqrt{p^2+q^2}}{8\pi^2 l} e^{-i(p\xi+q\eta)} \frac{\sin\frac{n\pi(\eta+l)}{2l}}{\sqrt{\frac{\pi^2 n^2}{4l^2}+p^2}}$$

$$:= \mathcal{K}_{m,n}(x,\xi),$$

or, with use of the Fourier transform $\mathcal{F}u(x) \to \tilde{u}(p)$,

$$\mathcal{K}_{m,n}(x,\xi) \tag{2.2}$$

$$= \mathcal{F}^{*} \int_{-\infty}^{\infty} dq \int_{-l}^{l} dy \int_{-l}^{l} d\eta \frac{\sin\frac{m\pi(y+l)}{2l}}{\sqrt{\frac{\pi^2 m^2}{4l^2}+p^2}} e^{iqy} \frac{\sqrt{p^2+q^2}}{4\pi l} e^{-iqn} \frac{\sin\frac{n\pi(\eta+l)}{2l}}{\sqrt{\frac{\pi^2 n^2}{4l^2}+p^2}} \mathcal{F}$$

$$:= \left\{\mathcal{F}^{*}\tilde{\mathcal{K}}(p)\mathcal{F}\right\}_{m,n},$$

where $\tilde{\mathcal{K}}(p)$ is the multiplication operator by the infinite matrix $\tilde{\mathcal{K}}_{m,n}(p)$. We will find the eigenvalues and eigenvectors of the matrix $\mathcal{K}(p)$. Then the spectral modes φ of the equation (1.13) are found by inverse Fourier transform $\varphi_\omega = \mathcal{F}\varphi$ from the eigenfunctions of the equation

$$\tilde{\mathcal{K}}(p)\varphi_\omega = \frac{2}{q}\Xi_0\varphi_\omega. \tag{2.3}$$

We will show that the matrix-function $\mathcal{K}$ is compact. Then denoting by $\kappa_1(p)$, $\kappa_2(p)$, $\kappa_3(p), \ldots$ the eigenvalues of $\tilde{\mathcal{K}}(p)$ and by $\varphi_1, \varphi_2, \varphi_3, \ldots$ the corresponding normalized eigenvectors, we form the eigenmode corresponding to $\kappa_m(s)$ as $\delta(p-s)\,\varphi_m = (2l)^{-1/2}\mathcal{F}^{+}e^{isx}\varphi_m$. The spectrum of the multiplication operator $\tilde{\mathcal{K}}(0)$ has a band-structure with thresholds at $\max_p \kappa_m(p) = \kappa_m$. It is more convenient to substitute now the exponential Fourier transform by the trigonometrical Fourier transform:

$$\delta(y-\eta) = \frac{1}{2\pi}\int_{\infty}^{\infty} e^{q(y-\eta)} dq = \frac{1}{2\pi}\int_{\infty}^{\infty} [\cos qy\, \cos q\eta + \sin qy\, \sin q\eta]\, dq.$$

Then the calculation of the matrix $\tilde{\mathcal{K}}(0)$ is reduced to calculation of the elementary integrals obtained via the change of variable: $y+l \to y$:

$$J_r^s(q) = \int_0^{2l} \sin q(y-l)\, \sin\frac{\pi r y}{2l}\, dy$$

$$= \cos ql \int_0^{2l} \sin qy\, \sin\frac{\pi r y}{2l}\, dy - \sin ql \int_0^{2l} \cos qy\, \sin\frac{\pi r y}{2l}\, dy$$

and

$$J_r^c(q) = \int_0^{2l} \cos q(y - l)\ \sin \frac{\pi r y}{2}\, dy$$

$$= \cos p\, l \int_0^{2l} \cos q\, y\ \sin \frac{\pi r y}{2l}\, dy + \sin q\, l \int_0^{2l} \sin q\, y\ \sin \frac{\pi r y}{2l}\, dy.$$

We have, with $y/l := \hat{y}$:

$$\int_0^{2l} \cos qy \sin \frac{\pi r y}{2l}\, dy = \frac{l}{2}\int_0^2 [\sin(ql + r\,\pi/2)\hat{y} - \sin(ql - r\,\pi/2)\hat{y}]\, d\hat{y}$$

$$= \frac{l}{2ql + r\,\pi}[-\cos(2q + r\,\pi) + 1] + \frac{l}{2ql - r\,\pi}[\cos(2q - r\,\pi) - 1]$$

$$= (-1)^r \cos 2ql \left(\frac{l}{2ql - r\,\pi} - \frac{1}{2ql + r\,\pi}\right) + \left(\frac{l}{2ql + r\,\pi} - \frac{l}{2ql - r\,\pi}\right)$$

$$= \left[(-1)^r \cos 2ql - 1\right] \frac{2\pi r l}{4q^2 l^2 - \pi^2 r^2}. \tag{2.4}$$

Similarly we obtain

$$\int_0^{2l} \sin qy \sin \frac{\pi r y}{2l}\, dx = \frac{l}{2}\int_0^2 [\cos(ql - r\,\pi/2)\hat{y} - \cos(ql + r\,\pi/2)\hat{y}]\, d\hat{y}$$

$$= \left[\sin 2ql\,(-1)^r\right] \frac{2\pi r l}{4q^2 l^2 - \pi^2 r^2}. \tag{2.5}$$

Substituting (2.4, 2.5) into $J_r^c(p)$, $J_r^s(p)$ we see, that all terms J_r^s with odd r and all terms J_r^c with even r are equal to zero, and all remaining terms are equal to

$$J_{2m}^s = \frac{2\pi m}{q^2 l^2 - \pi^2 m^2}\sin q\, l, \quad J_{2m+1}^c = -\frac{\pi(2m+1)}{q^2 l^2 - \pi^2(m+1/2)^2}\cos q\, l. \tag{2.6}$$

Then we have:

$$\tilde{K}_{rt}(p) \tag{2.7}$$

$$= \frac{1}{2\pi\, l}\int_{-\infty}^{\infty} J_r^s(q)\, \frac{\sqrt{p^2 + q^2}}{\sqrt{\frac{\pi^2 r^2}{4l^2} + p^2}\,\sqrt{\frac{\pi^2 t^2}{4l^2} + p^2}}\, J_t^s(q)\, dq$$

$$+ \frac{1}{2\pi\, l}\int_{-\infty}^{\infty} \frac{\sqrt{p^2 + q^2}}{\sqrt{\frac{\pi^2 r^2}{4l^2} + p^2}\,\sqrt{\frac{\pi^2 t^2}{4l^2} + p^2}}\, J_r^c(q)\, p\, J_t^c(q)\, dq$$

$$= \begin{cases} \dfrac{2\pi\, m\, n}{\sqrt{\frac{\pi^2 r^2}{4l^2} + p^2}\,\sqrt{\frac{\pi^2 t^2}{4l^2} + p^2}} \displaystyle\int_{-\infty}^{\infty} \dfrac{\sqrt{p^2 + q^2}\,\sin^2 q\, l}{l\,(q^2 l^2 - \pi^2 m^2)(q^2 l^2 - \pi^2 n^2)}\, dq, & \text{if } \begin{array}{l} r = 2m, \\ t = 2n \end{array} \\[2.5em] \dfrac{4\pi(m+1/2)(n+1/2)}{\sqrt{\frac{\pi^2 r^2}{4l^2} + p^2}\,\sqrt{\frac{\pi^2 t^2}{4l^2} + p^2}} \displaystyle\int_{-\infty}^{\infty} \dfrac{\sqrt{p^2 + q^2}\,\cos^2 q\, l}{l\,(q^2 l^2 - \pi^2(m+1/2)^2)(p^2 - \pi^2(n+1/2)^2)}\, dq, & \text{if } \begin{array}{l} r = 2m+1, \\ t = 2n+1, \end{array} \end{cases}$$

and 0 for complementary sets of indices. Essential features of the spectral structure of $\tilde{\mathcal{K}}(p)$ for $p \neq 0$ are the same as for the matrix $\tilde{\mathcal{K}}(0)$:

$$\tilde{\mathcal{K}}_{r,t}(0) = \begin{cases} \dfrac{4}{\pi l} \displaystyle\int_0^\infty s \dfrac{\sin^2 s}{(s^2 - \pi^2 m^2)(s^2 - \pi^2 n^2)} \, ds, \\ \qquad\qquad\qquad\qquad \text{if } r = 2m, \, t = 2n \\[2mm] \dfrac{4}{\pi l} \displaystyle\int_0^\infty s \dfrac{\cos^2 s}{(s^2 - \pi^2 (m+2^{-1})^2)(s^2 - \pi^2 (n+2^{-1})^2)} \, ds, \\ \qquad\qquad\qquad\qquad \text{if } r = 2m+1, \, t = 2n+1, \end{cases} \tag{2.8}$$

with integrals convergent due to compensation of the zeros of the denominators by the zeros of the numerators. One can see from (2.8) that the matrix $\tilde{\mathcal{K}}$ is a sum of two matrixes acting in invariant subspaces spanned by vectors with even and odd r respectively. Thus the problem of spectral analysis splits into two parts in corresponding subspaces on the slot $-l < \eta < l$:

$$\bigvee_m \sin \frac{m\pi\eta}{l} = E_{\text{odd}}, \quad \bigvee_m \cos \frac{(2m+1)\,\eta\pi}{2\,l} = E_{\text{even}}.$$

The subspace E_{odd}, for odd $r = 2m+1$, is spanned by even (symmetric) functions on the slot, and the subspace E_{even}, for even $r = 2m$, is spanned by odd (antisymmetric) functions on the slot. The spectral analysis of $\tilde{\mathcal{K}}$ can be accomplished in these invariant subspaces separately.

Based on matrix representation (2.8) we can prove that the operator $\mathcal{K}$ belongs to Hilbert-Schmidt class, hence is has discrete spectrum, and its square has a finite trace, hence the infinite determinant can be approximated by determinants of finite cut-off matrices. We derive these facts from asymptotic behavior of elements of $\mathcal{K}_{rt}$ for large (r, t).

Theorem 2.1. *Elements of the matrix $\tilde{\mathcal{K}}(0)$ have the following asymptotic for large r, t:*

$$\frac{\pi}{4\,l} \tilde{\mathcal{K}}_{rt}(0) = \text{Const} \frac{\ln r\,t^{-1}}{(r-t)(r+t)}, \, r > t > 0. > 0. \tag{2.9}$$

The *Proof* will be given for the part of the operator $\tilde{\mathcal{K}}(0)$ in the subspace of antisymmetric modes, $r = 2m$, $t = 2n$. The asymptotic of elements of the part of $\mathcal{K}$ in the symmetric subspace $r = 2m + 1$, $t = 2n + 1$ is derived similarly.

We present the integrand in the first integral (2.8) the following way:

$$s \frac{\sin^2 s}{(s^2 - \pi^2 m^2)(s^2 - \pi^2 n^2)} = \frac{1}{\pi^2(m^2 - n^2)} \left[\frac{s}{s^2(s^2 - \pi^2 m^2)} - \frac{s}{s^2(s^2 - \pi^2 n^2)} \right].$$

Then the corresponding integral is represented as

$$\frac{2}{\pi^3(m^2 - n^2)} \left[\int_0^\infty \frac{s \sin^2 s \, ds}{s^2(s^2 - \pi^2 m^2)} - \int_0^\infty \frac{s \sin^2 s \, ds}{s^2(s^2 - \pi^2 n^2)} \right]$$

$$:= \frac{2}{\pi^3(m^2 - n^2)} \left[\mathcal{J}_m - \mathcal{J}_n \right]. \tag{2.10}$$

Each of integrals in the right side can be calculated, due to Jordan lemma, as an integral on the imaginary axis p, e.g.:

$$\mathcal{J}_m = \frac{1}{2} \int_0^{i\infty} \frac{1 - e^{2is}}{s\left(s^2 - \pi^2 m^2\right)}\, ds = -\frac{1}{2}\int_0^{\infty} \frac{1 - e^{-2t}}{t\left(t^2 - \pi^2 m^2\right)}\, dt$$

The last integral is a sum of two integrals $\int_0^A + \int_A^\infty := \mathcal{J}_m^A + \mathcal{J}_m^\infty$. The first of them is estimated for large m by $\mathrm{Const}\, m^{-2}$, and the second is calculated explicitly after neglecting the exponential e^{-t}:

$$\mathcal{J}_m^\infty \approx \frac{1}{4}\ln \frac{A^2 + \pi^2 m^2}{A^2} \approx \frac{1}{2}\ln m. \tag{2.11}$$

Taking into account only the dominating term for large m we obtain, due to (2.10) the following asymptotic for the integral (2.8) for $m, n \to \infty$

$$\frac{\pi l}{4}\, \tilde{\mathcal{K}}_{2m,2n}(0) = \int_0^\infty s\, \frac{\sin^2 s}{(s^2 - \pi^2 m^2)(s^2 - \pi^2 n^2)}\, ds \approx \frac{\ln m/n}{\pi^3(m^2 - n^2)}, \quad m > n > 0. \tag{2.12}$$

End of the proof. $\qquad\qquad\qquad\qquad\qquad\qquad\qquad\qquad\qquad\qquad\square$

Corollary. The operator $\tilde{\mathcal{K}}(0)$ belongs to the Hilbert-Schmidt class because the series $\sum_{rt} |\tilde{\mathcal{K}}_{rt}(0)|^2 = \mathrm{Trace}\, \tilde{\mathcal{K}}^+ \tilde{\mathcal{K}}$ is convergent. Convergence of this series, due to smoothness of the asymptotic (2.12), is equivalent to the convergence of the corresponding integral on the first quadrant outside the unit disc:

$$\frac{1}{\pi^6}\int_{m^2+n^2 \geq 1} \frac{|\ln m/n|^2}{(m^2 - n^2)^2}\, dm\, dn = \frac{1}{\pi^6}\int_{\rho \geq 1}\frac{d\rho}{\rho^3}\int_{\theta=0}^{\pi/2} \frac{|\ln\tan\theta|^2}{|\cos 2\theta|^2}\, d\theta.$$

It is convergent because the integrand is a bounded continuous function of θ. Similar statement is true for $\tilde{\mathcal{K}}(p)$, $-\infty < p < \infty$, as well. This statement allows us to calculate the eigenvalues of the operator $\tilde{\mathcal{K}}(p)$ approximating $\tilde{\mathcal{K}}(p)$ by finite cut-off matrices, see Appendix B.

Summarizing above results we conclude that in case of simplest geometry of the device, with only two basic electrodes, the spectrum of the problem (1.13) has band-structure with thresholds defined by maxima κ_m of the eigenvalues $\kappa_m(p)$ of the operator $\tilde{\mathcal{K}}(p)$. One can guess that these maximal values are achieved at $p = 0$, then the upper thresholds of the lowest spectral bands can be calculated from the data given in Appendix B.

3. The slot-device with governing electrodes

Additional governing electrodes make the geometry of the slot-device non-trivial. Spectral properties of the corresponding matrices can't be revealed via separation of variables. The governing electrodes define new important properties of the device, in particular they permit the resonance manipulation of the plasma-waves in

88   A.B. Mikhailova, B. Pavlov and V.I. Ryzhii

the slot. The corresponding device with periodic array of governing electrodes can possess even more interesting spectral properties defined by the resonance band-gaps, see for instance [23, 24, 25, 26, 27, 28], where, in particular, the resonance effects caused by the "decoration" at the nodes are discussed.

In this section we consider the device with few governing electrodes γ_1, $\gamma_2, \ldots,$ two basic electrodes Γ_-, Γ_+ and one flat plasma-channel Γ squeezed between Γ_-, Γ_+. We will use DN-maps for the domains $\Omega_\pm = R_\pm \setminus \{\gamma_s \ldots\}$.

Begin with Dirichlet problem with boundary data on the plane

$$S = \{\Gamma \cup \Gamma_+ \cup \Gamma_-\} = \{z = 0\},$$

and on the governing electrodes γ_s, $s = 1, 2, 3, \ldots$. In [10] we suggested an iterative construction for calculation of the Poisson map in Ω based on the Poisson-maps in $\Omega_1 = R_3 \setminus \gamma_1$, $\Omega_2 = R_3 \setminus \gamma_2 \ldots$ In special case when γ_s, $s = 1, 2, \ldots$ are circular cylinders (rods), the corresponding Poisson maps are known, see [11]. The Poisson-kernel for the half-space $z > 0$ is:

$$\mathcal{P}_0(x, y, z; \xi, \eta, 0) = \frac{1}{4\pi^2} \int \int dp\, dq\, e^{-\sqrt{p^2 + q^2}\, z}\, e^{ip(x-\xi) + iq(y-\eta)}, \tag{3.1}$$

and the Poisson-kernel of the complement $R_3 \setminus \gamma_s$ of the circular cylinder γ_s radius ρ_s is

$$\mathcal{P}_s(\varphi, \rho, y; \theta, \eta) = \int_{-\infty}^{\infty} dq \sum_{k=-\infty}^{\infty} e^{ik(\varphi - \theta)}\, e^{iq(y-\eta)}\, \frac{\mathcal{H}_k^1(q\rho)}{\mathcal{H}_k^1(q\rho_s)}, \tag{3.2}$$

where $\mathcal{H}_k^1$ is the conventional Hankel function of the first kind.

Note that the potentials on the governing electrodes can be chosen such that the plasma-current is developed only in the slot between governing electrodes. Really, assume that the slot is a straight strip $\Gamma \subset S$ of the constant width $2l$ on the plane S between the basic electrodes $\Gamma_\pm$, with the voltages $V_\pm$ on them. Then the plasma-current will develop on Γ if the electric field on the slot is strong enough:

$$E_- < \frac{d_-}{2l}\left[V_+ - V_-\right]. \tag{3.3}$$

Here E_- is the ionization thresholds (the electron's "exit work") on Γ_-, $2l$ is the distance between Γ_+, Γ_- and d_- the thickness of the surface layer of the dimensional quantization near the edge of Γ_-. Physically the plasma-current can develop also between the governing electrodes and the basic electrode Γ_+. We assume now that it is not the case, choosing the potentials V_s between $V_\pm$ and the ionization thresholds E_s, E_- (the electron's exit work) on the governing electrodes and the negative electrode is large enough, compared with the voltage between the basic and governing electrodes,

$$E_s > \frac{d_s}{l_{s,+}}\left[V_+ - V_s\right], \quad E_- > \frac{d_-}{l_{-,s}}\left[V_s - V_-\right]. \tag{3.4}$$

Here $l_{s,+}$, $l_{-,s}$ are distances from γ_s to Γ_+ and from Γ_- to γ_s, and d_s is the thickness of the surface layer on γ_s.

We postpone to the forthcoming publication the discussion of the above physical limitations (3.3,3.4) for typical materials, but will concentrate now on derivation of equations for calculation of the amplitudes Ξ_ω, u_ω, φ_ω of oscillations of the plasma-current near the equilibrium values Ξ_0, φ_0, u_0.

The Dirichlet-to-Neumann map in the complement of the device is obtained from the Poisson map, see Appendix A. Hence the central problem of mathematical design of the slot-device, with additional electrodes γ_s for manipulating running waves on the slot, is the construction of Poisson map (or, equivalently, the Dirichlet Green function) for Laplacian on the basic domain $\Omega = R_3 \setminus \{\Gamma_+ \cup \Gamma_- \cup \gamma_1 \cup \gamma_2 \cup \cdots \}$. We will use the notations $\mathbf{P}_0$, $\mathbf{P}_s$, $s = 1, 2, \ldots$ for Poisson maps of the half-space R_+ $(z > 0)$ and the Poisson maps on the complements $\Omega_s = R_3 \setminus \gamma_s$ of the governing electrodes in R_3.

Denote by $\mathcal{P}_{st}$ the restriction onto $\partial\gamma_s$ of the Poisson map $\mathbf{P}_t$. This map transfers the space $C_{\partial\gamma_s}$ of continuous functions on $\partial\gamma_t$ into the space $C_{\partial\gamma_s}$ of continuous functions on $\partial\gamma_s$:

$$C_{\partial\gamma_s} \xleftarrow{\mathcal{P}_{st}} C_{\partial\gamma_t}.$$

The following statement (3.1) shows that the Poisson map $\mathbf{P}_{st}$ of the domain $\Omega_{(st)} = \Omega_s \cap \Omega_t$ can be constructed of the "partial" Poisson maps $\mathbf{P}_s$, $\mathbf{P}_t$ of the domains $\Omega_s = R_3 \setminus \gamma_s$, $\Omega_t = R_3 \setminus \gamma_t$, if the domains γ_s, γ_t are "separated" in certain sense. More precise,

We say that the domains are separated if there exist a domain $\hat{\gamma}_s \supset \gamma_s$, $\hat{\gamma}_s \cap \gamma_t = \emptyset$ such that the solution of the exterior Dirichlet problem

$$\Delta \hat{u}_s = 0, \ \hat{u}_s \Big|_{\partial\hat{\gamma}_s} = 1, \ \hat{u}_s(x) \to 0 \ \textit{if } x \to \infty$$

allows the estimation $\sup \hat{u}_s(x)\Big|_{\partial\gamma_t} < 1$, *and there exist a domain $\hat{\gamma}_t \supset \gamma_t$, with a similar property with respect γ_s.*

In particular, the domains are *separated* if each of them is contained in a ball, a cylinder of a half-space $B_s = \hat{\gamma}_s$ with the described property, since the corresponding harmonic function $\hat{u}_s$ can be constructed in that case explicitly. In all typical constructions of devices the condition of separation is fulfilled.

Theorem 3.1. *If the domains γ_s, γ_t are separated then the operators $I - \mathcal{P}_{st}\mathcal{P}_{ts}$, $I - \mathcal{P}_{ts}\mathcal{P}_{st}$ are invertible in $C_{\partial\gamma_s}$, $C_{\partial\gamma_t}$ respectively and the Poisson map $\mathbf{P}_{(st)}$ in the domain $\Omega_{(st)} = \Omega_s \cap \Omega_t$ is presented as*

$$\{\mathbf{P}_s, \ \mathbf{P}_t\} \left(\begin{array}{cc} \dfrac{I}{I - \mathcal{P}_{st}\mathcal{P}_{ts}} & -\dfrac{I}{I - \mathcal{P}_{st}\mathcal{P}_{ts}}\mathcal{P}_{st} \\[4mm] -\dfrac{I}{I - \mathcal{P}_{ts}\mathcal{P}_{st}}\mathcal{P}_{ts} & \dfrac{I}{I - \mathcal{P}_{ts}\mathcal{P}_{st}} \end{array} \right) := \left\{ \mathbf{P}^s_{(st)}, \ \mathbf{P}^t_{(st)} \right\},$$

so that the solution of the Dirichlet problem with the boundary data u_s, u_t on $\partial\gamma_s$, $\partial\gamma_t$ is presented as

$$u(x) = \int_{\partial\gamma_s} \mathbf{P}^s_{(st)}(x,\xi_s)u_s(\xi_s)d\xi_s + \int_{\partial\gamma_t} \mathbf{P}^t_{(st)}(x,\xi_t)u_t(\xi_t)d\xi_t.$$

Proof. Note that the maximum principle can be applied to the bounded harmonic functions on the domains obtained from R_+ via removing γ_s. Assuming that $(s,t) = (1,2)$ and the Poisson maps $\mathbf{P}_1$, $\mathbf{P}_2$ in Ω_1, Ω_2 are known, construct the solution of the Laplace equation in $R_3 \setminus (\gamma_1 \cup \gamma_2)$ with data u_s on $\partial\gamma_s$ in form

$$u = \mathbf{P}_1 \hat{u}_1 + \mathbf{P}_2 \hat{u}_2, \tag{3.5}$$

with still non defined "re-normalized" boundary data $\hat{u}_1$, $\hat{u}_2$. Then we obtain the following linear system for $\hat{u}_1$, $\hat{u}_2$:

$$\hat{u}_1 + \mathcal{P}_{12}\hat{u}_2 = u_1$$
$$\mathcal{P}_{21}\hat{u}_1 + \hat{u}_2 = u_2. \tag{3.6}$$

The operators $\mathcal{P}_{st}$ for $s \neq t$ are contracting in $C_{\partial\gamma_s} \times C_{\partial\gamma_t}$, due to maximum principle, hence the system (3.6) has unique solution

$$\hat{u}_1 = \frac{I}{I - \mathcal{P}_{12}\mathcal{P}_{21}}[u_1 - \mathcal{P}_{12}\,u_2]$$

$$\hat{u}_2 = \frac{I}{I - \mathcal{P}_{21}\mathcal{P}_{12}}[u_2 - \mathcal{P}_{21}\,u_1] \tag{3.7}$$

defined by the renorm-matrix corresponding to $\gamma_{12} = \gamma_1 \cup \gamma_2$:

$$\begin{pmatrix} \frac{I}{I - \mathcal{P}_{12}\mathcal{P}_{21}} & -\frac{I}{I - \mathcal{P}_{12}\mathcal{P}_{21}}\mathcal{P}_{12} \\ -\frac{I}{I - \mathcal{P}_{21}\mathcal{P}_{12}}\mathcal{P}_{21} & \frac{I}{I - \mathcal{P}_{21}\mathcal{P}_{12}} \end{pmatrix} := \mathcal{R}_{\gamma_{12}}. \tag{3.8}$$

This matrix transforms the boundary data u_1, u_2 into re-normalized data $\hat{u}_1$, $\hat{u}_2$ which can be used for construction of the solution of the original boundary problem by the formula (3.5) based on partial Poisson maps $\mathbf{P}_1$, $\mathbf{P}_2$. Then the Poisson map $\mathbf{P}_{(12)}$ in the complement $R_3 \setminus (\gamma_1 \cup \gamma_1)$ of the electrodes is obtained as the matrix product row by column:

$$\{\mathbf{P}_1, \mathbf{P}_2\}\,\mathcal{R}_{\gamma_1\gamma_2} := \left\{\mathbf{P}^1_{(12)}, \mathbf{P}^2_{(12)}\right\} := \mathbf{P}_{(12)}$$

so that we obtain the solution of the Dirichlet problem in $\Omega_{12} = R_3 \setminus (\gamma_1 \cup \gamma_2)$ in form:

$$u = \mathbf{P}_{(12)}\{u_1, u_2\}, \tag{3.9}$$

as announced. End of the proof. $\square$

Remark. The corresponding DN-map is obtained via differentiation of the constructed Poisson map with respect to the outward normal on each component $\partial\Omega_1$, $\partial\Omega_2$ of $\partial\Omega_{12}$. This can be presented symbolically as:

$$\Lambda_{(12)} = \{\Lambda_1, \Lambda_2\}\,\mathcal{R}_{\gamma_1\gamma_2}, \tag{3.10}$$

where Λ_s acts on $\partial\gamma_s$ and Λ_t acts on $\partial\gamma_t$.

Now the introduction of the electrodes $\gamma_3, \gamma_4, \ldots, \gamma_0$ into the scheme can be done by induction, if the separation conditions are fulfilled on each step: first we construct the corresponding renorm-matrix of restrictions $\mathcal{P}_{(12)3}$ and $\mathcal{P}_{3(12)}$ of the Poisson maps $\mathbf{P}_3$ onto $\partial\gamma_{12}$ and $\mathbf{P}_{12}$ onto $\partial\gamma_3$, respectively:

$$\mathcal{R}_{\gamma_3\,\gamma_{(12)}} = \begin{pmatrix} \dfrac{I}{I-\mathcal{P}_{3(12)}\,\mathcal{P}_{(21)3}} & -\dfrac{I}{I-\mathcal{P}_{3(12)}\,\mathcal{P}_{(21)3}}\mathcal{P}_{3(12)} \\[2ex] -\dfrac{I}{I-\mathcal{P}_{(12)3}\,\mathcal{P}_{3(12)}}\mathcal{P}_{(12)3} & \dfrac{I}{I-\mathcal{P}_{(12)3}\,\mathcal{P}_{3(12)}} \end{pmatrix},$$

then the corresponding Poisson map is obtained as the matrix product

$$\mathbf{P}_{(123)} = \{\mathbf{P}_3,\,\mathbf{P}_{(12)}\}\,\mathcal{R}_{\gamma_3\,\gamma_{(12)}}.$$

The corresponding Dirichlet-to-Neumann map is obtained as:

$$\Lambda_{(312)} = \{\Lambda_3,\,\Lambda_{12}\}\,\mathcal{R}_{\gamma_3\,\gamma_{(12)}}. \tag{3.11}$$

Convenient approximate formulae are obtained via replacement inverse operators by a finite sum of the corresponding Neumann series, for instance: $\left[I - \mathcal{P}_{12}\,\mathcal{P}_{21}\right]^{-1} = I + \mathcal{P}_{12}\,\mathcal{P}_{21} + \mathcal{P}_{12}\,\mathcal{P}_{21}\mathcal{P}_{12}\,\mathcal{P}_{21}\cdots$.

In particular case when the electrodes γ_s, γ_t are cylinders parallel to Γ and to each other, each of exterior Dirichlet problem with electrodes admits separation of the variable along the electrode. Then the Poisson map and the DN-map of the 3-d problem can be represented by Fourier transform based on the Poisson map and DN-map of the 2-d problem on the orthogonal section. Then we are able to summarize the algorithm of derivation of the equation (1.12) in case of two governing electrodes, based on formulae obtained for the DN-map $\Lambda_{(0,1,2)}$. Assume that $\gamma_1 \cup \gamma_2 \in \Omega^+$, and $R_3 \backslash \Omega^+ := \Omega^-$, $\gamma_s \in \Omega^+$, $\gamma_s \cap \Omega^- = \emptyset$, $s = 1, 2$. Due to the translation symmetry of $\Omega^+ = \Omega_0 \backslash [\gamma_1 \cup \gamma_2]$, the kernel $\Lambda_{(0,1,2)}(x,y,z)$ is connected to the kernel of the DN-map $\Lambda_{(0,1,2)}^+$ of the Helmholtz equation $-\triangle_2 u = \rho^2 u$ on the orthogonal cross-section (x,z) of $R_2 \cap \Omega^+$ by the formula

$$\Lambda_{(0,1,2)}^+(x,y,z;\xi,\zeta,\eta) = \int_{-\infty}^{\infty} \Lambda_{(0,1,2)}^+(x,z;\xi,\zeta;\rho)e^{i\rho(y-\eta)}\,d\rho. \tag{3.12}$$

Here (x,y,z), $(\xi,\zeta,\eta) \in \partial\Omega^+$, $(0,y,z)$, $(0,\eta,\zeta) \in R_2 \cap \partial\Omega$. Similarly the DN-map $\Lambda_{(0,1,2)}^-$ is defined by formula in Ω_- similar to (2.1). Then the equation for the restriction amplitude φ_ω of the electrostatic potential onto the slot is presented in form:

$$\left[\Lambda_+ + \Lambda_-\right]\varphi_\omega\Big|_\Gamma = \frac{2}{q}\mathrm{div}_2\,\Xi_0(x,y)\nabla_2\varphi_\omega\Big|_\Gamma. \tag{3.13}$$

Here we assume that the equilibrium concentration and equilibrium values of other parameters are obtained via solution of the system (1.5).

4. Plasma waves in the rigged channel

Consider the perturbed Laplacian on the slot

$$L_\Gamma = -\nabla_2 \Xi_0(x,y)\nabla_2,$$

with zero boundary conditions on the border $y = \pm l$. The electron's concentration $\Xi_0(x,y)$ on the slot is a function of two variables x, y which has, in case of two governing electrodes, the asymptotic Ξ_0 at infinity, $x \to \pm\infty$. If the additional electrodes are absent, then the electron concentration Ξ_0 is constant. If the additional electrons are present, with negative potentials V_s on them, we obtain a *rigged plasma channel* on the slot. On the compact part of the slot $\Xi_0(x,y)$ is defined by the configuration of the governing electrodes, and can be found in course of solution of the auxiliary stationary problem (1.4). We are not going to solve this problem now, but we may guess that, under the above conditions on stationary potentials $V_s < 0$, the equilibrium concentration is suppressed on the slot near to the governing electrodes due to Coulomb interaction with the negative potentials V_s on the electrodes.

4.1. Scattering of plasma waves in the rigged channel

In practice the stationary electron concentration Ξ_0 depends essentially only on the variable x along the channel. Then the spectral problem on the slot admits separation of variables

$$L_\Gamma \Psi = -\frac{\partial}{\partial x}\Xi_0(x)\frac{\partial \Psi}{\partial x} - \Xi_0(x)\frac{\partial^2}{\partial y^2}\Psi = \lambda\Psi. \tag{4.1}$$

For positive rapidly stabilizing concentration $\Xi_0(x) \to \Xi_0$ when $x \to \pm\infty$, the spectrum of the problem (4.1) is pure continuous. It has band-structure with step-wise growing multiplicity:

$$\sigma(L_\Gamma) = \cup_{r=1}^{\infty}\sigma_r,$$

of branches $\sigma_r = \left[\Xi_0\frac{\pi^2 r^2}{4\,l^2}, \infty\right)$. The corresponding scattered waves $\Psi(x,y) = \Psi_r(x,y,\lambda) = \frac{1}{\sqrt{l}}\sin\frac{\pi r(y+l)}{2l}\psi_r^+(x)$ fulfil (4.1), and the amplitude $\psi_r^+(x)$ of the scattered wave in the open channel, $\lambda > \Xi_0\frac{\pi^2 r^2}{4\,l^2}$ is a bounded solution of the spectral problem on the channel

$$-\frac{d}{dx}\Xi(x)\frac{d\psi_r(x)}{dx} + \Xi(x)\frac{\pi^2 r^2}{4\,l^2}\psi_r(x) = \lambda\psi_r(x)$$

with appropriate asymptotic at infinity. For the plasma waves incoming from $+\infty$ of x-axis, in open channels $\lambda\,\Xi_0^{-1} - \frac{\pi^2 r^2}{4l^2} > 0$

$$\overleftarrow{\psi}_r \approx e^{iQ_0^r x} + \overrightarrow{R}e^{-iQ_0^r x} \quad \text{when } x \to +\infty,$$

and

$$\overleftarrow{\psi}_r \approx \overleftarrow{T}e^{iQ_0^r x} \quad \text{when } x \to -\infty,$$

were $Q_0^r(\lambda) = \sqrt{\lambda\,\Xi_0^{-1} - \frac{\pi^2 r^2}{4l^2}}$. For plasma-waves initiated from $-\infty$ the asymptotic are

$$\overrightarrow{\psi}_r \approx e^{-iQ_0^r x} + \overleftarrow{R}e^{iQ_0^r x} \quad \text{when } x \to -\infty,$$

$$\overrightarrow{\psi}_r \approx \overrightarrow{T}e^{-iQ_0^r x} \quad \text{when } x \to +\infty.$$

The system of all scattered waves $\overleftarrow{\psi}_r$, $\overrightarrow{\psi}_r$, $\Xi_0 \pi^2\, r^2\, (2l)^{-2} < \lambda < \infty$ is complete and orthogonal in each channel (for each r). The whole system of eigenfunctions

$$\overleftarrow{\Psi}_r(x, y, \lambda) = \frac{1}{\sqrt{l}} \sin \frac{\pi\, r\,(y+l)}{2l} \overleftarrow{\Psi}_r, \quad r = 1, 2, \ldots$$

$$\overrightarrow{\Psi}_r(x, y, \lambda) = \frac{1}{\sqrt{l}} \sin \frac{\pi\, r\,(y+l)}{2l} \overrightarrow{\Psi}_r, \quad r = 1, 2, \ldots \tag{4.2}$$

in all open channels, $r = 1, 2, \ldots$ is complete and orthogonal in the space $L_2(\Gamma)$ of all square-integrable functions on the slot. Then the Green function of L_Γ is presented in spectral form as

$$[L_\Gamma - \mu I]^{-1}(x, \xi)$$
$$= \sum_r \int_0^\infty \frac{1}{\lambda - \mu} \left[\overrightarrow{\Psi}_r(x, \lambda)\rangle\langle\overrightarrow{\Psi}_r(\xi, \lambda) + \overleftarrow{\Psi}_r(x, \lambda)\rangle\langle\overleftarrow{\Psi}_r(\xi, \lambda) \right] \frac{dQ_0^r(\lambda)}{2\pi}.$$

We will use this formula for the regular point $\mu = 0$. It is convenient, following the previous section, to re-write the spectral problem (3.13) in form of equation similar to (2.3):

$$L_\Gamma^{-1/2} \left[\Lambda_- + \Lambda_+ \right] L_\Gamma^{-1/2} u = \frac{2}{q} u. \tag{4.3}$$

If the governing electrodes are absent, then the operator $\Lambda_- + \Lambda_+$, being reduced onto the slot, commutes with L_Γ, and the operator in the left side of the equation (4.3) in Fourier representation is just a multiplication by the 2×2 matrix $\mathcal{K}$ investigated in Section 2. If the governing electrodes are present, we generally obtain a sophisticated analytical problem. We postpone discussion of the corresponding general problem to a forthcoming publication, but consider a model of the above scattering process parametrized by Weyl functions of the restriction of the differential operator L_Γ onto the part of the slot near to the governing electrode.

4.2. Solvable model of the simplest rigged channel

Consider the special case when two cylindrical governing electrodes $\gamma_\pm$ are placed in upper and lower half-spaces $\Omega_\pm$ respectively on equal distances from the slot and parallel to each other and to the horizontal plane $S : z = 0$. We assume that the electrodes are skew-orthogonal to the slot Γ situated between the electrodes $\Gamma_\pm \subset S$ on the horizontal plane S. We do not calculate the electron's concentration $\Xi_0(x)$ via solution of the equations (1.4), but just assume that it depends only on the variable x along the slot:

$$\Xi_0(x) = \begin{cases} \sigma(x), & \text{if } -L < x < L, \\ \Xi_0, & \text{if } |x| > L. \end{cases} \tag{4.4}$$

We assume that Ξ_0 coincides with equilibrium electron concentration on the slot without governing electrodes, and is suppressed near the governing electrodes: $0 < \sigma(x) = \sigma(-x) < \Xi_0$. The scattered waves of the spectral problem

$$-\frac{d}{dx}\Xi_0(x)\frac{du}{dx} + \Xi_0(x)\frac{d^2u}{dy^2} = \lambda u \tag{4.5}$$

are found via separation of variables

$$-\frac{d}{dx}\Xi_0(x)\frac{d\varphi^r}{dx} + \Xi_0(x)\frac{\pi^2 r^2}{4L^2}\varphi^r = \lambda\varphi^r \tag{4.6}$$

and matching of exponentials with solutions of the homogeneous equation (4.6). Due to the symmetry of the equilibrium concentration on $(-L, L)$ it is convenient to parametrize the partial scattering matrices S^r in the channels by the Weyl functions of the operator (4.6) on $(0, L)$. Denote by $\varphi^r(x, \lambda)$, $\theta^r(x, \lambda)$ the solutions of (4.6) with the initial conditions at the origin $x = 0$

$$\varphi^r(0, \lambda) = 0,\ \frac{d\varphi^r}{dx}(0, \lambda) = 1;\ \theta^r(0, \lambda) = 1,\ \frac{d\theta^r}{dx}(0, \lambda) = 0.$$

Then the reflection coefficients of the spectral problems with Dirichlet and Neumann boundary conditions at the origin are found based on the Ansatz:

$$\psi^r_D(x, \lambda) = \begin{cases} \alpha\,\varphi^r(x, \lambda) & \text{if } 0 < x < L, \\ e^{iQ^r_0(x-L)} + S^D_r(\lambda)e^{-iQ^r_0(x-L)} & \text{if } L < x < \infty, \end{cases}$$

$$\psi^r_N(x, \lambda) = \begin{cases} \beta\,\theta^r(x, \lambda), & \text{if } 0 < x < L, \\ e^{iQ^r_0(x-L)} + S^N_r(\lambda)e^{-iQ^r_0(x-L)} & \text{if } L < x < \infty. \end{cases} \tag{4.7}$$

Denoting by

$$\mathcal{M}^D_r(\lambda) = \frac{d\varphi^r}{dx}(L, \lambda)\left[\varphi^r(L, \lambda)\right]^{-1},\ \mathcal{M}^N_r(\lambda) = \frac{d\theta^r}{dx}(L, \lambda)\left[\theta^r(L, \lambda)\right]^{-1}$$

the Weyl functions, see [12], at the point $x = L$, of the equation (4.6) with the Dirichlet and Neumann boundary conditions at the origin, we obtain:

$$S^D_r(\lambda) = \frac{iQ^r_0 - \mathcal{M}^D_r(\lambda)}{iQ^r_0 + \mathcal{M}^D_r}(\lambda),\ S^N_r(\lambda) = \frac{iQ^r_0 - \mathcal{M}^N_r(\lambda)}{iQ^r_0 + \mathcal{M}^N_r(\lambda)}. \tag{4.8}$$

We will assume that the scattered waves $\psi^r_D(x, \lambda)$, $\psi^r_N(x, \lambda)$ are continued on the whole x-axis as odd and even functions respectively. Then we may construct of them the scattered waves of the Schrödinger operator on the whole x-axis $\overleftarrow{\Psi}_r$, $\overrightarrow{\Psi}_r$ initiated from the right and from the left infinity respectively. Introducing the partial scattered waves and reflection/transmissoin coefficients by the Ansatz

$$\overleftarrow{\Psi}_r(x, \lambda) = \begin{cases} e^{iQ^r_0 x} + \overrightarrow{R}_r(\lambda)e^{iQ^r_0 x}, & \text{if } L < x < \infty, \\ \overleftarrow{T}_r(\lambda)e^{iQ^r_0 x}, & \text{if } -\infty < x < L. \end{cases}$$

we obtain $\overleftarrow{\Psi}_r(x,\lambda) = \frac{1}{2}[\psi^r_D(x) + \psi^r_N(x)]$ and hence

$$\overrightarrow{R}_r(\lambda) = \frac{1}{2}\left[S^D_r + S^N_r\right] e^{2iQ^r_0 L}, \quad \overleftarrow{T}_r(\lambda) = \frac{1}{2}\left[S^N_r - S^D_r\right] e^{2iQ^r_0 L}.$$

Similarly $\overleftarrow{\Psi}_r(x,\lambda) = \frac{1}{2}[\psi^r_N(x) - \psi^r_D(x)]$ and

$$\overleftarrow{\Psi}_r(x,\lambda) = \begin{cases} e^{-iQ^r_0 x} + \overleftarrow{R}_r(\lambda)\, e^{iQ^r_0 x}, & \text{if} \quad -\infty < x < -L, \\ \overrightarrow{T}_r(\lambda)\, e^{iQ^r_0 x}, & \text{if} \quad L < x < \infty, \end{cases}$$

where

$$\overleftarrow{R}_r(\lambda) = \frac{1}{2}\left[S^N_r + S^D_r\right] e^{2iQ^r_0 L}, \quad \overrightarrow{T}_r(\lambda) = \frac{1}{2}\left[S^N_r - S^D_r\right] e^{2iQ^r_0 L}.$$

If the spectrum of L_Γ is pure absolutely continuous, then the standard expansion by the corresponding scattered waves is given by the formula

$$u_r = \frac{1}{2\pi}\int_0^\infty \left[\overleftarrow{\Psi}_r(\lambda)\rangle\langle\overleftarrow{\Psi}_r(\lambda), u\rangle + \overrightarrow{\Psi}_r(\lambda)\rangle\langle\overrightarrow{\Psi}_r(\lambda), u\rangle\right] dQ^r_0 = \mathcal{F}^+_r\, \mathcal{F}_r\, u_r.$$

To derive a formula for the corresponding operator K, similar to (2.2), we represent the generalized kernel of the formal integral operator in the left part of the equation (1.12) using the translation invariance of the system of electrodes:

$$\left[\Lambda_- + \Lambda_+\right](x, y, 0; \xi, \eta, 0) := \frac{1}{\sqrt{2\pi}}\int_{-\infty}^{+\infty} e^{iq(y-\eta)}\left[\Lambda_-(q, x, \xi) + \Lambda_+(q, x, \xi)\right]\, dq.$$

The kernels $\Lambda_\pm(q, x, \xi)$ are Fourier transforms of the kernels of the DN-maps $\Lambda_\pm$. Multiplying the left side of (1.12) by $L_\Gamma^{-1/2}$ from both sides, and using the notations introduced in Section 3

$$\frac{1}{\sqrt{2l}}\int_{-l}^l e^{iqy}\sin\frac{\pi r(y+l)}{2l}\, dy = \mathcal{J}^c_r(q) + i\mathcal{J}^s_r(q) := \mathbf{J}_r(q),$$

we obtain the operator $L_\Gamma^{-1/2}\left[\Lambda_- - \Lambda_+\right]L_\Gamma^{-1/2} := \mathcal{K}$ in form of an integral operator K with the generalized matrix kernel with respect to the basis $\left\{\frac{1}{\sqrt{2l}}\sin\frac{\pi r(y+l)}{2l}\right\}_{r=1}^\infty$:

$$\frac{1}{8\pi^3}\int_{-\infty}^\infty dq \int_0^\infty\int_0^\infty \frac{dQ^r_0(\lambda)\, dQ^t_0(\mu)}{\sqrt{\lambda\,\mu}}\, \mathbf{J}_r(q)\,\overleftarrow{\Psi}_r(\lambda, x)\rangle\langle\overrightarrow{\Psi}_r(\lambda, x)$$

$$\begin{pmatrix} \langle\overleftarrow{\Psi}_r(\lambda)\left[\Lambda_-(q) + \Lambda_+(q)\right]\overleftarrow{\Psi}_t(\mu)\rangle, & \langle\overleftarrow{\Psi}_r(\lambda)\left[\Lambda_-(q) + \Lambda_+(q)\right]\overrightarrow{\Psi}_t(\mu)\rangle \\ \langle\overrightarrow{\Psi}_r(\lambda)\left[\Lambda_-(q) + \Lambda_+(q)\right]\overleftarrow{\Psi}_t(\mu)\rangle, & \langle\overrightarrow{\Psi}_r(\lambda)\left[\Lambda_-(q) + \Lambda_+(q)\right]\overrightarrow{\Psi}_t(\mu)\rangle \end{pmatrix}$$

$$\times\begin{pmatrix} \langle\overleftarrow{\Psi}_t(\lambda, \xi) \\ \langle\overrightarrow{\Psi}_t(\lambda, \xi) \end{pmatrix}\overline{\mathbf{J}}_t := K^L_{r,t}(x, \xi)$$

$$:= \mathcal{F}^+_r\, \mathcal{K}^L_{rs}(p)\,\mathcal{F}_s. \tag{4.9}$$

The matrix $\mathcal{K}^L(p)$ is similar to (2.2), but the role of Fourier transform is played now by the spectral transformations $\mathcal{F}_s$ defined by the scattered waves of the "partial" operators L_s on the slot. One may guess that the matrix $K^L(p)$ for each p belongs to Hilbert-Schmidt class, similarly to the above operator $\mathcal{K}(p)$. If this conjecture is right, the the eigenvalues of $\mathcal{K}(p)$ can be calculated via finite-dimensional approximation.

One can see from (4.9) that the operator $\mathcal{K}^L$ contains Weyl functions $\mathcal{M}_{r,t}^{D,N}$ which define the scattering matrix of the spectral problem (4.1) and hence the resonance and transport properties of the plasma channel. These transport properties may be manipulated via varying the potential(s) on the governing electrodes which affect the "bumps" of the stationary electron concentration and hence the resonance transmission of the plasma waves in the channel.

5. Appendix A: Dirichlet-to-Neumann map – basic facts

We describe here general features of the DN-map, see also [14, 16, 17], for Laplace operator defined in the space $L_2(\Omega)$ of square-integrable functions by the differential expression

$$L_D v = -\triangle v$$

on the class of twice differentiable functions $-\triangle v \in L_2(\Omega)$ vanishing on the piecewise smooth boundary $\Gamma = \partial\Omega$ of the domain $\Omega \subset R_3$. In this section $x = (x_1, x_2, x_3)$ and $y = (y_1, y_2, y_3)$ are three-dimensional variables. If the boundary of the domain has inner corners, then we assume that functions from the domain are submitted to the additional Meixner condition in form $\int_\Omega |\nabla v|^2 dx < \infty$. This condition guarantees uniqueness of solution of the non-homogeneous equation $L^D v - \lambda v = f \in L_2(\Omega)$ for complex values of the spectral parameter λ. Together with the operator $L := L^D$ we may consider the operator L^N defined by the same differential expression L with homogeneous Neumann conditions on the boundary

$$\left.\frac{\partial v}{\partial n}\right|_{\partial\Omega} = 0,$$

Both $L := L^D$ and L^N are *self-adjoint operators* in $L_2(\Omega)$. Corresponding resolvent kernels $G^{N,D}(x, y, \lambda)$ and the Poisson kernel

$$\mathcal{P}_\lambda(x, y) = -\frac{\partial G^D(x, y, \lambda)}{\partial n_y}, \ y \in \Gamma,$$

for regular values of the spectral parameter λ are locally smooth if $x \neq y$ and square integrable in Ω with boundary values $G^{N,D}(x, y, \lambda)$, $\mathcal{P}(x, y, \lambda)$ from appropriate Sobolev classes. Behavior of $G^N(x, y, \lambda)$, when both x, y are smooth points of the boundary $\Gamma = \partial\Omega$, is described by the following asymptotic which may be derived from the integral equations of potential theory:

$$G^{N}(x, x_\Gamma, \lambda) = \frac{1}{2\pi} \frac{1}{|x - x_\Gamma|} + \mathbf{Q}_\lambda + o(1). \tag{5.1}$$

Here the term $\mathbf{Q}_\lambda$ contains a local geometrical information on Ω near x_Γ and the spectral information, [19]. If the domain is compact, then the spectra $\sigma_{N,D}$ of operators $L^{N,D}$ are discrete and real. Solutions of classical boundary problems for operators $L^{N,D}$ may be represented for regular λ (from the complement of the spectrum) by the "re-normalized" simple layer potentials – for the Neumann problem

$$Lu = \lambda u, \ \left. \frac{\partial u}{\partial n} \right|_{\partial \Omega} = \rho,$$

$$u(x) = \int_{\partial \Omega} G^{N}(x, y, \lambda)\rho(y)d\Gamma, \tag{5.2}$$

and by the re-normalized double-layer potentials – for Dirichlet problem:

$$Lu = \lambda u, \ u\Big|_{\partial \Omega} = \hat{u}, \ u(x) = \int_{\partial \Omega} \mathcal{P}_D(x, y, \lambda)\hat{u}(y)d\Gamma. \tag{5.3}$$

Generally the DN-map is represented for regular points λ of the operator L_D as the derivative of the solution of the Dirichlet problem in the direction of the outer normal on the boundary of the domain Ω:

$$(\Lambda(\lambda)\hat{u})(x_\Gamma) = \left. \frac{\partial}{\partial n} \right|_{x=x_\Gamma} \int_{\partial \Omega} \mathcal{P}_D(x, y, \lambda)\hat{u}(y)d\Gamma. \tag{5.4}$$

The inverse map may be presented at the regular points of the operators $L_N^{\text{in,out}}$:

$$\left(Q^{\text{in,out}}(\lambda)\rho^{\text{in,out}}\right)(x_\Gamma) = \pm \int_{\Gamma} G^{N}_{\text{in,out}}(x, y, \lambda)\rho^{\text{in,out}}(y)d\Gamma. \tag{5.5}$$

The following statement, see [17], shows, that DN-map contains essential spectral information:

Theorem 5.1. *Consider the Laplace operator $L = -\triangle$ in $L_2(\Omega)$ with homogeneous Dirichlet boundary condition at the C_2-smooth boundary Γ of Ω. Then the DN-map Λ of L has the following representation on the complement of the spectrum Ξ_L in complex plane λ, $M > 0$:*

$$\Lambda_{\text{in}}(\lambda) = \Lambda_{\text{in}}(-M) - (\lambda + M)\mathcal{P}^{+}_{-M}\mathcal{P}_{-M} - (\lambda + M)^2\mathcal{P}^{+}_{-M}R_\lambda\mathcal{P}_{-M}, \tag{5.6}$$

where R_λ is the resolvent of L, and $\mathcal{P}_\lambda$ is the Poisson kernel of it. The operator $\mathcal{P}^{+}_{-M}\mathcal{P}_{-M}(x_\Gamma, y_\Gamma)$ is bounded in Sobolev class $W_2^{3/2}(\Gamma)$ of boundary values of twice differentiable functions $\{u : Lu \in L_2(\Omega)\}$ and the operator

$$\left(\mathcal{P}^{+}_{-M}R_\lambda\mathcal{P}_{-M}\right)(x_\Gamma, y_\Gamma) = \sum_{\lambda_s \in \Xi_L} \frac{\frac{\partial \varphi_s}{\partial n}(x_\Gamma)\frac{\partial \varphi_s}{\partial n}(y_\Gamma)}{(\lambda_s + M)^2(\lambda_s - \lambda)}$$

is compact.

Similar statement is true for DN-map in of the Schrödinger operator with rapidly decreasing potential in exterior domain

$$\Lambda_{\rm out}(\lambda) = \Lambda_{\rm out}(-M) + (\lambda + M)\mathcal{P}^{+}_{-M}\mathcal{P}_{-M} + (\lambda + M)^2 \mathcal{P}^{+}_{-M} R_\lambda \mathcal{P}_{-M}, \qquad (5.7)$$

with only difference that first terms of the decomposition contain the DN-map and Poisson kernel for the exterior domain and the last term may contain both the sum over discrete spectrum and the the integral over the absolutely continuous spectrum $\sigma_L^a = [0, \infty)$ of L, with the integrand combined of the normal derivatives of the corresponding scattered waves $\psi(x, |k|, \nu)$, $k = |k|\nu$, $|\nu| = 1$:

$$\mathcal{P}^{+}_{-M} R_\lambda \mathcal{P}_{-M}(x_\Gamma, y_\Gamma) = \frac{1}{(2\pi)^3} \int_{\Xi_1} \int_0^\infty \frac{\frac{\partial \psi}{\partial n}(x_\Gamma, |k|, \nu)\frac{\partial \bar{\psi}_s}{\partial n}(y_\Gamma |k|, \nu)}{(|k|^2 + M)^2(|k|^2 - \lambda)} |k|^2 d\nu \, dk.$$

Example. The Poisson map of Laplacian in the upper subspace $R_3^+ : z > 0$ is defined as

$$\mathcal{P}_\pm \varphi(x) = \frac{1}{2\pi} \int e^{i\langle p, x \rangle} e^{-|p|z} \tilde{\varphi}(p) \, d^2 p,$$

where $\tilde{\varphi} = \mathcal{F}\varphi$. The DN-map in the upper half-plane is given by the formula:

$$\Lambda_+ \varphi(x) = -\frac{\partial \varphi}{\partial z} = \frac{1}{2\pi} \int e^{i\langle p, x \rangle} |p| \, \tilde{\varphi}(p) \, d^2 p.$$

Here we use the notations (x_1, x_2, z) for the Cartesian coordinates $x = (x_1, x_2)$, $\langle p, x \rangle = p_1 x_1 + p_2 x_2$. It is a pseudo-differential operators degree one with the symbol $|p|$. The corresponding jump of the normal derivatives is also calculated as a positive pseudo-differential operator degree one:

$$-\Lambda_+ \varphi(x) - \Lambda_- \varphi(x) = \left(\frac{\partial \varphi}{\partial z}\bigg|_{0^+} - \frac{\partial \varphi}{\partial z}\bigg|_{0^-} \right)\bigg|_\Gamma = -\frac{1}{2\pi} \int e^{i\langle p, x \rangle} 2|p| \, \tilde{\varphi}(p) \, d^2 p.$$

$$(5.8)$$

In Section 2 we use the jump of the normal derivative framed by the projection P_Γ onto $L_2(\Gamma)$ on the slot $-l < y < l$, $-\infty < x < \infty$:

$$P_\Gamma \varphi(x, y) = \sum_{r=1}^\infty \frac{1}{2l} \sin \frac{\pi r(y + l)}{2l} \int_{-l}^l \sin \frac{\pi r(\eta + l)}{2l} \varphi(x, \eta) \, d\eta$$

$$= \sum_{r=1}^\infty \frac{1}{2l \, 2\pi} \sin \frac{\pi r(y + l)}{2l} \int_{-l}^l d\eta \int_{-\infty}^\infty dp \sin \frac{\pi r(\eta + l)}{2l} e^{-ip\eta} \mathcal{F}_1^+ \tilde{\varphi}(p, \eta).$$

This implies the spectral matrix representation for the framed DN-map and the framed jump of the normal derivative on the slot:

$$\left(P_\Gamma \left[-\Lambda_+ + \Lambda_- \right] P_\Gamma \right)_{rs}(p) = \frac{1}{2\pi} \int_{-\infty}^\infty dq \, \mathcal{J}_r(q) \sqrt{p^2 + q^2} \, \mathcal{J}_s(q), \qquad (5.9)$$

where

$$\mathcal{J}_r(p) = \frac{1}{\sqrt{2l}} \int_{-l}^l d\eta \sin \frac{\pi r \xi_2}{2l} e^{-iq\eta}.$$

6. Appendix B: cross-section eigenfunctions in the straight horizontal slot

In the paper [3] the cross-section eigenfunctions on the slot are found from the ordinary differential equation obtained via replacement the non-trivial left side Λ in the equation (1.12) by the constant. Then the eigenfunctions are found in explicit form of trigonometric functions. In this paper we developed DN- machinery to construct realistic equation for the non-equilibrium part of the quantum current and were able to prove, see Section 2, that the spectral problem for cross-section component on the slot is reduced to spectral analysis of a Hilbert-Schmidt operator. Nevertheless, it appeared that the eigenfunctions of that operator look very much the same as the eigenfunctions of the corresponding differential equation in [3], where they coincide with classical trigonometric functions. A minor difference may be noticed in the behavior of the first eigenfunctions near the electrodes, see first three eigenfunctions of the odd series below. At the moment we can't suggest any qualitative explanation of this phenomenon.

Here are first 5 eigenvalues of the "odd" series of the operator K: 0.3914815726, 0.1532098038, 0.09504925144, 0.06886402733, 0.05397919120, and first 5 eigenfunctions of the odd series:

$$
\begin{aligned}
\mathrm{fiod1}(x) := {}& -.9910874586\sin(.5\,(x+1)\,\pi) + .1212875218\sin(1.5\,(x+1)\,\pi) \\
&+ .04436147780\sin(2.5\,(x+1)\,\pi) + .02399915669\sin(3.5\,(x+1)\,\pi) \\
&+ .01530090916\sin(4.5\,(x+1)\,\pi) + .01070626954\sin(5.5\,(x+1)\,\pi) \\
&+ .007958603246\sin(6.5\,(x+1)\,\pi) + .006172535946\sin(7.5\,(x+1)\,\pi) \\
&+ .004945885162\sin(8.5\,(x+1)\,\pi) + .004054924722\sin(9.5\,(x+1)\,\pi)
\end{aligned}
$$

$$
\begin{aligned}
\mathrm{fiod2}(x) := {}& .1117423454\sin(.5\,(x+1)\,\pi) + .9818943229\sin(1.5\,(x+1)\,\pi) \\
&- .1343729349\sin(2.5\,(x+1)\,\pi) - .05608113124\sin(3.5\,(x+1)\,\pi) \\
&- .03307212574\sin(4.5\,(x+1)\,\pi) - .02241035233\sin(5.5\,(x+1)\,\pi) \\
&- .01641399421\sin(6.5\,(x+1)\,\pi) - .01263518518\sin(7.5\,(x+1)\,\pi) \\
&- .01012904963\sin(8.5\,(x+1)\,\pi) - .008267623797\sin(9.5\,(x+1)\,\pi)
\end{aligned}
$$

$$
\begin{aligned}
\mathrm{fiod3}(x) := {}& .05292785974\sin(.5\,(x+1)\,\pi) + .1162634038\sin(1.5\,(x+1)\,\pi) \\
&+ .9789142759\sin(2.5\,(x+1)\,\pi) - .1380115902\sin(3.5\,(x+1)\,\pi) \\
&- .06011078796\sin(4.5\,(x+1)\,\pi) - .03670653101\sin(5.5\,(x+1)\,\pi) \\
&- .02559615342\sin(6.5\,(x+1)\,\pi) - .01916493668\sin(7.5\,(x+1)\,\pi) \\
&- .01510111946\sin(8.5\,(x+1)\,\pi) - .01224449592\sin(9.5\,(x+1)\,\pi)
\end{aligned}
$$

$$
\begin{aligned}
\mathrm{fiod4}(x) := {}& .03274896712\sin(.5\,(x+1)\,\pi) + .06043162781\sin(1.5\,(x+1)\,\pi) \\
&+ .1157994972\sin(2.5\,(x+1)\,\pi) + .9773738676\sin(3.5\,(x+1)\,\pi) \\
&- .1405447629\sin(4.5\,(x+1)\,\pi) - .06244408352\sin(5.5\,(x+1)\,\pi) \\
&- .03885164416\sin(6.5\,(x+1)\,\pi) - .02744174349\sin(7.5\,(x+1)\,\pi) \\
&- .02092526245\sin(8.5\,(x+1)\,\pi) - .01593307543\sin(9.5\,(x+1)\,\pi)
\end{aligned}
$$

$$\begin{aligned}
\mathrm{fiod5}(x) := \ & .02298961699\,\sin(.5\,(x+1)\,\pi) + .03999406558\,\sin(1.5\,(x+1)\,\pi) \\
& + .06174483906\,\sin(2.5\,(x+1)\,\pi) + .1160041782\,\sin(3.5\,(x+1)\,\pi) \\
& + .9762227109\,\sin(4.5\,(x+1)\,\pi) - .1431966138\,\sin(5.5\,(x+1)\,\pi) \\
& - .06436782790\,\sin(6.5\,(x+1)\,\pi) - .04031008795\,\sin(7.5\,(x+1)\,\pi) \\
& - .02893430576\,\sin(8.5\,(x+1)\,\pi) - .02193605556\,\sin(9.5\,(x+1)\,\pi)
\end{aligned}$$

>plot(fiod1(x),x=-1..1,y=-1.5..1.5)

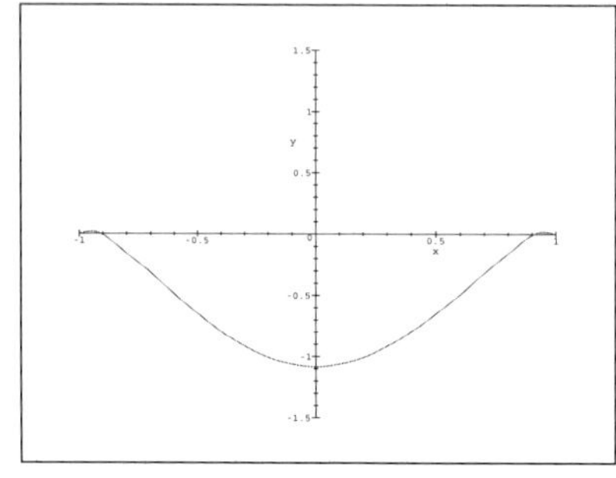

>plot(fiod2(x),x=-1..1,y=-1.5..1.5)

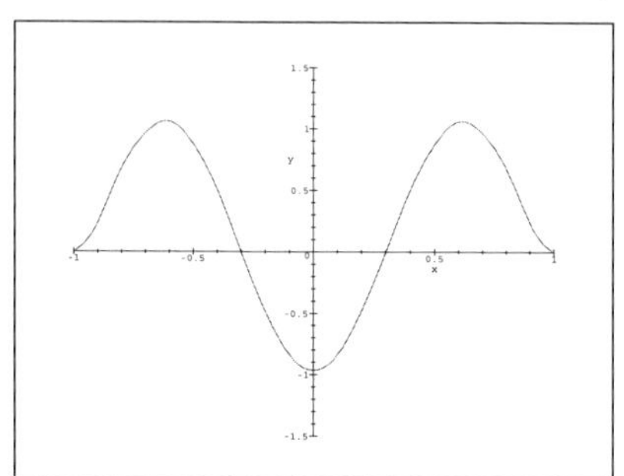

>plot(fiod3(x),x=-1..1,y=-1.5..1.5)

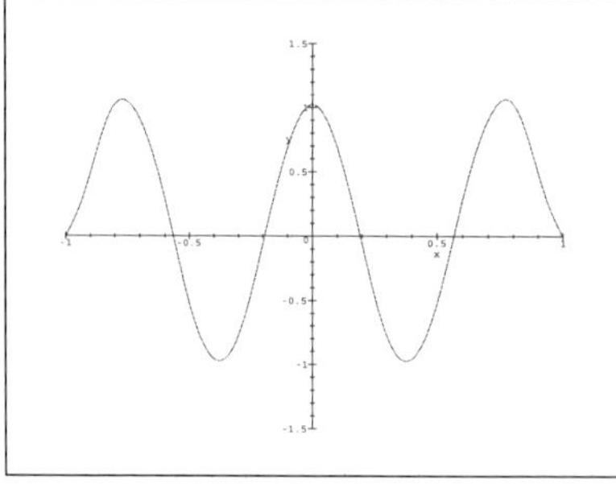

>plot(fiod4(x),x=-1..1,y=-1.5..1.5)

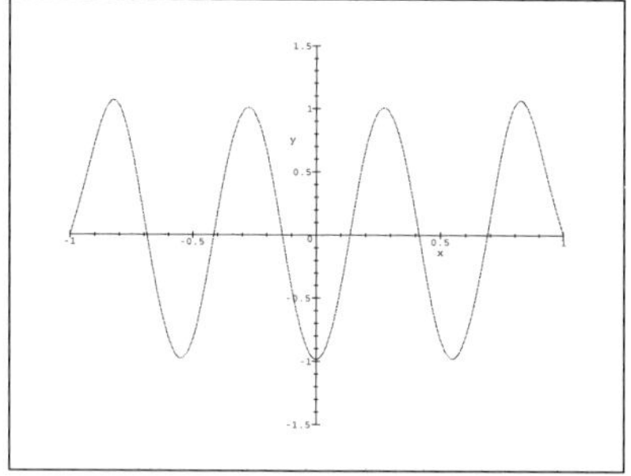

>plot(fiod5(x),x=-1..1,y=-1.5..1.5)

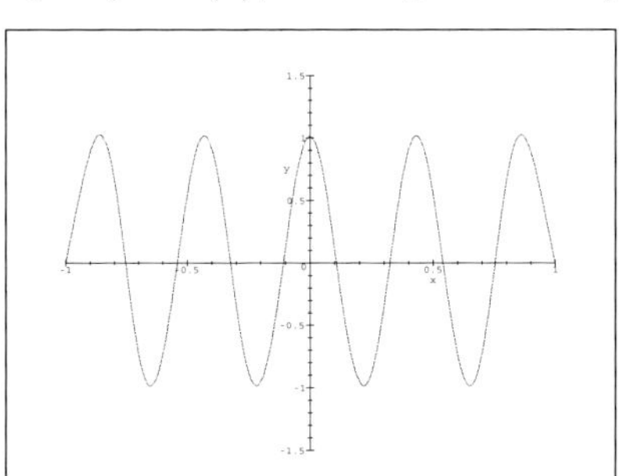

Acknowledgement

The authors are grateful to the Japanese Society of Promotion of Sciences which made our collaboration possible by awarding to B.P. the grant for short-time visit to Computer Solid State Laboratory at Aizu University, Fukushima, Japan. B.P. is also grateful to Gaven Martin and the referee for useful remarks.

References

[1] A. Chaplik, *Possible crystallization of charge carriers in low-density inversion layers*, Sov. Phys. JETP, **32** (1972), 395–398.

[2] M. Dyakonov, M. Shur, *Shallow water analogy for a ballistic field effect transistor: new mechanizm of plasma wave generation by dc current*, Phys. Rev. Letters, **71** (1993), 2465–2468.

[3] V. Ryzhii, I. Khmyrova, A. Satou, it Plazma mechanism of Teraherz photomixing in high-electron mobility transistor under interband photoexitation, Journ. Appl. Physics, **92** (2002), 10.

[4] V. Ryzhii, A. Satou, M.S. Shur, *Admittance of a slot diod with a two-dimensional electron channel*, Journal of Applied Physics, **93** (2003), 10041–10045.

[5] T. Otsuji, M. Hanabe, J. Shigenobu, S. Takahashi, E. Sano, *A Nobel Teraherz Plazma-Wave Photomixer with Resonant-Cavity Enhanced Structure*, Digest of the technical papers of the 12th IEEE Int. Conf. Teraherz. Electron., Sep. 2004

[6] W. Knap, F. Teppe, Y. Mezani, N. Dyakonova, J. Lusakovski, F. Boeuf, T. Shotnicki, D. Maude, S. Rumyantsev, M. Shur, *Plazma wave detection of subteraherz and teraherz radiation by silicon field-effect transistors*, Appl. Phys. Rev., **85** (2004), 675–677.

[7] N. Levinson, *The first boundary value problem for $\varepsilon \triangle u + A(x,y)u_x + B(x,y)u_y + C(x,y)u = D(x,y)$ for small ε*, Ann. of Math. 2 **51** (1950), 428–445.

[8] T. Otsuji, M. Hanabe, O. Ogawara, *Teraherz plasma wave resonance of two-dimensional electrons in INGaP/InGaAs/GaAs high electron mobility transistors*, Appl. Phys. Letters, **85** (2004), 2119–2121.

[9] M. Hanabe, T. Otsuji, V. Ryzhii, *Effect of Photogenerated Electrons on the Teraherz Plazma-Wave Resonance in HMET's under Interband Photoexcitation*, Proc. SPIE Int. Soc. Opt. Eng., **5466** (2004), 218–225.

[10] B. Pavlov, V. Ryzhii, *Quantum dot and anti-dot infrared photo-detectors: iterative methods for solving the Laplace equations in domains with involved geometry* , Theor. Math. Phys, **141**(2)(2004), 1469–1481.

[11] P.M. Morse, H. Feshbach, *Methods of theoretical physics* 2 volumes. McGraw-Hill Book Co., Inc., New York-Toronto-London, 1953.

[12] E.C. Titchmarsh, *Eigenfunctions expansions associated with second order differential equations*, Vol. 1, Clarendon Press, Oxford, 1962.

[13] M. Abramovitz, I.A. Stegun, *Handbook of mathematical functions* Dover, NY, 1965.

[14] A.P. Calderón *On an inverse boundary value problem*, Seminar on Numerical Analysis and its Applications to Continuum Physics, Soc. Brasiliera de Mathematica, Rio de Janeiro, (1980), 65–73.

[15] V. Kondratjev, E. Landis, *Qualitative theory of second order linear partial differential equations*, in: Partial Differential Equations 3, ITOGI NAUKI SSSR, VINITI, Moscow 1988, 99–215.

[16] J. Sylvester, G. Uhlmann, *The Dirichlet to Neumann map and applications*, In: Proceedings of the Conference "Inverse problems in partial differential equations (Arcata,1989)", SIAM, Philadelphia, 1990, 101–139.

[17] B. Pavlov, *S-Matrix and Dirichlet-to-Neumann Operators*, in: *Scattering* (Encyclopedia of Scattering), ed. R. Pike, P. Sabatier, Academic Press, Harcourt Science and Tech. Company (2001), 1678–1688

[18] V. Ryzhii, G. Khrenov, *High-Frequency operation of lateral Hot-Electron Transistors*, in: Trans. Electron Devices, **42**, 166 (1995).

[19] M.D. Faddeev, *Asymptotic Behaviour of the Green's function for the Neumann problem at the boundary point*, Zapiski Nauchn, Seminarov LOMI **131** (1983), 142–147 (English Translation: J. Sov. Math. **30** (1985), 2336–2340).

[20] L. Landau, E. Livshitz, *Fluid Mechanics* Second edition, Volume **6** of Course of Theoretical Physics Second English Edition, Revised. Translated from Russian by J. Sykes and W. Reid. Pergamon press Oxford, 1959.

[21] B. Pavlov, S. Fedorov, *The group of shifts and harmonic analysis on a Riemann surface of genus one*, Algebra i Analiz **1** (1989), 132–168 (English translation: Leningrad Math. J. **1**(1990), 447–489).

[22] V. Ryzhii, A. Satou, I. Khmyrova and A. Chaplik, *Plazma oscillations in a diode structure with a two-dimensional electron channel*, Journal of applied Physics, **96** (2004), 7625–7628.

[23] B. Pavlov, N. Smirnov, *A model of a crystal from potentials of zero-radius with internal structure*, in: Wave propagation. Scattering theory (Russian), 155–164, 258, Probl. Mat. Fiz., N12, Leningrad. Univ., Leningrad, 1987.

[24] B. Pavlov, *The theory of extensions and explicitly-solvable models* (In Russian), Uspekhi Mat. Nauk, **42** (1987), 99–131 (English translation in: Russian Math. Surveys 42:6 (1987), 127–168).

[25] J. Schenker, M. Aizenman, *The creation of spectral gaps by decoration*, Lett. Math. Phys. **53**,3 (2000), 253–262.

[26] V. Oleinik, N. Sibirev, Asymptotics of localized spectral bands of the periodic waveguide, in: Proceedings POMI, The day for Diffraction 2002, St. Petersburg, Russia, June 2002, 94–102.

[27] V. Oleinik, N. Sibirev, B. Pavlov, *Analysis of the dispersion equation for the Schrödinger operator on periodic metric graphs*, Waves Random media, **14** (2004), 157–183.

[28] C. Fox, V. Oleinik, B. Pavlov, *Dirichlet-to-Neumann map machinery for resonance gap and bands of periodic structures*, Department of Mathematics of the University of Auckland report series, 518, March 2004, 26 p. Sbm. to "Contemp. Mathematics", Proceedings of the Conference on Math. Phys. Birmingham, Alabama, 30 March–3 April 2005.

Anna B. Mikhailova
V.A. Fock Institute for Physics
of St.Petersburg University
Russia
e-mail: eddy@bk.ru

Boris Pavlov
V.A. Fock Institute for Physics
of St.Petersburg University,
Russia

on leave from the

Department of Mathematics
the University of Auckland
Private Bag 92019
Auckland, New Zealand
e-mail: pavlov@math.auckland.ac.nz

Victor I. Ryzhii
Solid State Computer Laboratory
Aizu University
Aizu-Wakamatsu City,Japan
e-mail: v-ryzhii@u-aizu.ac.jp

Operator Theory:
Advances and Applications, Vol. 174, 105–116

Inverse Spectral Problem for Quantum Graphs with Rationally Dependent Edges

Marlena Nowaczyk

Abstract. In this paper we study the problem of unique reconstruction of the quantum graphs. The idea is based on the trace formula which establishes the relation between the spectrum of Laplace operator and the set of periodic orbits, the number of edges and the total length of the graph. We analyse conditions under which is it possible to reconstruct simple graphs containing edges with rationally dependent lengths.

1. Introduction

Differential operators on metric graphs (quantum graphs) is a rather new and rapidly developing area of modern mathematical physics. Such operators can be used to model the motion of quantum particles confined to certain low dimensional structures. This has many possible applications to quantum computing and design of nanoelectronic devices [1], which explains recent interest in the area.

The main mathematical tool used in this article is the trace formula, which establishes the connection between the spectrum of the Laplace operator on a metric graph and *the length spectrum* (the set of all periodic orbits on the graph), the number of edges and the total length of the graph.

J.P. Roth [12] proved trace formula for quantum graphs using the heat kernel approach. An independent way to derive trace formula using scattering approach was suggested by B. Gutkin, T. Kottos and U. Smilansky [6, 8] and mathematically rigorous proof of this result was provided by P. Kurasov and M. Nowaczyk [10]. The trace formula is applied in order to reconstruct the graph from the spectrum of the corresponding Laplace operator. It has been proven that this procedure can be carried out in the case when the lengths of the edges are rationally independent and the graph has no vertices of valence 2. In current paper we go further and consider graphs with trivially and weakly rationally dependent edges. We have decided to restrict our considerations to the case of the so-called Laplace operator on metric graphs – the second derivative operator with natural (free, standard, Kirchhoff) boundary conditions at vertices.

Explicit examples constructed in [6, 11, 2] show that the inverse spectral and scattering problems for quantum graphs do not have, in general, unique solutions.

For a historical background on quantum graphs, their applications and theory development see Introduction and References in our previous paper [10].

2. Basic definitions

All notations and definitions in this paper will follow those used in [10]. We are not going to repeat the rigorous derivation of the trace formula presented there, but in this section we will introduce the definitions which we are going to use.

Consider arbitrary finite metric graph Γ consisting of N edges. The edges will be identified with the intervals of the real line $\Delta_j = [x_{2j-1}, x_{2j}] \subset \mathbb{R}$, $j = 1, 2, \ldots, N$ and the set of all edges will be denoted by $E = \{\Delta_j\}_{j=1}^N$. Their lengths will be denoted by $d_j = |x_{2j} - x_{2j-1}|$ and corresponding set of all lengths by $D = \{d_j\}$. Let us denote by M the number of vertices in the graph Γ. Vertices can be obtained by dividing the set $\{x_k\}_{k=1}^{2N}$ of endpoints into equivalence classes $V_m, m = 1, 2, \ldots, M$. The coordinate parameterization of the edges does not play any important role, therefore we are going to identify metric graphs having the same topological structure and the same lengths of the edges. This equivalence is more precisely described in [11, 2].

Consider the Hilbert space of square integrable functions on Γ

$$\mathcal{H} \equiv L^2(\Gamma) = \oplus \sum_{j=1}^N L^2(\Delta_j) = \oplus \sum_{n=1}^N L^2[x_{2j-1}, x_{2j}]. \tag{1}$$

The Laplace operator H on Γ is the sum of second derivative operators acting in each space $L^2(\Delta_j)$,

$$H = \oplus \sum_{j=1}^N \left(-\frac{d^2}{dx^2} \right). \tag{2}$$

This differential expression does not uniquely determine the self-adjoint operator. Two differential operators in $L^2(\Gamma)$ are naturally associated with the differential expression (2), namely the minimal operator with the domain $\mathrm{Dom}\,(H_{\min}) = \oplus \sum_{j=1}^N C_0^\infty(\Delta_j)$ and the maximal operator $H_{\max}$ with the domain $\mathrm{Dom}\,(H_{\max}) = \oplus \sum_{j=1}^N W_2^2(\Delta_j)$, where W_2^2 denotes the Sobolev space.

The Hilbert space $\mathcal{H}$ introduced above does not reflect the connectivity of the graph. It is the boundary conditions that connect values of the function on different edges. Therefore these conditions have to be chosen in a special way so that they reflect the connectivity of the graph. See [11] for the discussion how the most general boundary conditions can be chosen. In the current paper we restrict our consideration to the case of natural (free, standard, Kirchhoff) boundary conditions given by

$$\begin{cases} f(x_j) = f(x_k), \quad x_j, x_k \in V_m, \\ \sum_{x_j \in V_m} \partial_n f(x_j) = 0, \end{cases} \qquad m = 1, 2, \ldots, M, \tag{3}$$

where $\partial_n f(x_j)$ denotes the normal derivative of the function f at the endpoint x_j. The functions satisfying these conditions are continuous at the vertices. In the case of the vertex with valence 2 conditions (3) imply that the function and its first derivative are continuous at the vertex, i.e., the vertex can be removed by substituting the two edges joined at the vertex by one edge with the length equal to the sum of the lengths of the two edges. This procedure is called *cleaning* [11] and a graph Γ with no vertices of valence 2 is called *clean*.

The *Laplace operator* $H(\Gamma)$ on the metric graph Γ is the operator $H_{\max}$ given by (2) restricted to the set of functions satisfying boundary conditions (3). This operator is self-adjoint [11] and uniquely determined by the graph Γ. The spectrum of the operator $H(\Gamma)$ is discrete and consists of positive eigenvalues accumulating at $+\infty$. Therefore the inverse spectral problem for $H(\Gamma)$ is to reconstruct the graph Γ from the set of eigenvalues.

3. Trace formula

Let us establish the secular equation determining all positive eigenvalues of the operator H. Suppose that ψ is an eigenfunction for the operator corresponding to the positive spectral parameter $E = k^2 > 0$. Then this function is a solution to the one-dimensional Schrödinger equation on the edges $-\frac{d^2 \psi}{dx^2} = k^2 \psi$. The general solution to the differential equation on the edge $\Delta_j = [x_{2j-1}, x_{2j}]$ with the length $d_j = |x_{2j} - x_{2j-1}|$ can be written in the basis of incoming waves as follows

$$\psi(x) = a_{2j-1} e^{ik|x - x_{2j-1}|} + a_{2j} e^{ik|x - x_{2j}|}, \tag{4}$$

where a_m is the amplitude of the wave coming in from the endpoint x_m.

Now let us introduce two matrices $\mathcal{E}$ and Σ corresponding to evaluation of amplitudes through edges and vertices respectively. First matrix

$$\mathcal{E} = \begin{pmatrix} e^1 & 0 & \cdots \\ \hline 0 & e^2 & \cdots \\ \hline \vdots & \vdots & \ddots \end{pmatrix}, \qquad \text{where} \quad e^j = \begin{pmatrix} 0 & e^{ikd_j} \\ e^{ikd_j} & 0 \end{pmatrix}. \tag{5}$$

The second matrix is formed by blocks of vertex scattering matrices

$$\Sigma = \begin{pmatrix} \sigma^1 & 0 & \cdots \\ \hline 0 & \sigma^2 & \cdots \\ \hline \vdots & \vdots & \ddots \end{pmatrix}, \tag{6}$$

where for natural boundary conditions the vertex scattering matrices do not depend on the energy and elements are given by

$$\sigma_{jk}^m = \begin{cases} \frac{2}{v_m}, & j \neq k, \\ \frac{2-v_m}{v_m}, & j = k, \end{cases} \quad \text{for } v_m \neq 1 \qquad \text{and} \qquad \sigma = 1 \text{ for } v_m = 1. \tag{7}$$

After evaluation of the amplitudes through edges and then through vertices we arrive to the same incoming amplitudes. Therefore the amplitudes $\mathbf{a}$ determine an eigenfunction of $H(\Gamma)$ for $E > 0$ if and only if $\mathbf{a} = \Sigma\mathcal{E}\mathbf{a}$, i.e., when the matrix

$$U(k) = \Sigma\mathcal{E}(k) \tag{8}$$

has eigenvalue 1 and $\mathbf{a}$ is the corresponding eigenvector.

Let us denote the eigenvalues of the Laplace operator H in nondecreasing order as follows

$$E_0 = k_0^2 = 0 < E_1 = k_1^2 \leq E_2 = k_2^2 \leq \cdots$$

and we will introduce the distribution u connected with the spectral measure

$$u \equiv \delta(k) + \sum_{n=1}^{\infty} \left(\delta(k - k_n) + \delta(k + k_n) \right).$$

Now we are going to present the relation between spectrum of Laplace operator H and lengths of periodic orbits, number of edges and total length of the graph. Before we do this, however, we need to give a few definitions related to periodic orbits of a graph.

By a *periodic orbit* we understand any oriented closed path on Γ. We do not allow to turn back at any internal point of the edge, but walking the same edge multiple times is allowed. Note that so defined orbit does not have any starting point. With any such (continuous) periodic orbit p one can associate the *discrete periodic orbit* consisting of all edges forming that orbit. Also let:

- $\mathcal{P}$ be the set of all periodic orbits for the graph Γ,
- $l(p)$ be the geometric length of a periodic orbit p,
- $\mathrm{prim}(p)$ denote a primitive periodic orbit, such that p is a multiple of $\mathrm{prim}(p)$,
- $\mathcal{L} = d_1 + d_2 + \cdots + d_N$ be the total length of the graph Γ,
- $\mathcal{T}(p)$ be the set of all scattering coefficients along the orbit p.

Let us introduce coefficients which are independent of the energy:

$$\mathcal{A}_p = l(\mathrm{prim}(p))\left(\prod_{\sigma_{ij}^m \in \mathcal{T}(p)} \sigma_{ij}^m \right), \qquad \mathcal{A}_p^* = l(\mathrm{prim}(p))\left(\prod_{\sigma_{ij}^m \in \mathcal{T}(p)} \overline{\sigma_{ij}^m} \right). \tag{9}$$

The following theorem has been proven in [10], following the ideas of B. Gutkin and U. Smilansky [6].

Proposition 1 (Theorem 1 from [10]). *Let $H(\Gamma)$ be the Laplace operator on a finite connected metric graph Γ, then the following two trace formulae establish* *the relation between the spectrum $\{k_j^2\}$ of $H(\Gamma)$ and the set of periodic orbits $\mathcal{P}$, the number of edges N and the total length $\mathcal{L}$ of the graph:*

$$u(k) \equiv \delta(k) + \sum_{n=1}^{\infty} \left(\delta(k - k_n) + \delta(k + k_n) \right) \tag{10}$$

$$= -(N - M + 1)\delta(k) + \frac{\mathcal{L}}{\pi} + \frac{1}{2\pi} \sum_{p \in \mathcal{P}} \left(\mathcal{A}_p\, e^{ikl(p)} + \mathcal{A}_p^*\, e^{-ikl(p)} \right),$$

and

$$\hat{u}(l) \equiv 1 + \sum_{n=1}^{\infty} \left(e^{-ik_n l} + e^{ik_n l} \right) \tag{11}$$

$$= -(N - M + 1) + 2\mathcal{L}\delta(l) + \sum_{p \in \mathcal{P}} \left(\mathcal{A}_p \delta(l - l(p)) + \mathcal{A}_p^* \delta(l + l(p)) \right)$$

where $\mathcal{A}_p$, $\mathcal{A}_p^*$ are independent of the energy complex numbers given by (9).

The formula (11) converges in the sense of distributions (see [10] pp. 4908–4909 for explicit calculations).

4. The inverse spectral problem

In this section we are going to apply formula (11) to prove that the inverse spectral problem has unique solution for certain simple (i.e., without loops or multiple edges), clean, finite connected metric graphs with rationally dependent lengths of edges.

The set L of lengths of all periodic orbits is usually called the length spectrum. In some cases, formula (11) allows us to recover the length spectrum (of periodic orbits) from the energy spectrum (of the Laplace operator H). On the other hand, there are known graphs for which some lengths of periodic orbits cannot be recovered. Formula (11) implies directly that the spectrum of a graph allows one to recover the lengths l of all periodic orbits from the *reduced length spectrum* $L' \subset L$ defined as

$$L' = \{ l : \Big(\sum_{\substack{p \in \mathcal{P} \\ l(p)=l}} \mathcal{A}_p \Big) \neq 0 \}. \tag{12}$$

Although for any periodic orbit p the coefficient $\mathcal{A}_p$ defined in (9) is non-zero it can happen that the sum of all coefficients in front of $\delta(l - l(p))$ is zero. This is the reason why we use reduced length spectrum instead of more common length spectrum.

4.1. Graphs with trivially rationally dependent edges

In this subsection we will discuss graphs where the set of all lengths of edges is rationally independent, while some edges can have equal lengths (we will call such case a graph with *trivially rationally dependent* edges). One can prove that such graphs can be uniquely reconstructed from length spectrum and total length of the graph — and, therefore, can be uniquely reconstructed from spectrum of Laplace operator on this graph.

We shall now remind Lemma 2 from paper [10] and we will re-state this lemma for graphs with trivially rationally dependent edges.

Lemma 2. *Let Γ be a graph with trivially rationally dependent lengths of edges. Assume that the edges of the same length are not neighbors to each other. Then the reduced length spectrum L' contains at least the following lengths:*

- *$4d_j$, for all $j = 1, \ldots, N$;*
- *$2d_j$ if there exist exactly one edge of length d_j;*
- *$2(d_j + d_k)$ iff the edges having lengths d_j and d_k are neighbors;*
- *$2(d_i + d_j + d_k)$ if Δ_i, Δ_j and Δ_k form a path but do not form a cycle.*

Proof. Consider any orbit p of the length $4d_j$. Then the coefficient $\mathcal{A}_p$ product consists of exactly two squared reflection coefficients and therefore is strictly positive. The coefficient in front of $\delta(l - 4d_j)$ in the sum (11): $\sum_{p:l(p)=4d_j} \mathcal{A}_p$ is also strictly positive. Thus $4d_j$ belongs to the reduced length spectrum L'.

The other three parts of this proof follow from the Lemma 2 and its proof in [10]. $\qquad\square$

Lemma 3. *Assume that the metric graph Γ is finite, clean, connected and simple. Let Γ have edges of trivially rationally dependent lengths. Let us denote number of edges of length d_1 by β_1, number of edges of length d_2 by β_2, ..., number of edges of length d_n by β_n (where $\beta_i \geq 1$ for $i = 1, \ldots, n$).*

Then the total length $\mathcal{L}$ of the graph and the reduced length spectrum L' determine the lengths of all edges (d_j), as well as the number of edges having these particular lengths (β_j).

Proof. Consider the finite subset L'' of $L' \subset L$, consisting of all lengths less than or equal to $4\mathcal{L}$

$$L'' = \{l \in L' : l \leq 4\mathcal{L}\}.$$

This finite set contains at least the numbers $4d_j$ and those numbers form a basis for a set of all lengths of periodic orbits, i.e., every length $l \in L''$ (as well as in L) can be written as a combination of $4d_j$

$$l = \frac{1}{4}\sum_{j=1}^{n} n_j 4d_j, \quad n_j \in \mathbb{N},$$

where n_j are the smallest possible non-negative integers. Since all d_j are rationally independent then this combination is unique. Such a basis is not unique but any two bases $\{4d_j\}$ and $\{4d'_j\}$ are equal with respect to a permutations of its elements.

The total length of the graph $\mathcal{L}$ can also be written as

$$\mathcal{L} = \frac{1}{4}\sum_{j=1}^{n} \beta_j 4d_j, \quad \beta_j \in \mathbb{N}. \tag{13}$$

Because the graph Γ is simple (i.e., without loops or multiple edges), the coefficients β_j indicate the total number of edges of length d_j. $\qquad\square$

Lemma 4. *Assume that the metric graph Γ is finite, clean, connected and simple. Let Γ have edges with trivially rationally dependent lengths. Also assume that any two edges Δ, Δ' with lengths d_i, d_j (where i can be equal j) for which $\beta_i \geq 2$ and $\beta_j \geq 2$ (i.e., they are both repeating edges) are separated by at least two non-repeating edges (i.e., edges for which $\beta_k = 1$).*

Then the graph Γ can be reconstructed from the set $D = \{d_j\}$ of the lengths of all edges and the reduced length spectrum L'.

Proof. At the beginning we are going to reconstruct the graph Γ without repeating edges. In order to do this, we shall use the idea of reconstructing the simple subgraph in the proof of Lemma 4 in the paper [10].

Let us denote by Γ^* the subgraph of Γ which can be obtained by deleting all edges with $\beta_j \geq 2$. Γ^* does not have to be a connected graph, so let us denote its components by $\Gamma^{(1)}, \Gamma^{(2)}, \ldots, \Gamma^{(s)}$. The reconstruction will be done iteratively and we will construct an increasing finite sequence of subgraphs such that $\Gamma_1 \subset \Gamma_2 \subset \cdots \subset \Gamma_{N^*} = \Gamma^*$. The corresponding subsets of edges will be denoted by E_k for $k = 1, \ldots, N^*$.

The reconstruction of any component $\Gamma^{(j)}$ is done in the following way. For $k = 1$ take the graph $\Gamma_1^{(j)}$, consisting of an arbitrary non-repeating edge, say Δ_1. In order to get $\Gamma_2^{(j)}$, pick any neighbor of Δ_1, say Δ_2, and attach it to any of the endpoints of Δ_1 (the set of neighbors of Δ_1 can be easily obtained from the reduced length spectrum L').

Suppose that connected subgraph $\Gamma_k^{(j)}$ consisting of k edges ($k \geq 2$) is already reconstructed. Pick any edge, say Δ_{k+1}, which is a neighbor of at least one of the edges in $\Gamma_k^{(j)}$. Let us denote by E_k^{nbh} the subset of E_k consisting of all edges which are neighbors of Δ_{k+1}. We have to identify (one or two) vertices in $\Gamma_k^{(j)}$ to which the new Δ_{k+1} is attached – every such vertex is uniquely determined by listing of the edges joined at this vertex (since the subgraph $\Gamma_k^{(j)}$ is simple, connected and contains at least two edges). Therefore we have to separate E_k^{nbh} into two classes of edges, each attached to one endpoint of Δ_{k+1}. Observe that one of the two sets can be empty, which corresponds to the case the edge Δ_{k+1} is attached to $\Gamma_k^{(j)}$ at one vertex only.

Take any two edges from E_k^{nbh}, say Δ' and Δ''. The edges Δ' and Δ'' belong to the same class if and only if:

- Δ' and Δ'' are neighbors themselves and
- $d' + d'' + d_{k+1} \notin L'$, i.e., the edges Δ', Δ'' and Δ_{k+1} do not form a cycle (note that if Δ', Δ'' and Δ_{k+1} form a cycle, then there are two periodic orbits of length $d' + d'' + d_{k+1}$ and the corresponding $\mathcal{A}$-coefficients are equal – which implies that $d' + d'' + d_{k+1} \in L'$).

In this way we either separate the set E_k^{nbh} into two classes of edges or E_k^{nbh} consists of edges joined at one vertex. In the first case, the new edge Δ_{k+1} connects the two vertices uniquely determined by those two subclasses. In the second case, the edge Δ_{k+1} is attached at one end point to $\Gamma_k^{(j)}$ at the vertex uniquely determined by E_k^{nbh}. It does not matter which of the two end points of Δ_{k+1} is attached to the chosen vertex of $\Gamma_k^{(j)}$, since the two possible resulting graphs are equivalent.

Denote the graph created this way by $\Gamma_{k+1}^{(j)}$.

When there are no more edges left which are neighbors of $\Gamma_k^{(j)}$, then pick any new non-repeating edge from E and start the reconstruction procedure for new component of graph Γ^*, say $\Gamma^{(j')}$. After a finite number of steps one arrives at the graph Γ^*.

It remains now to add the repeating edges. Since each repeating edge of length d_n is separated from any other repeating edge of length d_m by at least two non-repeating edges, then there is no interference between adding edges d_n and d_m to Γ^*. Following previous lemma, from reduced length spectrum L' and total length of the graph $\mathcal{L}$ we know that we have exactly β_n edges of length d_n.

As the first step we want to split all neighbors of all d_n edges into $2\beta_n$ classes (some of which can be empty). The set of all neighbors of d_n from graph Γ^* will be denoted by $\mathbb{E}_n$. We say that Δ_j and Δ_k from $\mathbb{E}_n$ are in the same class if:

- Δ_j and Δ_k are neighbors to each other,
- they do not build a cycle of length $d_n + d_j + d_k$,
- if there is an edge Δ_m which is a neighbor to Δ_j and to Δ_k but is not a neighbor to any edge of length d_n, then there is a cycle of length $d_m + d_j + d_k$.

In that way we obtain non-empty sets $\mathbb{E}_n^1, \mathbb{E}_n^2, \ldots, \mathbb{E}_n^{\alpha_n}$ which correspond to vertices $v_1, v_2, \ldots, v_{\alpha_n}$ where $\alpha_n \leq 2\beta_n$.

As the second step we have to identify, for each edge of length d_n, two vertices (or only one) to which this particular edge is attached. We are going to check all pairs of vertices v_i and v_j from the list above. An edge of length d_n is attached to those two vertices if

- v_i and v_j are connected by a path of two edges d' and d'' where $d' \in \mathbb{E}_n^i$ and $d'' \in \mathbb{E}_n^j$ and there exist a periodic orbit of length $d' + d'' + d_n$ in L', or
- v_i and v_j are not connected by any path of two edges and for each pair $d' \in \mathbb{E}_n^i$ and $d'' \in \mathbb{E}_n^j$ there exist a periodic orbits of length $2(d' + d'' + d_n)$ in L'.

For each of those vertices $v_1, v_2, \ldots, v_{\alpha_n}$ for which neither of the above conditions are satisfied, we attach a loose edge of length d_n.

We repeat this procedure for all edges of repeating lengths. Since the graph is finite, after finite number of steps we reconstruct the whole graph Γ. $\qquad\square$

Theorem 5. *The spectrum of a Laplace operator on a metric graph determines the graph uniquely, provided that:*

- *the graph is clean, finite, simple and connected,*
- *the edges are trivially rationally dependent,*

- *any two repeating edges are separated by at least two non-repeating edges (having rationally independent lengths).*

Proof. The spectrum of the operator determines the left-hand side of the trace formula (10). Formula (11) shows that the spectrum of the graph determines the total length of the graph and the reduced length spectrum. Lemma 3 implies that the lengths of all edges and their multiplicities can be extracted from this quantities under the conditions of the theorem. It follows from Lemma 4 that the whole graph can be reconstructed. $\qquad\square$

4.2. Graphs with weakly rationally dependent edges

In the last part of this paper we shall consider some special kind of graph with rationally dependent edges and we will prove that for those graphs the unique reconstruction from the spectrum of Laplace operator is still possible. We shall use, as before, the trace formula and some properties of mutually prime numbers.

Definition 6. Assume that the metric graph Γ is finite, clean, connected and simple. We say that the edge lengths are weakly rationally dependent if the lengths of edges belong to the set

$$\left\{ d_1, \frac{p_{12}}{q_{12}}d_1, \frac{p_{13}}{q_{13}}d_1, \ldots, \frac{p_{1r_1}}{q_{1r_1}}d_1, d_2, \frac{p_{22}}{q_{22}}d_2, \ldots, \frac{p_{2r_2}}{q_{2r_2}}d_2, \ldots, d_n, \frac{p_{n2}}{q_{n2}}d_n, \ldots, \frac{p_{nr_n}}{q_{nr_n}}d_n \right\},$$

where $p_{ij}/q_{ij} > 1$ are proper fractions, $q_{j2}, q_{j3}, \ldots, q_{jr_j}$ are mutually prime for all $j = 1, \ldots, n$ and $d_1, d_2, \ldots, d_n$ are rationally independent.

Observe that if $n = 1$ then all edges in the graph are rationally dependent. On the other hand, if all $p_{ij} = 0$ for $j \geq 2$ and all i then all edges in the graph are rationally independent. Note that the denominators q_{ij} are mutually prime but it doesn't immediately indicate that they are prime numbers.

Lemma 7. *Assume that the metric graph Γ has weakly rationally dependent edges. Then the total length $\mathcal{L}$ of the graph and the reduced length spectrum L' determine the lengths of all edges.*

Proof. As in Lemma 3 we will use an approach of finding a basis for all periodic orbits. We claim that the set $\{2s_j\}$, where s_j is length of any edge in the graph, is a basis for all periodic orbits. Consider as before the finite subset L'' of $L' \subset L$ consisting of all lengths less than or equal to $2\mathcal{L}$

$$L'' = \{l \in L' : l \leq 2\mathcal{L}\}.$$

It is obvious that any periodic orbit can be written as a half-integer combination of $2s_j$ elements

$$l = \frac{1}{2}\sum_{j=1}^{N} \alpha_j 2s_j, \quad \alpha_j \in \mathbb{N}.$$

We shall prove that for graph with weakly rationally dependent edges this combination is unique.

Among all periodic orbits there exist periodic orbits of length $2s_j$. Assume that for some arbitrary j such orbit is a linear combination of other edges and since $d_1, d_2, \ldots, d_n$ are rationally independent it is enough to consider only rationally dependent edges. For sake of notation clearness we will omit the first index in numbers p_{ij} and q_{ij} as well as index at d_i. Thus we have the following equation

$$2\frac{p_j}{q_j}d = \alpha_1 \frac{p_1}{q_1}d + \alpha_2 \frac{p_2}{q_2}d + \cdots + \alpha_{j-1}\frac{p_{j-1}}{q_{j-1}}d + \alpha_{j+1}\frac{p_{j+1}}{q_{j+1}}d + \cdots + \alpha_n\frac{p_n}{q_n}d \quad (14)$$

$$2\frac{p_j}{q_j} = \frac{\alpha_1 p_1 q_2 \cdots q_{j-1}q_{j+1} \cdots q_n + \cdots + \alpha_n q_1 q_2 \cdots q_{j-1}q_{j+1} \cdots q_{n-1}p_n}{q_1 q_2 \cdots q_{j-1}q_{j+1} \cdots q_n}$$

$$2p_j q_1 \cdots q_{j-1}q_{j+1} \cdots q_n = \alpha_1 p_1 q_2 \cdots q_n + \cdots + \alpha_{j-1}q_1 q_2 \cdots p_{j-1}q_j \cdots q_n$$
$$+ \alpha_{j+1}q_1 q_2 \cdots q_j p_{j+1} \cdots q_n + \cdots + \alpha_n q_1 q_2 \cdots q_{n-1}p_n.$$

Let us compare both sides of the previous equation, one by one, modulo each of $q_1, q_2, \ldots, q_{j-1}, q_{j+1}, \ldots, q_n$, thus giving the following system of equations

$$\begin{cases} 0 &= \alpha_1 p_1 q_2 \cdots q_n \quad &(\mathrm{mod}\ q_1) \\ &\vdots \\ 0 &= \alpha_{j-1}q_1 q_2 \cdots p_{j-1}q_j \cdots q_n \quad &(\mathrm{mod}\ q_{j-1}) \\ 0 &= \alpha_{j+1}q_1 q_2 \cdots q_j p_{j+1} \cdots q_n \quad &(\mathrm{mod}\ q_{j+1}) \\ &\vdots \\ 0 &= \alpha_n q_1 q_2 \cdots q_{n-1}p_n \quad &(\mathrm{mod}\ q_n) \end{cases}$$

Since all q_i are mutually prime and p_i/q_i are proper fractions, the only solution to this system of equations is $\alpha_i = 0 \pmod{q_i}$ for all $i = 1, 2, \ldots, j-1, j+1, \ldots, n$. It means that all elements on the right-hand side of (14) are nonnegative integers, while the left-hand side of the same equation is an integer if and only if $j = 1$ or $j = 2$ (then $p_1 = q_1 = 1$ or, respectively, $q_2 = 2$ and $p_2 = 3$).

In the first case, the left-hand side equals 2, while at the same time the right-hand side is either 0 or is strictly greater than 2. In the second case, the left-hand side is equal to 3, while the right-hand side is equal to $\alpha_1 + r$, where r is either 0 or is strictly greater than 3. Thus, to fulfill equation (14), r has to be 0 and α_1 has to be 3. This is, however, impossible – since there is exactly one periodic orbit of length 3 (consisting of double edge of length $\frac{p_2}{q_2} = \frac{3}{2}$).

Thus we have proven that the set $\{2s_j\}$ where s_j are lengths of all edges in the graph Γ form the basis for all lengths of periodic orbits.

Hence we have determined all lengths of edges if these edges are weakly rationally dependent. $\qquad\square$

Lemma 8. *Assume that the metric graph Γ has weakly rationally dependent edges. Then the graph Γ can be reconstructed from the sets $D = \{d_j\}$ and the reduced length spectrum L'.*

Proof. As we have just shown in Lemma 7, from reduced length spectrum L' one can obtain lengths of all edges in graph Γ with weakly rationally dependent edges.

Following Lemma 2 we can deduce that the reduced length spectrum L' contains at least the shortest orbit formed by any two neighboring edges Δ_j and Δ_k, i.e., $2(d_j + d_k)$. Thus we can identify all neighbors of each edge. The algorithm of reconstruction the graph Γ will be the same as in proof of Lemma 4 in part where we reconstruct components of Γ^*. $\qquad\square$

Theorem 9. *The spectrum of a Laplace operator on a metric graph determines the graph uniquely, provided that:*

- *the graph is clean, finite, simple and connected,*
- *the edges are weakly rationally dependent.*

Proof. The spectrum of the operator determines the left-hand side of the trace formula (10). Formula (11) shows that the spectrum of the graph determines the total length of the graph and the reduced length spectrum. Lemma 7 implies that the lengths of all edges can be extracted from this quantities under the conditions of the theorem. It follows from Lemma 8 that the whole graph can be reconstructed. $\quad\square$

Acknowledgment

The author would like to thank Pavel Kurasov for fruitful and important discussions.

References

[1] V. Adamyan, *Scattering matrices for microschemes,* Operator Theory: Advances and Applications **59** (1992), 1–10.

[2] J. Boman and P. Kurasov, *Symmetries of quantum graphs and the inverse scattering problem,* Adv. Appl. Math. **35** (2005), 58–70.

[3] L. Friedlander, *Genericity of simple eigenvalues for a metric graph,* Israel Journal of Mathematics **146** (2005), 149–56.

[4] N.I. Gerasimenko and B.S. Pavlov, *Scattering problems on noncompact graphs,* Teoret. Mat. Fiz. **74** (1988) 345–59 (Eng. transl. Theoret. and Math. Phys. **74** (1988) 230–40).

[5] V. Guillemin and R. Melrose, *An inverse spectral result for elliptical regions in R^2,* Adv. in Math. **32** (1979), 128–48.

[6] B. Gutkin and U. Smilansky, *Can one hear the shape of a graph?* J. Phys. A. Math. Gen. **34** (2001), 6061–6068.

[7] V. Kostrykin and R. Schrader, *Kirchoff's rule for quantum wires,* J. Phys A:Math. Gen. **32** (1999), 595–630.

[8] T. Kottos and U. Smilansky, *Periodic orbit theory and spectral statistics for quantum graphs,* Ann. Physics **274** (1999), 76–124.

[9] P. Kuchment, *Quantum graphs. I. Some basic structures,* Special section on quantum graphs, Waves Random Media **14** (2004), S107–28.

[10] P. Kurasov and M. Nowaczyk, *Inverse spectral problem for quantum graphs,* J. Phys. A. Math. Gen, **38** (2005), 4901–4915.

[11] P. Kurasov and F. Stenberg, *On the inverse scattering problem on branching graphs*, J. Phys. A: Math. Gen. **35** (2002), 101–121.

[12] J.-P. Roth, *Le spectre du laplacien sur un graphe* Lectures Notes in Mathematics: Theorie du Potentiel **1096** (1984), 521–539.

[13] A. V. Sobolev and M. Solomyak, *Schrödinger operators on homogeneous metric trees: spectrum in gaps*, Rev. Math. Phys. **14** (2002), 421–467.

[14] M. Solomyak, *On the spectrum of the Laplacian on regular metric trees*, Special section on quantum graphs Waves Random Media **14** (2004), 155–171.

Marlena Nowaczyk
Department of Mathematics
Lund Institute of Technology
Box 118
210 00 Lund, Sweden
e-mail: `marlena@maths.lth.se`

Operator Theory:
Advances and Applications, Vol. 174, 117–158

Functional Model of a Class of Non-selfadjoint Extensions of Symmetric Operators

Vladimir Ryzhov

Dedicated to the memory of Vladimir Veselov

Abstract. This paper offers the functional model of a class of non-selfadjoint extensions of a symmetric operator with equal deficiency indices. The explicit form of dilation of a dissipative extension is offered and the Sz.-Nagy-Foiaş model as developed by B. Pavlov is constructed. A variant of functional model for a non-selfadjoint non-dissipative extension is formulated. We illustrate the theory by two examples: singular perturbations of the Laplace operator in $L_2(\mathbb{R}^3)$ by a finite number of point interactions, and the Schrödinger operator on the half-axis $(0, \infty)$ in the Weyl limit circle case at infinity.

Mathematics Subject Classification (2000). Primary 47A45; Secondary 47A20, 47B25.

Keywords. Non-selfadjoint operators, functional model, Green formula, Weyl function, almost solvable extensions.

Introduction

Functional model approach plays a prominent role in the study of non-selfadjoint and non-unitary operators on a Hilbert space. The rich and comprehensive theory has been developed since pioneering works of M. Brodskiĭ, M. Livŝic, B. Szökefalvi-Nagy, C. Foiaş, L. de Branges, and J. Rovnyak, see [33], [34] and references therein. The functional model techniques are based on the fundamental theorem of B. Szökefalvi-Nagy and C. Foiaş stating that each linear contraction T, $\|T\| \leq 1$ on a separable Hilbert space H can be extended to a unitary operator U on a wider Hilbert space $\mathscr{H} \supset H$ such that $T^n = P_H U^n|_H$, $n \geq 0$, where P_H is the orthogonal projection from the space $\mathscr{H}$ onto its subspace H. Operator U is called **dilation** of the contraction T. A unitary operator U with such properties is not unique, but if the contraction T does not have reducing unitary parts (such operators are called completely non-unitary, or simple) and if U is **minimal** in the sense that the linear set $\{U^k H : k \in \mathbb{Z}\}$ is dense in the dilation space $\mathscr{H}$, then the unitary dilation U is unique up to a unitary equivalence. B. Szökefalvi-Nagy and C. Foiaş proved as well that the spectrum of the minimal unitary dilation of a simple contraction is absolutely continuous and coincides with the unit circle $\mathbb{T} := \{z \in \mathbb{C} : |z| = 1\}$. In the spectral representation of the unitary operator U, when U becomes a multiplication $f \mapsto k * f$, $k \in \mathbb{T}$ on some L_2 space of vector-functions f, the contraction $T = P_H U|_H$ takes the form of its functional model $T \cong P_H k * |_H$.

Originating in the specific problems of physics of the time, the initial research on functional model quickly shifted into the realm of "pure mathematics" and most of the model results are now commonly regarded as "abstract". One of the few exceptions is the scattering theory developed by P. Lax and R. Phillips [26]. The theory was originally devised for the analysis of the scattering of electromagnetic and acoustic waves off compact obstacles. The research, however, not only resulted in important discoveries in the scattering theory, but deeply influenced the subsequent developments of the operator model techniques as well.

The connection between the Lax-Phillips approach and the Sz.-Nagy-Foiaş dilation theory is established by means of the Cayley transform that maps a bounded operator T such that $\mathcal{R}(T - I)$ is dense in H into a possibly unbounded operator $A := -i(T + I)(T - I)^{-1}$, $\mathcal{D}(A) := \mathcal{R}(T - I)$. If T is unitary, then A is selfadjoint, and when T is contractive, the imaginary part of the operator A (properly understood, if needed, in the sense of sesquilinear forms) is positive. The latter operators A are called **dissipative**. By definition, the selfadjoint dilation $\mathscr{A} = \mathscr{A}^*$ of a dissipative operator $A = -i(T + I)(T - I)^{-1}$ is the Cayley transform of the unitary dilation of T. Correspondingly, the dilation $\mathscr{A}$ is called **minimal** if the set $\{(\mathscr{A} - zI)^{-1}H : \operatorname{Im} z \neq 0\}$ is dense in $\mathscr{H}$.

The main object of the Lax-Phillips scattering theory is a strongly continuous contractive group of operators on a Hilbert space. The generator of this group is a dissipative operator that describes the geometry of the scatterer. Its selfadjoint dilation is present in the problem statement from the very beginning, and as all other mathematical objects of the theory, allows a clear physical interpretation.

Another line of examples of the fruitful interplay between the functional model theory and mathematical physics originates in the works [35], [37], [38], [39] of B. Pavlov on dissipative Schrödinger operators with a complex potential on $L_2(\mathbb{R}^3)$ and with a dissipative boundary condition on $L_2(0, \infty)$. In comparison with the Lax-Phillips theory these studies are distinguished by the absence of the "natural" selfadjoint dilation known upfront. In both cases the selfadjoint dilations have to be "guessed" and explicitly assembled from the objects given in the initial problem statement. This approach eventually evolved into a recipe that not only allows to recover the selfadjoint dilation (see [24]), but also to build its spectral representation, obtaining the eigenfunction expansion of the original dissipative operator. The dilation and the model space used by B. Pavlov are well suited for the study of differential operators, and as in the case of the Lax-Phillips theory, the objects emerging from the model considerations have clear physical meaning.

(See [40].) The technique of expansion by the dilation's eigenfunctions of absolutely continuous spectrum in order to pass to the spectral representation is well known in the physical literature, where this otherwise formal procedure is properly rectified by the distribution theory. In application to the setting of a generic dissipative operator, this approach requires a certain adaptation of the rigged Hilbert spaces technique. (See [40] for an example.)

The next step in the development was made by S. Naboko, who offered a "direct" method of passing on to the functional model representation for the dissipative operators with the relatively bounded imaginary part [28], [29], [30]. The approach is based on the preceding works of B. Pavlov, but without resorting to the dilation's eigenfunctions of continuous spectrum, the spectral mapping is expressed in terms of boundary values of certain operator- and vector-valued functions analytic in the upper and lower half-planes. In a sense, this is exactly what one should expect trying to justify the distributions by methods of the analytic functions theory [15]. As an immediate benefit, this direct approach opened up the opportunity to include non-selfadjoint relatively bounded perturbations of a selfadjoint operator with the relative bound lesser than 1 in the model-based considerations. It turned out that for an operator of this class there exists a model space where the action of the operator can be expressed in a simple and precise form. The ability to abandon the dissipativity restriction imposed on the operator class suitable for the model-based study allowed S. Naboko to conduct the profound spectral analysis of additive perturbations of the selfadjoint operators, to develop the scattering theory for such perturbations, and to introduce valuable definitions of spectral subspaces of a non-selfadjoint non-dissipative operator. The idea of utilization of the functional model of a "close" operator for the study of the operator under consideration was adopted by N. Makarov and V. Vasyunin in [27], who offered the analogue of S. Naboko's construction for an arbitrary bounded operator considered as a perturbation of a unitary. It comes quite naturally that the relationship between these two settings is established by the Cayley transform.

Although the question of model representation of a bounded operator became settled on the abstract level with the work [27], the challenges with various applications to the physical problems remain to be addressed. (See [43] for valuable details on dissipative case.) Speaking of two basic examples of non-selfadjoint Schrödinger operators tracked back to the original works of B. Pavlov, it has to be noted that the example of the Schrödinger operator with a complex-valued potential can be studied from the more general point of view of relative bounded perturbations developed in [28].[1] At the same time the second example, non-selfadjoint extensions of a symmetric differential operator, mostly remains outside of the general theory since these operators could not be divided into a selfadjoint one, plus a relatively bounded additive perturbation. Consequently, in order to utilize the functional model approach for the study of extensions of symmetric operators arising in the

---

[1] The functional model of additive perturbations has been applied to the spectral analysis of the transport operator in [31], [25].

physical applications, one is left solely with the recipe of B. Pavlov. In other words, one has to "guess" the selfadjoint dilation and to prove the eigenfunction expansion theorem.

The present paper concerns the functional model construction for a wide class of extensions of symmetric operators known in the literature as **almost solvable extensions**. Our approach is identical to that of S. Naboko and as such does not involve the eigenfunctions expansion at all. All considerations are carried out in the general setting of the model for non-dissipative non-selfadjoint operators. Although results obtained here are applicable to many interesting physical and mathematical problems, the limitations of almost solvable extensions theory hamper the study of the most interesting case of a multi-dimensional boundary value problem for the partial differential operators. (See Remark 1.5 for more details.) Dissipative extensions of symmetric operators with finite deficiency indices are much easier to analyze. A few successful attempts that utilize the B. Pavlov schema to examine operators of this class encountered in applications were published recently. In particular, Pavlov's approach to the model construction of dissipative extensions of symmetric operators was followed by B. Allahverdiev in his works [4], [5], [6], [7], and some others, and by the group of authors [20], [10], [11], [12], where the theory of the dissipative Schrödinger operator on a finite interval was applied to the problems arising in the semiconductor physics. In comparison with these results, Section 2 below offers an abstract perspective on the selfadjoint dilation and its resolvent for a dissipative almost solvable extension, and more importantly, verifies correctness of many underlying arguments needed for the further development in the general situation. These abstract results are immediately applicable to any dissipative almost solvable extension, thereby relieve of the burden to prove them in each particular case. Since the eigenfunction expansion is not used in the model construction, all the objects are well defined and there is no need for special considerations with regard to formal procedures dealing with "generalized" vectors. Finally, the paper proposes a model of an almost solvable extension with no assumption of its dissipativity.

The paper is organized as follows. In Section 1 we briefly review some definitions and results pertinent to our study. The section culminates with the calculation of the characteristic function of a non-selfadjoint almost solvable extension of a symmetric operator expressed in terms of the extension's "parameter" and the Weyl function. (See the definitions below.) The relationship of these three objects is believed to be first obtained in the paper [35] for a symmetric operator with the deficiency indices $(1,1)$, but seems to remain unnoticed. We take an opportunity and formulate this result in the more general setting of almost solvable extensions. In Section 2 we show how to build the functional model of a non-selfadjoint almost solvable extension of a symmetric operator following the approach of [28]. All the results are accompanied with the full proofs, starting from the exact form of dilation of a dissipative almost solvable extension and ending in the main model theorem for a general non-selfadjoint non-dissipative extension. In Section 3 the theory is illustrated by two examples of symmetric operators with finite deficiency indices. We refrain from giving the model construction of non-selfadjoint extensions of these operators, because all such results are easily derived from the theory developed in Section 2.

The author would like to thank Prof. Serguei Naboko for his attention to the work and encouragement. As well, the author is grateful to the referee for numerous useful suggestions with regard to the paper's readability.

We use symbol $\mathscr{B}(H_1, H_2)$ where H_1, H_2 are separable Hilbert spaces, for the Banach algebra of bounded operators, defined everywhere in H_1 with values in H_2. The notation $A : H_1 \to H_2$ is equivalent to $A \in \mathscr{B}(H_1, H_2)$. Also, $\mathscr{B}(H) := \mathscr{B}(H, H)$. The real axis, complex plane are denoted as $\mathbb{R}$, $\mathbb{C}$, respectively. Further, $\mathbb{C}_\pm := \{z \in \mathbb{C} : \pm \operatorname{Im} z > 0\}$, $\mathbb{R}_\pm := \{x \in \mathbb{R} : \pm x > 0\}$, where Im stands for the imaginary part of a complex number. The domain, range and kernel of a linear operator A are denoted as $\mathcal{D}(A)$, $\mathcal{R}(A)$, and $\ker(A)$; the symbol $\rho(A)$ is used for the resolvent set of A.

1. Preliminaries

Let us recall a few basic facts about unbounded linear operators.

For a closed linear operator L with dense domain $\mathcal{D}(L)$ on a separable Hilbert space H a sesquilinear form $\Psi_L(\cdot, \cdot)$ defined on the domain $\mathcal{D}(L) \times \mathcal{D}(L)$:

$$\Psi_L(f, g) = \frac{1}{i}[(Lf, g)_H - (f, Lg)_H], \quad f, g \in \mathcal{D}(L) \tag{1.1}$$

plays a role of the imaginary part of L in the sense that $2\operatorname{Im}(Lf, f) = \Psi_L(f, f)$, $f \in \mathcal{D}(L)$.

Definition 1.1. The operator L is called **dissipative** if

$$\operatorname{Im}(Lf, f) \geq 0, \quad f \in \mathcal{D}(L). \tag{1.2}$$

Definition 1.2. The operator L is called **maximal dissipative** if (1.2) holds and the resolvent $(L - zI)^{-1} \in \mathscr{B}(H)$ exists for any $z \in \mathbb{C}_-$.

In what follows A denotes a closed and densely defined symmetric operator on the separable Hilbert space H with equal deficiency indices $0 < n_+(A) = n_-(A) \leq \infty$. We will assume that A is simple, i.e., it has no reducing subspaces where it induces a self-adjoint operator. The adjoint operator A^* is closed and $A \subseteq A^*$ in a sense that $\mathcal{D}(A) \subseteq \mathcal{D}(A^*)$ and $Ax = A^*x$ for $x \in \mathcal{D}(A)$.

1.1. Boundary triples and almost solvable extensions

An extension $\mathcal{A}$ of the operator A is called **proper**, if $A \subseteq \mathcal{A} \subseteq A^*$. The following definition, see [19], [16], [21], may be considered as an abstract version of the second Green formula.

Definition 1.3. A triple $\{\mathcal{H}, \Gamma_0, \Gamma_1\}$ consisting of an auxiliary Hilbert space $\mathcal{H}$ and linear mappings Γ_0, Γ_1 defined on the set $\mathcal{D}(A^*)$, is called a **boundary triple** for the operator A^* if the following conditions are satisfied:

1. Green's formula is valid

$$(A^*f, g)_H - (f, A^*g)_H = (\Gamma_1 f, \Gamma_0 g)_{\mathcal{H}} - (\Gamma_0 f, \Gamma_1 g)_{\mathcal{H}}, \quad f, g \in \mathcal{D}(A^*). \tag{1.3}$$

2. For any $Y_0, Y_1 \in \mathcal{H}$ there exist $f \in \mathcal{D}(A^*)$, such that $\Gamma_0 f = Y_0$, $\Gamma_1 f = Y_1$. In other words, the mapping $f \mapsto \Gamma_0 f \oplus \Gamma_1 f$, $f \in \mathcal{D}(A^*)$ into $\mathcal{H} \oplus \mathcal{H}$ is surjective.

The boundary triple can be constructed for any closed densely defined symmetric operator with equal deficiency indices. Moreover, the space $\mathcal{H}$ can be chosen so that $\dim \mathcal{H} = n_+(A) = n_-(A)$. (See references above for further details.)

Definition 1.4. A proper extension $\mathcal{A}$ of the symmetric operator A is called **almost solvable (a.s.)** if there exist a boundary triple $\{\mathcal{H}, \Gamma_0, \Gamma_1\}$ for A^* and an operator $B \in \mathscr{B}(\mathcal{H})$ such that

$$f \in \mathcal{D}(\mathcal{A}) \iff \Gamma_1 f = B\Gamma_0 f. \tag{1.4}$$

Note that this definition implies the inclusion $\mathcal{D}(\mathcal{A}) \subset \mathcal{D}(A^*)$ and in fact the operator $\mathcal{A}$ is a restriction of A^* to the linear set $\{f \in \mathcal{D}(A^*) : \Gamma_1 f = B\Gamma_0 f\}$.

It can be shown (see [18]) that if a proper extension $\mathcal{A}$ has regular points in both the upper and lower half-planes, then this extension is almost solvable. In other words, there exist a boundary triple $\{\mathcal{H}, \Gamma_0, \Gamma_1\}$ and an operator $B \in \mathscr{B}(\mathcal{H})$ such that $\mathcal{A} = A_B$. We will refer to the operator B as a "parameter" of the extension A_B.

The next theorem summarizes some facts concerning a.s. extensions needed for the purpose of the paper.

Theorem 1.1. *Let A be a closed symmetric operator with dense domain on a separable Hilbert space H with equal (finite or infinite) deficiency indices and $\{\mathcal{H}, \Gamma_0, \Gamma_1\}$ be the boundary triple for its adjoint A^*. Let $B \in \mathscr{B}(\mathcal{H})$ and A_B be the corresponding a.s. extension of A. Then*

1. *$A \subset A_B \subset A^*$.*
2. *$(A_B)^* \subset A^*$, $(A_B)^* = A_{B^*}$.*
3. *A_B is maximal, i.e., $\rho(A_B) \neq \emptyset$.*
4. *B is dissipative $\iff$ A_B is maximal dissipative.*
5. *$B = B^* \iff A_B = (A_B)^*$.*

Proof. The proof can be found in [19], [18]. Note that the last two assertions can easily be verified using equality

$$\Psi_{A_B}(f, g) = \frac{1}{i}[(A_B f, g) - (f, A_B g)] = \frac{1}{i}((B - B^*)\Gamma_0 f, \Gamma_0 g), \tag{1.5}$$
$$f, g \in \mathcal{D}(A_B)$$

which directly follows from (1.3), (1.4). $\qquad\square$

Remark 1.5. In many cases of operators associated with partial differential equations, the boundary triple constructed according to the results cited in Definition 1.3 could not be easily linked to the Green formula as traditionally understood in a sense of differential expressions. For example, let Ω be a smooth bounded domain in $\mathbb{R}^3$, and A be a minimal symmetric operator in $L_2(\Omega)$ associated with the Laplace differential expression $-\Delta$ in Ω. Then A^* is defined on the set of functions $u \in L_2(\Omega)$ such that $\Delta u \in L_2(\Omega)$. The well-known Green formula (see [1], for example) suggest the "natural" definition of mappings Γ_0, Γ_1 as $\Gamma_0 : u \mapsto u|_{\partial\Omega}$, $\Gamma_1 : u \mapsto \frac{\partial u}{\partial n}|_{\partial\Omega}$, $u \in D(A^*)$ with the boundary space $\mathcal{H} = L_2(\partial\Omega)$. However, because there exist functions in $\mathcal{D}(A^*)$ that do not possess boundary values on $\partial\Omega$, operators Γ_0 and Γ_1 are not defined on the whole of $\mathcal{D}(A^*)$, and the theory of almost solvable extensions is inapplicable to this choice of the triple $\{\mathcal{H}, \Gamma_0, \Gamma_1\}$.

1.2. Weyl function

For a given boundary triple $\{\mathcal{H}, \Gamma_0, \Gamma_1\}$ for the operator A^* introduce an operator A_∞ as a restriction of A^* on the set of elements $y \in \mathcal{D}(A^*)$ satisfying the condition $\Gamma_0 y = 0$:

$$A_\infty := A^*|_{\mathcal{D}(A_\infty)}, \quad \mathcal{D}(A_\infty) := \{y \in \mathcal{D}(A^*) : \Gamma_0 y = 0\}. \tag{1.6}$$

Formally, the operator A_∞ is an almost solvable selfadjoint extension of A corresponding to the choice $B = \infty$. (See (1.4).) This justifies the notation. It turns out ([19], [18]), that the operator A_∞ is selfadjoint indeed. Further, for any $z \in \mathbb{C}_-\cup\mathbb{C}+$ the domain $\mathcal{D}(A^*)$ can be represented in the form of the direct sum:

$$\mathcal{D}(A^*) = \mathcal{D}(A_\infty)\dotplus \ker(A^* - zI) \tag{1.7}$$

according to the decomposition $f = y + h$ with $f \in \mathcal{D}(A^*)$, $y \in \mathcal{D}(A_\infty)$, and $h \in \ker(A^* - zI)$, where

$$y := (A_\infty - zI)^{-1}(A^* - zI)f, \quad h := f - y.$$

Taking into account equality $\mathcal{D}(A_\infty) = \ker(\Gamma_0)$ and the surjective property of Γ_0, it follows from the formula (1.7) that for each $e \in \mathcal{H}$ and $z \in \mathbb{C}_- \cup \mathbb{C}_+$ the equation $\Gamma_0 h = e$ has a unique solution that belongs to $\ker(A^* - zI)$. In other words, a restriction of operator Γ_0 on the set $\ker(A^* - zI)$ is invertible. Denote $\gamma(z)$ the corresponding inverse operator:

$$\gamma(z) = \left[\Gamma_0|_{\ker(A^* - zI)}\right]^{-1}, \quad z \in \mathbb{C}_- \cup \mathbb{C}_+. \tag{1.8}$$

By a simple computation we deduce from (1.3) with $f \in \mathcal{D}(A_\infty)$, $g \in \ker(A^* - zI)$ that

$$\gamma^*(\bar{z}) = \Gamma_1(A_\infty - zI)^{-1}, \quad z \in \mathbb{C}_- \cup \mathbb{C}_+. \tag{1.9}$$

The Weyl function $M(\cdot)$ corresponding to the boundary triple $\{\mathcal{H}, \Gamma_0, \Gamma_1\}$ is defined as an operator-function with values in $\mathscr{B}(\mathcal{H})$, such that for each $z \in \mathbb{C}_- \cup \mathbb{C}_+$, and $f_z \in \ker(A^* - zI)$

$$M(z)\Gamma_0 f_z = \Gamma_1 f_z. \tag{1.10}$$

Another representation of $M(\cdot)$ easily follows from (1.8) and (1.10)

$$M(z) = \Gamma_1 \gamma(z), \quad z \in \mathbb{C}_+ \cup \mathbb{C}_-. \tag{1.11}$$

The next theorem sums up a few properties of the Weyl function.

Theorem 1.2. *Let $M(\cdot)$ be the Weyl function (1.10), $z \in \mathbb{C}_- \cup \mathbb{C}_+$ and an operator $B \in \mathscr{B}(\mathcal{H})$ be a parameter of a.s. extension A_B of A. The following assertions hold:*

1. *$M(z)$ is analytic,*
2. *$\operatorname{Im} M(z) \cdot \operatorname{Im} z > 0$,*
3. *$[M(z)]^* = M(\bar{z})$,*
4. *$M(z) - M(\zeta) = (z - \zeta)\gamma^*(\bar{\zeta})\gamma(z), \quad z, \zeta \in \mathbb{C}_+ \cup \mathbb{C}_-,$*
5. *$z \in \rho(A_B) \iff (B - M(z))$ is boundedly invertible in $\mathcal{H}$,*
6. *$(A_B - zI)^{-1} - (A_\infty - z)^{-1} = \gamma(z)(B - M(z))^{-1}\gamma^*(\bar{z}), \quad z \in \rho(A_B).$*

Proof. The proof of the theorem can be found in [18]. $\qquad\qquad\qquad \square$

It follows from Theorem 1.2 that the Weyl function $M(\cdot)$ is a Herglotz function. It is analytic in the upper half-plane, with positive imaginary part.

1.3. Characteristic function of an almost solvable extension

As before, let A be a simple densely defined symmetric operator with equal deficiency indices and $\{\mathcal{H}, \Gamma_0, \Gamma_1\}$ be the boundary triple for A^*. Let $M(\cdot)$ be the Weyl function corresponding to that triple. According to Theorem 1.1, for any $B \in \mathscr{B}(\mathcal{H})$ the extension A_B is selfadjoint if $B = B^*$. We shall assume that $B \neq B^*$ and calculate the characteristic function of the non-selfadjoint operator A_B. (See the definition below.) For simplicity sake we assume that the operator A_B is simple. In other words, A_B has no non-trivial selfadjoint parts. It turns out that there exists an elegant formula which ties together the characteristic function of A_B, Weyl function $M(\cdot)$ and the extension "parameter" B. In the particular case of the one-dimensional Schrödinger operator on $\mathbb{R}_+$, this formula was obtained in [35].

Let us recall the definition of the characteristic function of a linear non-selfadjoint operator. In our narrative we follow the abstract approach developed by Štraus [44].

For a closed linear operator L with dense domain $\mathcal{D}(L)$ introduce a linear set $\mathcal{G}(L)$:

$$\mathcal{G}(L) = \big\{ g \in \mathcal{D}(L) : \Psi_L(f, g) = 0, \quad \forall f \in \mathcal{D}(L) \big\},$$

and a linear space $\mathfrak{L}(L)$ defined as closure of the quotient $\mathcal{D}(L)/\mathcal{G}(L)$ endowed with an inner product $[\xi, \eta]_{\mathfrak{L}} = \Psi_L(f, g)$, $\xi, \eta \in \mathfrak{L}(L)$, $f \in \xi$, $g \in \eta$, where $\Psi_L(f, g)$ is defined in (1.1). The inner product $[\cdot, \cdot]_{\mathfrak{L}}$ is symmetric and non-degenerate, but not necessarily positive. The non-degeneracy means the implication

$$[\xi, \eta]_{\mathfrak{L}} = 0, \forall \eta \in \mathfrak{L} \quad \Rightarrow \quad \xi = 0.$$

Definition 1.6. A **boundary space** for the operator L is any linear space $\mathfrak{L}$ which is isomorphic to $\mathfrak{L}(L)$. A **boundary operator** for the operator L is the linear operator Γ with the domain $\mathcal{D}(L)$ and the range in the boundary space $\mathfrak{L}$ such that

$$[\Gamma f, \Gamma g]_{\mathfrak{L}} = \Psi_L(f, g), \quad f, g \in \mathcal{D}(L). \tag{1.12}$$

We shall assume that the operator L is non-selfadjoint and its resolvent set is non-empty: $\rho(L) \neq \emptyset$. Let $\mathfrak{L}$ endowed with an inner product $[\cdot, \cdot]$ be a boundary space for L with boundary operator Γ, and let $\mathfrak{L}'$ with an inner product $[\cdot, \cdot]'$ be a boundary space for $-L^*$ with boundary operator Γ' mapping $\mathcal{D}(L^*)$ onto $\mathfrak{L}'$.

Definition 1.7. A **characteristic function** of the operator L is an operator-valued function Θ_L defined on the set $\rho(L^*)$ whose values $\Theta_L(z)$ map $\mathfrak{L}$ into $\mathfrak{L}'$ according to the equality

$$\Theta_L(z)\Gamma f = \Gamma'(L^* - zI)^{-1}(L - zI)f, \quad f \in \mathcal{D}(L). \tag{1.13}$$

Since the right-hand side of (1.13) is analytic with regard to $z \in \rho(L^*)$, the function Θ_L is analytic on $\rho(L^*)$.

Let us carry out the calculation of the characteristic function of an a.s. extension A_B of the symmetric operator A parameterized by the bounded operator $B \in \mathcal{B}(\mathcal{H})$.

Let $B = B_R + iB_I$ where $B_R = \frac{1}{2}(B + B^*)$ and $B_I = \frac{1}{2i}(B - B^*)$ be the real and the imaginary parts of operator B, and

$$E = \mathrm{clos}\,\mathcal{R}(B_I), \quad \alpha = |2B_I|^{1/2}, \quad J = \mathrm{sign}(B_I|_E). \tag{1.14}$$

Obviously, operators α and J commute as functions of the selfadjoint operator B_I. Note as well the involutional properties of the mapping J acting on the space E, namely, the equalities $J = J^* = J^{-1}$. If the operator B is dissipative (i.e., $B_I \geq 0$), then $J = I_E$ and $\alpha = (2B_I)^{1/2}$.

Using notation (1.14) the equality (1.5) can be rewritten in the form

$$\Psi_{A_B}(f, g) = 2(B_I \Gamma_0 f, \Gamma_0 g)_E = (J\alpha\Gamma_0 f, \alpha\Gamma_0 g)_E, \quad f, g \in \mathcal{D}(A_B)$$

where equality $2B_I|_E = \alpha J\alpha|_E$ holds due to the spectral theorem. According to the definition (1.12) we can choose the boundary space of the operator A_B to be the space E with the metric $[\cdot, \cdot] = (J\cdot, \cdot)_{\mathcal{H}} = (J\cdot, \cdot)_E$ and define the boundary operator Γ as the map

$$\Gamma : f \mapsto J\alpha\Gamma_0 f, \quad f \in \mathcal{D}(\Gamma), \quad \mathcal{D}(\Gamma) = \mathcal{D}(A_B). \tag{1.15}$$

Since $-A_B^* = -A_{B^*}$, see Theorem 1.1, we can repeat the arguments above and choose the boundary space of $-A_B^*$ to be the same Hilbert space E with the same metric $[\cdot, \cdot]' = [\cdot, \cdot] = (J\cdot, \cdot)_E$, and the boundary operator Γ' to be equal to the operator $\Gamma = J\alpha\Gamma_0$. Note that the metric $[\cdot, \cdot] = [\cdot, \cdot]'$ is positive if the operator B is dissipative.

Now we are ready to calculate the characteristic function of the operator A_B that corresponds to the chosen boundary spaces and operators. Let $z \in \rho(A_B^*)$ be a complex number and $f \in \mathcal{D}(A_B)$. Then from the equality $g_z = (A_B^* - zI)^{-1}(A_B - zI)f$ we obtain

$$A_B f - A_B^* g_z = z(f - g_z)$$

which due to inclusions $A_B \subset A^*$, $A_B^* \subset A^*$ shows that the vector $f - g_z$ belongs to the linear set $\ker(A^* - zI)$. By the Weyl function definition (1.10) the following equality holds for each $z \in \rho(A_B^*)$, $f \in \mathcal{D}(A_B)$

$$M(z)\Gamma_0(f - g_z) = \Gamma_1(f - g_z).$$

Since $f \in \mathcal{D}(A_B)$ and $g_z \in \mathcal{D}(A_B^*)$, the right-hand side here can be rewritten in the form $B\Gamma_0 f - B^*\Gamma_0 g_z$, and after elementary regrouping we obtain

$$(M(z) - B)\Gamma_0 f = (M(z) - B^*)\Gamma_0 g_z.$$

By virtue of Theorem 1.2 the operator $(M(z) - B^*)$ is boundedly invertible for $z \in \rho(A_B^*)$. Therefore,

$$\Gamma_0 g_z = (B^* - M(z))^{-1}(B - M(z))\Gamma_0 f$$

and due to (1.15),

$$
\begin{aligned}
\Gamma' g_z = J\alpha\Gamma_0 g_z &= J\alpha(B^* - M(z))^{-1}(B - M(z))\Gamma_0 f \\
&= J\alpha(B^* - M(z))^{-1} \times \big[B^* - M(z) + (B - B^*)\big]\Gamma_0 f \\
&= J\alpha\big[I + 2i(B^* - M(z))^{-1}B_I\big]\Gamma_0 f \\
&= J\alpha\big[I + i(B^* - M(z))^{-1}\alpha J\alpha\big]\Gamma_0 f \\
&= \big[I_E + iJ\alpha(B^* - M(z))^{-1}\alpha\big]J\alpha\Gamma_0 f \\
&= \big[I_E + iJ\alpha(B^* - M(z))^{-1}\alpha\big]\Gamma f
\end{aligned}
$$

so that finally for any $f \in \mathcal{D}(A_B)$ and $z \in \rho(A_B^*)$ the following equality holds

$$\Gamma'(A_B^* - zI)^{-1}(A_B - zI)f = \big[I_E + iJ\alpha(B^* - M(z))^{-1}\alpha\big]\Gamma f.$$

Now the comparison with the definition (1.13) yields that the characteristic function $\Theta_{A_B}(\cdot) : E \to E$ corresponding to the boundary operators and spaces chosen above is given by the formula

$$\Theta_{A_B}(z) = I_E + iJ\alpha(B^* - M(z))^{-1}\alpha\big|_E, \quad z \in \rho(A_B^*). \qquad (1.16)$$

Similar calculations can be found in [22].

A few remarks are in order. Following the schema followed above, it is easy to compute the characteristic function $\Theta_B(\cdot)$ of the operator B. Indeed, for $x, y \in \mathcal{H}$

$$
\begin{aligned}
\Psi_B(x, y) = \frac{1}{i}\big[(Bx, y)_{\mathcal{H}} - (x, By)_{\mathcal{H}}\big] &= \frac{1}{i}\big((B - B^*)x, y\big)_{\mathcal{H}} \\
&= 2(B_I x, y)_{\mathcal{H}} = (J\alpha x, \alpha y)_E,
\end{aligned}
$$
$$\Psi_{-B^*}(x, y) = \Psi_B(x, y)$$

so that we can choose the space $E = \operatorname{clos}\mathcal{R}(B_I)$ as a boundary space of the operators B and $-B^*$, see (1.14), and assume the boundary operators for B and $-B^*$ to be the mapping of the vector $x \in \mathcal{H}$ into $J\alpha x \in E$. Computations, similar to those conducted above, lead to the following expression for the characteristic function $\Theta_B(\cdot)$ of the operator B:

$$\Theta_B(z) = I_E + iJ\alpha(B^* - zI)^{-1}\alpha\big|_E.$$

Remark 1.8. Comparison with (1.16) shows that the characteristic function Θ_{A_B} of the extension A_B can be formally obtained by the substitution of zI in the expression for the characteristic function Θ_B of the "parameter" operator B with the Weyl function $M(z)$ of the operator A. Or more formally,

$$\Theta_{A_B}(z) = \Theta_B(M(z)), \quad z \in \rho(B^*) \cap \rho(A_B^*).$$

This interesting formula can be traced back to the paper of B. Pavlov [35].

Remark 1.9. Values of the characteristic operator function $\Theta_{A_B}(\cdot)$ in the upper half-plane $\mathbb{C}_+$ are J-contractive operators in E, i.e., for $\varphi \in E$

$$\big(J\Theta_{A_B}(z)\varphi, \Theta_{A_B}(z)\varphi\big)_E \le (J\varphi, \varphi)_E, \quad z \in \rho(A_B^*) \cap \mathbb{C}_+. \tag{1.17}$$

This result follows from the general contractive property of characteristic functions of linear operators obtained in [44]. It is remarkable that the proof cited below does not require the knowledge of the characteristic function itself. Its contractiveness follows directly from its definition.

Theorem 1.3. *[44] Let $\mathfrak{L}$, $\mathfrak{L}'$, Γ, Γ' be the boundary spaces and boundary operators for the operators L and $-L^*$ respectively as described in Definition 1.6, $[\cdot, \cdot]$, $[\cdot, \cdot]'$ be the metrics in $\mathfrak{L}$, $\mathfrak{L}'$, and $\Theta_L(\cdot)$ be the characteristic function of L, see Definition 1.7. Then the following equality holds*

$$[\varphi, \varphi_1] - [\Theta_L(z)\varphi, \Theta_L(\zeta)\varphi_1]' = (1/i)(z - \bar{\zeta})(\Omega_z\varphi, \Omega_\zeta\varphi_1)_H \tag{1.18}$$

where $z, \zeta \in \rho(L^*)$, $\varphi, \varphi_1 \in \mathfrak{L}$, and the operator Ω_z, $z \in \rho(L^*)$ is uniquely defined as the map $\Omega_z : \Gamma f \mapsto f - (L^* - zI)^{-1}(L - zI)f$, $f \in \mathcal{D}(L)$.

Proof. By the polarization identity it is sufficient to show that (1.18) is valid for $z = \zeta$, $\varphi = \varphi_1$. Standard density arguments allow us to prove the statement of the theorem only for the dense set of vectors $\{\varphi\}$ in $\mathfrak{L}$, for which there exist $f \in \mathcal{D}(L)$, such that $\varphi = \Gamma f$. Let φ be such a vector and $f \in \mathcal{D}(L)$ satisfies the condition $\Gamma f = \varphi$. Let g_z be the vector $(L^* - zI)^{-1}(L - zI)f$. Note that $g_z \in \mathcal{D}(L^*)$, $(Lf, g_z) = (f, L^*g_z)$, and $Lf - L^*g_z = z(f - g_z)$. Then

$$(z - \bar{z})(\Omega_z\varphi, \Omega_z\varphi)_H = (z - \bar{z})(\Omega_z\Gamma f, \Omega_z\Gamma f)$$
$$= (z - \bar{z})(f - g_z, f - g_z)$$
$$= \big(z(f - g_z), f - g_z\big) - \big(f - g_z, z(f - g_z)\big)$$
$$= \big(Lf - L^*g_z, f - g_z\big) - \big(f - g_z, Lf - L^*g_z\big)$$
$$= (Lf, f) + (L^*g_z, g_z) - (f, Lf) - (g_z, L^*g_z)$$
$$= \big((Lf, f) - (f, Lf)\big) - \big((-L^*g_z, g_z) - (g_z, -L^*g_z)\big)$$

so that

$$(1/i)(z - \bar{z})(\Omega_z\varphi, \Omega_z\varphi)_H = \Psi_L(f, f) - \Psi_{-L^*}(g_z, g_z)$$
$$= [\Gamma f, \Gamma f] - [\Gamma'g_z, \Gamma'g_z]' = [\Gamma f, \Gamma f] - [\Theta_L(z)\Gamma f, \Theta_L(z)\Gamma f]'$$
$$= [\varphi, \varphi] - [\Theta_L(z)\varphi, \Theta_L(z)\varphi]'.$$

The proof is complete. $\qquad\qquad\square$

128   V. Ryzhov

From (1.18) and the representation of the boundary spaces of operators A_B, $-A_B^*$ described above we conclude

$$(J\varphi, \varphi)_E - (J\Theta_{A_B}(z)\varphi, \Theta_{A_B}(z)\varphi)_E = (2\operatorname{Im} z)\|\Omega_z\varphi\|^2$$

which proves (1.17).

Remark 1.10. If the extension parameter B is a dissipative operator from $\mathscr{B}(\mathcal{H})$, then the corresponding extension A_B is a closed dissipative operator and its resolvent set $\rho(A_B)$ includes the lower half-plane $\mathbb{C}_-$. The conjugate operator $(A_B)^*$ is an extension of A corresponding to the operator B^*, see Theorem 1.1, so that the upper half-plane $\mathbb{C}_+$ consists of the regular points of A_B^*. Since B is dissipative, the involution operator J defined in (1.14) is in fact the identity operator on E. Moreover, the metric of the boundary spaces $\mathfrak{L}$, $\mathfrak{L}'$ is positively defined, hence the space $\mathfrak{L}$ is a Hilbert space. It follows from (1.17) that in this case values of the characteristic function $\Theta_{A_B}(z)$, $z \in \mathbb{C}_+$ are contractive operators from $\mathscr{B}(E)$:

$$\|\Theta_{A_B}(z)\varphi\|_E \le \|\varphi\|_E, \quad z \in \mathbb{C}_+, \, B \in \mathscr{B}(\mathcal{H}) \text{ is dissipative.} \tag{1.19}$$

Remark 1.11. Let $z \in \mathbb{C}_+$. According to Theorem 1.2, the imaginary part of the operator $M(z)$ is positive. In other words, values of the function $M(z)$ are dissipative operators in $\mathscr{B}(E)$. Assume $B = iI_\mathcal{H}$. This operator is dissipative, so is the corresponding extension A_{iI}. Since $B^* = -iI_\mathcal{H}$, $E = \mathcal{H}$, $J = I_\mathcal{H}$, $\alpha = (2|B_I|)^{1/2} = \sqrt{2}I_\mathcal{H}$, the characteristic function (1.16) can be written as

$$\Theta_{A_{iI}}(z) = I_\mathcal{H} + 2i\big(-iI_\mathcal{H} - M(z)\big)^{-1} = \big(M(z) - iI\big)\big(M(z) + iI\big)^{-1}.$$

We see that for $z \in \mathbb{C}_+$ the contractive function $\Theta_{iI}(z)$ and the Herglotz function $M(z)$ are related to each other via Cayley transform. In fact, the operator function $\Theta_{A_{iI}}$ is a characteristic function of the symmetric operator A as defined in [45].

2. Functional model

In this section a variant of model for a non-selfadjoint non-dissipative a.s. extension of the symmetric operator A is constructed.

Let $B \in \mathscr{B}(\mathcal{H})$ and A_B be the corresponding a.s. extension of A. The question of simultaneous simplicity of operators B and A_B was formulated in [43], and the author is unaware of any results which would shed light on the intricate relationship between selfadjoint parts of B and A_+. In the following it is always assumed that both B and A_B are simple operators. Further, by virtue of Theorem 1.1, A_B is maximal and the resolvent set of A_B is non-empty: $\rho(A_B) \ne \emptyset$. The conjugate operator $(A_B)^*$ is simple and maximal as well. It coincides with the extension of A parametrized by $B^* : (A_B)^* = A_{B^*}$. The characteristic function $\Theta_{A_B}(\cdot)$ is analytic on $\rho(A_B^*)$ with values in $\mathscr{B}(E)$ and J-contractive on $\rho(A_B^*) \cap \mathbb{C}_+$, see (1.16), (1.17) and (1.14) for the notation.

Assume that the operator $B = B_R + iB_I$, where $B_R := (1/2)(B + B^*)$, $B_I := (1/2i)(B - B^*)$, is not dissipative so that $J \ne I_E$. Along with the operator $B =$

$B_R + iJ\frac{\alpha^2}{2}$ consider a dissipative operator $B_+ := B_R + i|B_I| = B_R + i\frac{\alpha^2}{2}$ and let A_{B_+} be the corresponding a.s. extension of A. Then the operators B_+ and A_{B_+} are both dissipative, B_+ is bounded, and as shown in [28], B_+ is simple. As mentioned above, these observations alone do not guarantee simplicity of A_{B_+}. Nevertheless, A_{B_+} is simple. This fact follows from Theorem 2.1 below and explicit relationship between Cayley transformations of A_B and A_{B_+} found in [27] in a more general setting. Namely, it follows from [27] that selfadjoint parts of A_B and A_{B_+} coincide. The same result can be obtained by methods developed in the system theory [8].

Finally, due to dissipativity the lower half-plane $\mathbb{C}_-$ consists of the regular points of A_{B_+} and similarly, $\mathbb{C}_+ \subset \rho(A_{B_+}^*)$.

According to Remarks 1.9 and 1.10, values of characteristic functions of two extensions A_B, A_{B_+} are J-contractive and contractive operators respectively in $\rho(A_B^*) \cap \mathbb{C}_+$. It turns out that these values are related via the so-called Potapov-Ginzburg transformation [9]. This observation was first made in [28] for additive perturbations of a selfadjoint operator and in [27] for the general case. (Cf. [8] for an alternative, but equivalent approach.) We formulate this relationship in the special situation of almost solvable extensions of a symmetric operator and sketch a simple proof based on findings of [28].

Theorem 2.1. *The characteristic functions* $\Theta := \Theta_{A_B}$, $S := \Theta_{A_{B_+}}$ *of two simple maximal a.s. extensions* A_B, A_{B_+} *of the symmetric operator* A *corresponding to the extension parameters* $B, B_+ \in \mathscr{B}(\mathcal{H})$, *where* $B = B_R + iB_I$, $B_+ = B_R + i|B_I|$ *are related to each other via following the Potapov-Ginzburg transformation.*

$$
\begin{aligned}
\Theta(z) &= \ \ (X^- + X^+ S(z)) \cdot (X^+ + X^- S(z))^{-1}, \\
\Theta(z) &= -(X^+ - S(z)X^-)^{-1} \cdot (X^- - S(z)X^+), \\
S(z) &= \ \ (X^- + X^+ \Theta(z)) \cdot (X^+ + X^- \Theta(z))^{-1}, \\
S(z) &= -(X^+ - \Theta(z)X^-)^{-1} \cdot (X^- \Theta(z)X^+)^{-1} \\
\Theta(\zeta) &= \ \ (X^+ + X^- [S(\bar{\zeta})]^*) \cdot (X^- + X^+ [S(\bar{\zeta})]^*)^{-1}, \\
\Theta(\zeta) &= -(X^- - [S(\bar{\zeta})]^* X^+)^{-1} \cdot (X^+ - [S(\bar{\zeta})]^* X^-), \\
[S(\bar{\zeta})]^* &= \ \ (X^+ + X^- \Theta(\zeta)) \cdot (X^- + X^+ \Theta(\zeta))^{-1}, \\
[S(\bar{\zeta})]^* &= -(X^- - \Theta(\zeta)X^+)^{-1} \cdot (X^+ - \Theta(\zeta)X^-).
\end{aligned}
\tag{2.1}
$$

Here $z \in \rho(A_B^*) \cap \mathbb{C}_+$, $\zeta \in \rho(A_B^*) \cap \mathbb{C}_-$ *and* $X^\pm := (I_E \pm J)/2$ *are two complementary orthogonal projections in the space* E.

Proof. The existence of the Potapov-Ginzburg transformation S of a J-contractive operator Θ and formulae (2.1) can be found in the literature ([9], [8]). On the other hand, it has been shown in the paper [28] that the characteristic functions of two bounded operators $B = B_R + iB_I$ and $B_+ = B_R + i|B_I|$ are related via Potapov-Ginzburg transformation. Taking into account Remark 1.8 we arrive at the theorem's assertion. $\qquad\square$

In what follows we will use the simplified notation Θ, S introduced in Theorem 2.1 for the characteristic functions Θ_{A_B} and $\Theta_{A_{B_+}}$, respectively. Note that due to Remark 1.10, the analytic operator functions $S(z)$ and $S^*(\zeta) := [S(\bar{\zeta})]^*$ are contractive if $z \in \mathbb{C}_+$, $\zeta \in \mathbb{C}_-$. Moreover, there exist non-tangential strong boundary values almost everywhere on the real axis: $S(k) := s - \lim_{\varepsilon \downarrow 0} S(k + i\varepsilon)$, $S^*(k) := s - \lim_{\varepsilon \downarrow 0} S^*(k - i\varepsilon)$, a.e. $k \in \mathbb{R}$. These boundary values are contractive and mutually conjugate operators for almost all $k \in \mathbb{R}$ ([32]).

2.1. B. Pavlov's form of Sz.-Nagy-Foiaş model

The functional model of a dissipative operator can be derived from the B. Sz.-Nagy-C.Foiaş model for the contraction, whose Cayley transform it represents [32]. An independent approach was given in the framework of acoustic scattering by P. Lax and R. Phillips [26]. In our narrative we will use an equivalent model construction given by B. Pavlov in [37], [38] and elaborated further in the paper [28] of S. Naboko.

Let $\mathscr{A}$ be the minimal selfadjoint dilation of the simple dissipative operator A_{B_+}. In other words, the operator $\mathscr{A} = \mathscr{A}^*$ is defined on a wider space $\mathscr{H} \supset H$ such that (cf. [32])

$$
\begin{aligned}
P_H(\mathscr{A} - zI)^{-1}|_H &= (A_{B_+} - zI)^{-1}, \quad z \in \mathbb{C}_- \\
P_H(\mathscr{A} - zI)^{-1}|_H &= (A_{B_+}^* - zI)^{-1}, \quad z \in \mathbb{C}_+
\end{aligned}
\tag{2.2}
$$

and $\mathscr{H} := \mathrm{span}\{(\mathscr{A} - zI)^{-1}H : z \in \mathbb{C}_\pm\}$. Here $P_H : \mathscr{H} \to H$ is the orthogonal projection from the dilation space $\mathscr{H}$ onto H. The dilation $\mathscr{A}$ can be chosen in many ways. Following [37], [38], we will use the dilation space in the form of the orthogonal sum $\mathscr{H} := D_- \oplus H \oplus D_+$, where $D_\pm := L_2(\mathbb{R}_\pm, E)$. The space H is naturally embedded into $\mathscr{H}$: $H \to 0 \oplus H \oplus 0$, whereas spaces $D_\pm$ are embedded into $L_2(E) = D_- \oplus D_+$. The dilation representation offered in the next theorem is a straightforward generalization of B. Pavlov's construction [39]. Its form was announced in [41] without a proof. (See [43], [24] for a more general approach.)

Define a linear operator $\mathscr{A}$ by formula

$$
\mathscr{A} \begin{pmatrix} v_- \\ u \\ v_+ \end{pmatrix} = \begin{pmatrix} iv'_- \\ A^*u \\ iv'_+ \end{pmatrix}, \qquad \begin{pmatrix} v_- \\ u \\ v_+ \end{pmatrix} \in \mathcal{D}(\mathscr{A}),
\tag{2.3}
$$

where the domain $\mathcal{D}(\mathscr{A})$ consists of vectors $(v_-, u, v_+) \in \mathscr{H}$, such that

$$
v_\pm \in W_2^1(\mathbb{R}_\pm, E) \quad \text{and} \quad u \in \mathcal{D}(A^*)
$$

satisfy two "boundary conditions":

$$
\left.\begin{aligned}
\Gamma_1 u - B_+ \Gamma_0 u &= \alpha v_-(0) \\
\Gamma_1 u - B_+^* \Gamma_0 u &= \alpha v_+(0)
\end{aligned}\right\}.
\tag{2.4}
$$

Here boundary values $v_\pm(0) \in E$ are well defined according to imbedding theorems for spaces $W_2^1(\mathbb{R}_\pm, E)$.

Remark 2.1. There is a certain "geometrical" aspect of conditions (2.4). Indeed, the left-hand side of relations (2.4) are vectors from $\mathcal{H}$, whereas vectors on the right-hand side belong to the potentially "smaller" space $E \subset \mathcal{H}$. Since the vector $v_\pm(0) \in E$ can be chosen arbitrarily, it means that for $(v_-, u, v_+) \in \mathcal{D}(\mathscr{A})$.

$$\overline{\mathcal{R}\left(\Gamma_1 u - B_+\Gamma_0 u\right)} = \overline{\mathcal{R}\left(\Gamma_1 u - B_+^*\Gamma_0 u\right)} = E.$$

Remark 2.2. By termwise substraction we obtain from (2.4):

$$(B_+ - B_+^*)\Gamma_0 u = i\alpha^2 \Gamma_0 u = \alpha(v_+(0) - v_-(0)).$$

Standard arguments based on the functional calculus for the bounded selfadjoint operator α combined with the facts that $\mathcal{R}(\alpha)$ is dense in E and $v_\pm(0) \in E$ yields:

$$i\alpha\Gamma_0 u = v_+(0) - v_-(0), \quad (v_-, u, v_+) \in \mathcal{D}(\mathscr{A}). \tag{2.5}$$

Remark 2.3. Let $\mathscr{G}$ be a set of vectors $u \in \mathcal{D}(A^*)$ such that $(v_-, u, v_+) \in \mathcal{D}(\mathscr{A})$ with some $v_\pm \in D_\pm$. It is clear that $\mathscr{G}$ includes $\mathcal{D}(A_{B_+}) \cup \mathcal{D}(A_{B_+}^*)$. Indeed, if for example $v_-(0) = 0$ in (2.4), then we conclude that $u \in \mathcal{D}(A_{B_+})$, whereas $v_+(0)$ can be chosen appropriately in order to satisfy the second condition (2.4). The same argument applied to the case $v_+(0) = 0$ shows that $\mathcal{D}(A_{B_+}^*) \subset \mathscr{G}$.

Now we can formulate the main theorem concerning the selfadjoint dilation of A_{B_+}. For notational convenience let us introduce the following four operators

$$Y_\pm : y_\pm \mapsto iy_\pm', \quad \mathcal{D}(Y_\pm) := W_2^1(\mathbb{R}_\pm, E)$$

$$Y_\pm^0 : y_\pm \mapsto iy_\pm', \quad \mathcal{D}(Y_\pm^0) := \overset{o}{W}{}_2^1(\mathbb{R}_\pm, E),$$

where W_2^1, $\overset{o}{W}{}_2^1$ are usual Sobolev spaces [1]. Direct computation shows that

$$(Y_\pm)^* = (Y_\pm^0) \quad \text{and} \quad \rho(Y_+) = \rho(Y_-^0) = \mathbb{C}_+, \quad \rho(Y_-) = \rho(Y_+^0) = \mathbb{C}_-.$$

Theorem 2.2. *The operator $\mathscr{A}$ is a minimal selfadjoint dilation of the dissipative operator A_{B_+}. The resolvent of $\mathscr{A}$ is given by the following formulae:*

$$(\mathscr{A} - zI)^{-1} \begin{pmatrix} h_- \\ h_0 \\ h_+ \end{pmatrix}$$

$$= \begin{pmatrix} \psi_-(\xi) \\ (A_{B_+} - z)^{-1}h_0 - \gamma(z)(B_+ - M(z))^{-1}\alpha\psi_-(0) \\ (Y_+^0 - z)^{-1}h_+ + e^{-iz\xi}\{i\alpha\Gamma_0(A_{B_+} - z)^{-1}h_0 + S^*(\bar{z})\psi_-(0)\} \end{pmatrix}, \quad z \in \mathbb{C}_-$$

$$(\mathscr{A} - zI)^{-1} \begin{pmatrix} h_- \\ h_0 \\ h_+ \end{pmatrix}$$

$$= \begin{pmatrix} (Y_-^0 - z)^{-1}h_- + e^{-iz\xi}\{-i\alpha\Gamma_0(A_{B_+}^* - z)^{-1}h_0 + S(z)\psi_+(0)\} \\ (A_{B_+}^* - z)^{-1}h_0 - \gamma(z)(B_+^* - M(z))^{-1}\alpha\psi_+(0) \\ \psi_+(\xi) \end{pmatrix}, \quad z \in \mathbb{C}_+$$

where $(h_-, h_0, h_+) \in \mathcal{H}$, $\psi_\pm := (Y_\pm - z)^{-1}h_\pm$, $z \in \mathbb{C}_\pm$.

132   V. Ryzhov

Proof. Let $\mathcal{U} := (v_-, u, v_+) \in \mathcal{D}(\mathscr{A})$. Then

$$
\begin{aligned}
(\mathscr{A}\mathcal{U}, \mathcal{U}) &- (\mathcal{U}, \mathscr{A}\mathcal{U}) \\
&= (iv'_-, v_-) + (A^*u, u) + (iv'_+, v_+) - (v_-, iv'_-) - (u, A^*u) - (v_+, iv'_+) \\
&= i \int_{-\infty}^0 (v'_- \bar{v}_- + v_- \bar{v}'_-) dk + i \int_0^{+\infty} (v'_+ \bar{v}_+ + v_+ \bar{v}'_+) dk + (A^*u, u) - (u, A^*u) \\
&= i\|v_-(0)\|^2 - i\|v_+(0)\|^2 + (\Gamma_1 u, \Gamma_0 u) - (\Gamma_0 u, \Gamma_1 u).
\end{aligned}
$$

By substitution $\Gamma_1 u$ from (2.4) and (2.5) we obtain for two last summands

$$
\begin{aligned}
(\Gamma_1 u, \Gamma_0 u) &- (\Gamma_0 u, \Gamma_1 u) \\
&= (\alpha v_-(0) + B_+ \Gamma_0 u, \Gamma_0 u) - (\Gamma_0 u, \alpha v_+(0) + B_+^* \Gamma_0 u) \\
&= (v_-(0), \alpha \Gamma_0 u) - (\alpha \Gamma_0 u, v_+(0)) \\
&= (v_-(0), (-i)[v_+(0) - v_-(0)]) - ((-i)[v_+(0) - v_-(0)], v_+(0)) \\
&= i(v_-(0), v_+(0)) - i\|v_-(0)\|^2 + i\|v_+(0)\|^2 - i(v_-(0), v_+(0)) \\
&= i\|v_+(0)\|^2 - i\|v_-(0)\|^2.
\end{aligned}
$$

Finally,

$$
(\mathscr{A}\mathcal{U}, \mathcal{U}) - (\mathcal{U}, \mathscr{A}\mathcal{U}) = 0, \quad \mathcal{U} \in \mathcal{D}(\mathscr{A}),
$$

therefore $\mathscr{A}$ is symmetric.

Further, it is easy to see on ground that $\|\psi_\pm(0)\|_E \leq C\|\psi_\pm\|_{W_2^1(\mathbb{R}_\pm, E)}$ that operators defined by the right-hand sides of formulae for resolvent of $\mathscr{A}$ in the theorem's statement are bounded for corresponding $z \in \mathbb{C}_\pm$. If we show that they yield vectors that belong to the domain of operator $\mathscr{A}$ and they indeed describe inverse operators for $\mathscr{A} - zI$, it would mean that the symmetric operator $\mathscr{A}$ is closed and its deficiency indices equal zero. Hence $\mathscr{A}$ is selfadjoint.

Let $z \in \mathbb{C}_-$ be a complex number and $\mathcal{V} := (\tilde{v}_-, \tilde{u}, \tilde{v}_+)$ be a vector from the right-hand side of the corresponding resolvent equality under consideration. The first and third component of $\mathcal{V}$ obviously belong to the Sobolev spaces $W_2^1(\mathbb{R}_\pm, E)$. We need to verify first that $\mathcal{V}$ satisfies the boundary conditions (2.4).

$$
\begin{aligned}
(\Gamma_1 - B_+ \Gamma_0)\tilde{u} &= (\Gamma_1 - B_+ \Gamma_0)\big[(A_{B_+} - z)^{-1} h_0 - \gamma(z)(B_+ - M(z))^{-1} \alpha \psi_-(0)\big] \\
&= -(\Gamma_1 - B_+ \Gamma_0)\gamma(z)(B_+ - M(z))^{-1} \alpha \psi_-(0) \\
&= -(M(z) - B_+)(B_+ - M(z))^{-1} \alpha \psi_-(0) \\
&= \alpha \psi_-(0) = \alpha \tilde{v}_-(0)
\end{aligned}
$$

where we used the equalities $\Gamma_1 \gamma(z) = M(z)$ and $\Gamma_0 \gamma(z) = I_{\mathcal{H}}$, see (1.8), (1.11).

Further,

$$(\Gamma_1 - B_+^* \Gamma_0)\tilde{u}$$
$$= (\Gamma_1 - B_+^* \Gamma_0)\big[(A_{B_+} - z)^{-1}h_0 - \gamma(z)(B_+ - M(z))^{-1}\alpha\psi_-(0)\big]$$
$$= (\Gamma_1 - B_+\Gamma_0)\tilde{u} + i\alpha^2\Gamma_0\big[(A_{B_+} - z)^{-1}h_0 - \gamma(z)(B_+ - M(z))^{-1}\alpha\psi_-(0)\big]$$
$$= \alpha\psi_-(0) + i\alpha^2\Gamma_0(A_{B_+} - z)^{-1}h_0 - i\alpha^2(B_+ - M(z))^{-1}\alpha\psi_-(0)$$
$$= i\alpha^2\Gamma_0(A_{B_+} - z)^{-1}h_0 + \alpha\big[I - i\alpha(B_+ - M(z))^{-1}\alpha\big]\psi_-(0)$$
$$= \alpha\big[i\alpha\Gamma_0(A_{B_+} - z)^{-1}h_0 + S^*(\bar{z})\psi_-(0)\big] = \alpha\tilde{v}_+(0).$$

Thus, both conditions (2.4) are satisfied. Now consider $(\mathscr{A} - zI)\mathcal{V}$ for $z \in \mathbb{C}_-$. Since $\tilde{v}_- = (\mathrm{Y}_- - z)^{-1}h_-$ and $\tilde{v}_+ = (\mathrm{Y}_+^0 - z)^{-1}h_+ + e^{-iz\xi}\tilde{v}_+(0)$ it is easy to see that $(\mathrm{Y}_\pm - z)\tilde{v}_\pm = h_\pm$. Inclusions $A_{B_+} \subset A^*$ and $\mathcal{R}(\gamma(z)) \subset \ker(A^* - zI)$ help to compute the middle component $(A^* - z)\tilde{u}$:

$$(A^* - z)\big[(A_{B_+} - z)^{-1}h_0 - \gamma(z)(B_+ - M(z))^{-1}\alpha\psi_-(0)\big] = h_0.$$

Thus, $(\mathscr{A} - zI)(\mathscr{A} - zI)^{-1} = I$.

In order to check correctness of the equality $(\mathscr{A} - zI)^{-1}(\mathscr{A} - zI) = I$, let $\mathcal{U} := (v_-, u, v_+) \in \mathcal{D}(\mathscr{A})$ and $z \in \mathbb{C}_-$ be a complex number. Then

$$(\mathscr{A} - zI)^{-1}(\mathscr{A} - zI)U = (\mathscr{A} - zI)^{-1}\begin{pmatrix}(\mathrm{Y}_- - zI)v_- \\ (A^* - zI)u \\ (\mathrm{Y}_+ - zI)v_+\end{pmatrix}$$

$$= \begin{pmatrix} v_-(\xi) \\ (A_{B_+} - z)^{-1}(A^* - z)u - \gamma(z)(B_+ - M(z))^{-1}\alpha v_-(0) \\ v_+^0(\xi) + e^{-iz\xi}\big\{i\alpha\Gamma_0(A_{B_+} - z)^{-1}(A^* - z)u + S^*(\bar{z})v_-(0)\big\}\end{pmatrix}$$

where $v_+^0(\xi) := (\mathrm{Y}_+^0 - zI)^{-1}(\mathrm{Y}_+ - zI)v_+$.

We need to show first that the middle component here coincides with u. Note that the vector $\Psi(z) := (A_{B_+} - zI)^{-1}(A^* - zI)u - u$ belongs to $\ker(A^* - zI)$, therefore the expression $[\gamma(z)]^{-1}\Psi(z)$ represents an element $\Gamma_0\Psi(z)$ from $\mathcal{H}$. Now we can rewrite the middle component as follows:

$$u + \gamma(z)\big[\Gamma_0\Psi(z) - (B_+ - M(z))^{-1}\alpha v_-(0)\big]$$
$$= u + \gamma(z)(B_+ - M(z))^{-1}\big[(B_+ - M(z))\Gamma_0\Psi(z) - \alpha v_-(0)\big].$$

By the definition (1.10) of the Weyl function $M(\cdot)$ and the first of conditions (2.4), the expression in square brackets can be rewritten as

$$(B_+\Gamma_0 - \Gamma_1)\Psi(z) - (\Gamma_1 - B_+\Gamma_0)u = B_+\Gamma_0(\Psi(z) + u) - \Gamma_1(\Psi(z) + u).$$

The only thing left is the observation that $\Psi(z) + u$ belongs to the domain $\mathcal{D}(A_{B_+})$, hence this expression equals zero.

Because $v_+(\xi) = v_+^0(\xi) + e^{-iz\xi}v_+(0)$, in order to check correctness of the expression for the third component in the computations above we only need to show that

$$i\alpha\Gamma_0(A_{B_+} - z)^{-1}(A^* - z)u + S^*(\bar{z})v_-(0) = v_+(0).$$

Recalling that $S^*(\bar{z}) = I - i\alpha(B_+ - M(z))^{-1}\alpha$, $v_-(0) = v_+(0) - i\alpha\Gamma_0 u$ (see (1.16), (2.5)) and utilizing notation $\Psi(z)$ once again, we obtain

$$i\alpha\Gamma_0(A_{B_+} - z)^{-1}(A^* - z)u + S^*(\bar{z})v_-(0)$$
$$= i\alpha\Gamma_0(\Psi(z) + u) + v_+(0) - i\alpha\Gamma_0 u - i\alpha(B_+ - M(z))^{-1}\alpha v_-(0)$$
$$= v_+(0) + i\alpha\Gamma_0\Psi(z) - i\alpha(B_+ - M(z))^{-1}\alpha v_-(0)$$
$$= v_+(0) + i\alpha(B_+ - M(z))^{-1}\big[(B_+ - M(z))\Gamma_0\Psi(z) - \alpha v_-(0)\big].$$

It was shown at the previous step that the expression in square brackets is equal to zero.

The resolvent formula in the case $z \in \mathbb{C}_+$ is verified analogously.

Finally, dilation equalities (2.2) are obvious for the operators $(\mathscr{A} - zI)^{-1}$. Minimality of dilation $\mathscr{A}$ follows from the relation

$$\mathrm{span}\{(\mathscr{A} - zI)^{-1}H \,:\, z \in \mathbb{C}_\pm\}$$
$$= \mathrm{span}\left\{ \begin{pmatrix} e^{-iz_+ + \xi}\alpha\Gamma_0(A_{B_+}^* - z_+)^{-1}H \\ (A_{B_+} - z_-)^{-1}H + (A_{B_+}^* - z_+)^{-1}H \\ e^{-iz_- - \xi}\alpha\Gamma_0(A_{B_+} - z_-)^{-1}H \end{pmatrix} \,:\, z_\pm \in \mathbb{C}_\pm \right\},$$

properties of exponents in $L_2(\mathbb{R}_\pm)$, and density of sets

$$\{\alpha\Gamma_0(A_{B_+}^* - z)^{-1}H \,:\, z \in \mathbb{C}_+\}, \quad \{\alpha\Gamma_0(A_{B_+} - z)^{-1}H \,:\, z \in \mathbb{C}_-\}$$

in E. This density is a simple consequence of the fact that E is a boundary space and $\alpha\Gamma_0$ is a boundary operator for A_{B_+}, $A_{B_+}^*$ as defined in Section 1.3.

The proof is complete. $\qquad\square$

The spectral mapping that maps dilation $\mathscr{A}$ into the multiplication operator $f \mapsto k \cdot f$ on some L_2-space gives the model representation of the dissipative operator A_{B_+}:

$$\left. \begin{array}{ll} P_H(k - z)^{-1}|_H \cong (A_{B_+} - zI)^{-1}, & z \in \mathbb{C}_-,\, k \in \mathbb{R} \\ P_H(k - z)^{-1}|_H \cong (A_{B_+}^* - zI)^{-1}, & z \in \mathbb{C}_+,\, k \in \mathbb{R} \end{array} \right\}. \tag{2.6}$$

Following [37], [38], [28] we arrive at the model Hilbert space $\mathbf{H} = L_2\begin{pmatrix} I & S^* \\ S & I \end{pmatrix}$ by the factorization against elements with zero norm and subsequent completion of the linear set $\{\begin{pmatrix} \widetilde{g} \\ g \end{pmatrix} \,:\, \widetilde{g}, g \in L_2(\mathbb{R}, E)\}$ of two-components E-valued vector functions with respect to the norm

$$\left\| \begin{pmatrix} \widetilde{g} \\ g \end{pmatrix} \right\|_{\mathbf{H}}^2 := \int_{\mathbb{R}} \left\langle \begin{pmatrix} I & S^* \\ S & I \end{pmatrix} \begin{pmatrix} \widetilde{g} \\ g \end{pmatrix}, \begin{pmatrix} \widetilde{g} \\ g \end{pmatrix} \right\rangle_{E \oplus E} dk. \tag{2.7}$$

Note that in general the completion operation makes it impossible to treat individual components $\widetilde{g}$, g of a vector $\begin{pmatrix} \widetilde{g} \\ g \end{pmatrix} \in \mathbf{H}$ as regular L_2-functions. However, two equivalent forms of the **H**-norm

$$\left\|\begin{pmatrix}\widetilde{g}\\ g\end{pmatrix}\right\|_{\mathbf{H}}^2 = \|S\widetilde{g}+g\|_{L_2(E)}^2 + \|\Delta_* g\|_{L_2(E)}^2 = \|\widetilde{g}+S^* g\|_{L_2(E)}^2 + \|\Delta\widetilde{g}\|_{L_2(E)}^2,$$

where $\Delta := \sqrt{I - S^* S}$ and $\Delta_* := \sqrt{I - SS^*}$ show that for each $\begin{pmatrix}\widetilde{g}\\ g\end{pmatrix} \in \mathbf{H}$ expressions $S\widetilde{g}+g$, $\widetilde{g}+S^* g$, $\Delta\widetilde{g}$, and $\Delta_* g$ are in fact usual square summable vector-functions from $L_2(E)$.

Subspaces in **H**

$$\mathfrak{D}_+ := \begin{pmatrix}H_2^+(E)\\ 0\end{pmatrix}, \quad \mathfrak{D}_- := \begin{pmatrix}0\\ H_2^-(E)\end{pmatrix}, \quad \mathfrak{H} := \mathbf{H} \ominus [\mathfrak{D}_+ \oplus \mathfrak{D}_-]$$

where $H_2^{\pm}(E)$ are Hardy classes of E-valued vector functions analytic in $\mathbb{C}_{\pm}$, are mutually orthogonal.[2] The subspace $\mathfrak{H}$ can be described explicitly:

$$\mathfrak{H} = \left\{\begin{pmatrix}\widetilde{g}\\ g\end{pmatrix} \in \mathbf{H} \ : \ \widetilde{g}+S^* g \in H_2^-(E), \ S\widetilde{g}+g \in H_2^+(E)\right\}.$$

The orthogonal projection $P_{\mathfrak{H}}$ from **H** onto $\mathfrak{H}$ is defined by the following formula

$$P_{\mathfrak{H}}\begin{pmatrix}\widetilde{g}\\ g\end{pmatrix} = \begin{pmatrix}\widetilde{g} - P_+(\widetilde{g}+S^* g)\\ g - P_-(S\widetilde{g}+g)\end{pmatrix}, \quad \widetilde{g}, g \in L_2(E)$$

where $P_{\pm}$ are the orthogonal projections from $L_2(E)$ onto Hardy classes $H_2^{\pm}(E)$.

The following lemma is a version of the corresponding result from [28].

Lemma 2.4. *Let $u \in H$. Linear mappings*

$$u \mapsto \alpha\Gamma_0(A_{B_+}^* - z)^{-1}u, \quad u \mapsto \alpha\Gamma_0(A_{B_+} - \zeta)^{-1}u$$

are bounded operators from H into classes $H_2^+(E)$, $H_2^-(E)$, respectively, with the norms less than $\sqrt{2\pi}$, i.e., for $u \in H$ the following estimates hold

$$\|\alpha\Gamma_0(A_{B_+}^* - z)^{-1}u\|_{H_2^+(E)} \leq \sqrt{2\pi}\|u\|,$$

$$\|\alpha\Gamma_0(A_{B_+} - \zeta)^{-1}u\|_{H_2^-(E)} \leq \sqrt{2\pi}\|u\|.$$

Proof. For a given vector $u \in H$ and $\zeta \in \mathbb{C}_-$, $\zeta = k - i\varepsilon$, $k \in \mathbb{R}$, $\varepsilon > 0$ denote $g_\zeta := (A_{B_+} - \zeta)^{-1}u$. Then since $B_+ = B_R + i\frac{\alpha^2}{2}$ and $g_\zeta \in \mathcal{D}(A_{B_+})$, so that

---

[2] Analytic functions from vector-valued Hardy classes $H_2^{\pm}(E)$ are equated with their boundary values existing almost everywhere on the real axis. These boundary values form two complementary orthogonal subspaces in $L_2(\mathbb{R}, E) = H_2^+(E) \oplus H_2^-(E)$. (See [42] for details.)

136 V. Ryzhov

$B_+\Gamma_0 g_\varsigma = \Gamma_1 g_\varsigma$, we obtain

$$i\|\alpha\Gamma_0(A_{B_+} - \zeta)^{-1}u\|^2 = i\|\alpha\Gamma_0 g_\varsigma\|^2 = i(\alpha^2\Gamma_0 g_\varsigma, \Gamma_0 g_\varsigma)$$

$$= \left(i\frac{\alpha^2}{2}\Gamma_0 g_\varsigma, \Gamma_0 g_\varsigma\right) - \left(\Gamma_0 g_\varsigma, i\frac{\alpha^2}{2}\Gamma_0 g_\varsigma\right)$$

$$= \left(\left(B_R + i\frac{\alpha^2}{2}\right)\Gamma_0 g_\varsigma, \Gamma_0 g_\varsigma\right) - \left(\Gamma_0 g_\varsigma, \left(B_R + i\frac{\alpha^2}{2}\right)\Gamma_0 g_\varsigma\right)$$

$$= (B_+\Gamma_0 g_\varsigma, \Gamma_0 g_\varsigma) - (\Gamma_0 g_\varsigma, B_+\Gamma_0 g_\varsigma) = (\Gamma_1 g_\varsigma, \Gamma_0 g_\varsigma) - (\Gamma_0 g_\varsigma, \Gamma_1 g_\varsigma)$$

$$= (A^* g_\varsigma, g_\varsigma) - (g_\varsigma, A^* g_\varsigma) = \left(A_{B_+} g_\varsigma, g_\varsigma\right) - \left(g_\varsigma, A_{B_+} g_\varsigma\right)$$

$$= \left(A_{B_+}(A_{B_+} - \zeta)^{-1}u, (A_{B_+} - \zeta)^{-1}u\right) - \left((A_{B_+} - \zeta)^{-1}u, A_{B_+}(A_{B_+} - \zeta)^{-1}u\right)$$

$$= \left(u, (A_{B_+} - \zeta)^{-1}u\right) - \left((A_{B_+} - \zeta)^{-1}u, u\right) + (\zeta - \bar{\zeta})\|g_\varsigma\|^2.$$

Here we used the inclusion $A_{B_+} \subset A^*$ and the Green formula (1.3). The remaining part of the proof reproduces corresponding reasoning of paper [28]. Let E_t, $t \in \mathbb{R}$ be the spectral measure of the selfadjoint dilation $\mathscr{A}$. Then

$$\frac{1}{2}\|\alpha\Gamma_0(A_{B_+} - \zeta)^{-1}u\|^2 = \frac{1}{2i}\left[\left(u, (\mathscr{A} - \zeta)^{-1}u\right) - \left((\mathscr{A} - \zeta)^{-1}u, u\right) + (\zeta - \bar{\zeta})\|g_\varsigma\|^2\right]$$

$$= \frac{1}{2i}\left(\left[(\mathscr{A} - \bar{\zeta})^{-1} - (\mathscr{A} - \zeta)^{-1}\right]u, u\right) - \varepsilon\|g_\varsigma\|^2$$

$$= \varepsilon\left\|(\mathscr{A} - \zeta)^{-1}u\right\|^2 - \varepsilon\|(A_{B_+} - \zeta)^{-1}u\|^2$$

$$= \varepsilon\left\|(\mathscr{A} - k + i\varepsilon)^{-1}u\right\|^2 - \varepsilon\|(A_{B_+} - k + i\varepsilon)^{-1}u\|^2$$

$$= \varepsilon\int_{\mathbb{R}} \frac{1}{(t-k)^2 + \varepsilon^2} d(E_t u, u) - \varepsilon\|(A_{B_+} - k + i\varepsilon)^{-1}u\|^2.$$

By the Fubini theorem,

$$\frac{1}{2}\int_{\mathbb{R}} \|\alpha\Gamma_0(A_{B_+} - k + i\varepsilon)^{-1}u\|^2 dk$$

$$= \int_{\mathbb{R}}\left\{\varepsilon\int_{\mathbb{R}} \frac{1}{(t-k)^2 + \varepsilon^2} d(E_t u, u)\right\} dk - \varepsilon\int_{\mathbb{R}} \|(A_{B_+} - k + i\varepsilon)^{-1}u\|^2 dk$$

$$= \int_{\mathbb{R}}\left\{\varepsilon\int_{\mathbb{R}} \frac{1}{(t-k)^2 + \varepsilon^2} dk\right\} d(E_t u, u) - \varepsilon\int_{\mathbb{R}} \|(A_{B_+} - k + i\varepsilon)^{-1}u\|^2 dk$$

$$= \pi\int_{\mathbb{R}} d(E_t u, u) - \varepsilon\int_{\mathbb{R}} \|(A_{B_+} - k + i\varepsilon)^{-1}u\|^2 dk$$

$$= \pi\|u\|^2 - \varepsilon\int_{\mathbb{R}} \|(A_{B_+} - k + i\varepsilon)^{-1}u\|^2 dk.$$

Hence,

$$\|\alpha\Gamma_0(A_{B_+} - \zeta)^{-1}u\|^2_{H_2^-(E)} = \sup_{\varepsilon > 0}\int_{\mathbb{R}} \|\alpha\Gamma_0(A_{B_+} - k + i\varepsilon)^{-1}u\|^2 dk \leq 2\pi\|u\|^2.$$

Another statement of the lemma is proven analogously. $\qquad\square$

It follows from the properties of Hardy classes $H_2^\pm$ that for each $u \in H$ there exist L_2-boundary values of the analytic vector-functions $\alpha\Gamma_0(A_{B_+}^* - zI)^{-1}u$ and $\alpha\Gamma_0(A_{B_+} - \zeta I)^{-1}u$ almost everywhere on the real axis. For these limits we will use the notation:

$$\alpha\Gamma_0(A_{B_+}^* - k - i0)^{-1}u := \lim_{\varepsilon\downarrow 0} \alpha\Gamma_0(A_{B_+}^* - (k + i\varepsilon))^{-1}u,$$

$$\alpha\Gamma_0(A_{B_+} - k + i0)^{-1}u := \lim_{\varepsilon\downarrow 0} \alpha\Gamma_0(A_{B_+} - (k - i\varepsilon))^{-1}u, \tag{2.8}$$

$$u \in H \text{ and almost all } k \in \mathbb{R}.$$

Note that the point set on the real axis where these limits exist depends on the vector $u \in H$. Moreover, the left-hand side in (2.8) does not define any operator functions on the real axis $\mathbb{R}$. These expressions can only be understood as formal symbols for the limits that appear on the right-hand side.

In accordance with [28], introduce two linear mappings $\mathscr{F}_\pm : \mathscr{H} \to L_2(\mathbb{R}, E)$

$$\mathscr{F}_+ : (v_-, u, v_+) \longmapsto -\frac{1}{\sqrt{2\pi}} \alpha\Gamma_0(A_{B_+} - k + i0)^{-1}u + S^*(k)\widehat{v}_-(k) + \widehat{v}_+(k)$$

$$\mathscr{F}_- : (v_-, u, v_+) \longmapsto -\frac{1}{\sqrt{2\pi}} \alpha\Gamma_0(A_{B_+}^* - k - i0)^{-1}u + \widehat{v}_-(k) + S(k)\widehat{v}_+(k)$$

where $(v_-, u, v_+) \in \mathscr{H}$, and $\widehat{v}_\pm$ are the Fourier transforms of functions $v_\pm \in D_\pm$. By virtue of the Paley-Wiener theorem, $\widehat{v}_\pm \in H_2^\pm(E)$, see [42]. The distinguished role of mappings $\mathscr{F}_\pm$ is revealed in the next theorem.

Theorem 2.3. *There exists a unique mapping Φ from the dilation space $\mathscr{H}$ onto the model space $\mathbf{H}$ with the properties:*

1. *Φ is an isometry.*
2. *$\widetilde{g} + S^*g = \mathscr{F}_+\mathfrak{h}$, $S\widetilde{g} + g = \mathscr{F}_-\mathfrak{h}$, where $\binom{\widetilde{g}}{g} = \Phi\mathfrak{h}$, $\mathfrak{h} \in \mathscr{H}$.*
3. *For $z \notin \mathbb{R}$*

$$\Phi \circ (\mathscr{A} - zI)^{-1} = (k - z)^{-1} \circ \Phi,$$

   where $\mathscr{A}$ is the minimal selfadjoint dilation of the operator A_{B_+}
4. *$\Phi H = \mathfrak{H}$, $\Phi D_\pm = \mathfrak{D}_\pm$.*

Property (3) means that Φ maps $\mathscr{A}$ into the multiplication operator on the space $\mathbf{H}$; therefore, the dissipative operator A_{B_+} is mapped into its model representation as required in (2.6).

The proof of the theorem is carried out at the end of this section.

Computation of functions $\mathscr{F}_\pm h$, $h \in H$ can be further simplified. More precisely, there exists a formula which allows one to avoid the calculation of the resolvent of the dissipative operator A_{B_+}. To that end we recall the definition (1.6) of the operator A_∞ given earlier. There exists a certain "resolvent identity" for A_∞ and A_{B_+}, which we will obtain next.

Let $\zeta \in \mathbb{C}_-$. Then the equation $(A_{B_+} - \zeta)\phi = h$ has a unique solution for each $h \in H$. We can represent this solution in the form of sum $\phi = f + g$, where

$g := (A_\infty - \zeta)^{-1}h$ and $f \in \ker(A^* - \zeta)$. Obviously, $f = [(A_{B_+} - \zeta)^{-1} - (A_\infty - \zeta)^{-1}]h$. Since $\phi \in \mathcal{D}(A_{B_+})$ and $\Gamma_0 g = 0$, we have

$$0 = (\Gamma_1 - B_+\Gamma_0)\phi = \Gamma_1(f + g) - B_+\Gamma_0 f = M(\zeta)\Gamma_0 f + \Gamma_1 g - B_+\Gamma_0 f.$$

Hence, $\Gamma_1 g = (B_+ - M(\zeta))\Gamma_0 f$ and since $0 \in \rho(B_+ - M(\zeta))$, we obtain

$$\Gamma_0 f = (B_+ - M(\zeta))^{-1}\Gamma_1 g.$$

The left-hand side can be rewritten in the form

$$\Gamma_0 f = \Gamma_0(f + g) = \Gamma_0 \phi = \Gamma_0(A_{B_+} - \zeta)^{-1}h.$$

Now, by the definition of g,

$$\Gamma_0(A_{B_+} - \zeta I)^{-1}h = (B_+ - M(\zeta))^{-1}\Gamma_1(A_\infty - \zeta I)^{-1}h.$$

Since vector $h \in H$ is arbitrary, it follows that

$$\Gamma_0(A_{B_+} - \zeta I)^{-1} = (B_+ - M(\zeta))^{-1}\Gamma_1(A_\infty - \zeta I)^{-1}, \quad \zeta \in \mathbb{C}_-. \tag{2.9}$$

Similar computations yield the formula for the conjugate operator $A^*_{B_+}$:

$$\Gamma_0(A^*_{B_+} - zI)^{-1} = (B^*_+ - M(z))^{-1}\Gamma_1(A_\infty - zI)^{-1}, \quad z \in \mathbb{C}_+. \tag{2.10}$$

Substituting (2.9) and (2.10) into the definitions of functions $\mathscr{F}_\pm h$, $h \in H$ we arrive at the result ($h \in H$, $k \in \mathbb{R}$):

$$\begin{aligned}
\mathscr{F}_+ h &= -\frac{1}{\sqrt{2\pi}}\lim_{\varepsilon\downarrow 0}\alpha(B_+ - M(k - i\varepsilon))^{-1}\Gamma_1(A_\infty - (k - i\varepsilon))^{-1}h \\
\mathscr{F}_- h &= -\frac{1}{\sqrt{2\pi}}\lim_{\varepsilon\downarrow 0}\alpha(B^*_+ - M(k + i\varepsilon))^{-1}\Gamma_1(A_\infty - (k + i\varepsilon))^{-1}h.
\end{aligned} \tag{2.11}$$

For each $h \in H$ these limits exist for almost any $k \in \mathbb{R}$ and represent two square integrable vector-functions.

The advantage of formulae (2.11) becomes apparent when, for example, the space $\mathcal{H}$ is finite-dimensional. In this case all computations are reduced to the calculation of the resolvent of the selfadjoint operator A_∞ and the matrix inversion problem for the matrix-valued function $(B_+ - M(z))$, $z \in \mathbb{C}_-$[3].

Taking into account that $\Gamma_1(A_\infty - zI)^{-1} = \gamma^*(\bar{z})$, we obtain from (2.9) and (2.10) following relations. They will be used in the proof of Theorem 2.3.

$$\left.\begin{aligned}
\Gamma_0(A_{B_+} - \zeta I)^{-1} &= (B_+ - M(\zeta))^{-1}\gamma^*(\bar\zeta), \quad \zeta \in \mathbb{C}_- \\
\Gamma_0(A^*_{B_+} - zI)^{-1} &= (B^*_+ - M(z))^{-1}\gamma^*(\bar{z}), \quad z \in \mathbb{C}_+
\end{aligned}\right\}. \tag{2.12}$$

At last, for the sake of completeness, we formulate the theorem that describes the resolvent of operator A_{B_+} in the upper half-plane. Its proof is based solely on the Hilbert resolvent identities and can be found in [2]. It is curious to notice that in contrast with similar results of the next section, the vectors on the right-hand side of these formulae already belong to space H, making application of projection P_H redundant. In the notation below we customarily identify initial and model

---

[3]Recall that $(B^*_+ - M(\bar{z}))^{-1} = [(B_+ - M(z))^{-1}]^*$, $z \in \mathbb{C}_-$.

spaces and operators whose unitary equivalence is established by the isometry Φ in hope that it will not lead to confusion.

Theorem 2.4. *For* $\binom{\widetilde{g}}{g} \in H$

$$(A_{B_+} - zI)^{-1}\binom{\widetilde{g}}{g} = (k - z)^{-1}\binom{\widetilde{g} - [S(z)]^{-1}(S\widetilde{g} + g)(z)}{g}, \quad z \in \mathbb{C}_+$$

$$(A_{B_+}^* - \zeta I)^{-1}\binom{\widetilde{g}}{g} = (k - \zeta)^{-1}\binom{\widetilde{g}}{g - [S^*(\bar{\zeta})]^{-1}(\widetilde{g} + S^*g)(\zeta)}, \quad \zeta \in \mathbb{C}_-.$$

Here $(S\widetilde{g} + g)(z)$ *and* $(\widetilde{g} + S^*g)(\zeta)$ *are values of the analytical continuation of the functions* $S\widetilde{g} + g \in H_2^+(E)$ *and* $\widetilde{g} + S^*g \in H_2^-(E)$ *into complex points* $z \in \mathbb{C}_+$, $\zeta \in \mathbb{C}_-$, *respectively.*

The remaining part of this section outlines principal steps of the proof of Theorem 2.3 in the form of a few propositions.

Introduce a linear set in $\mathscr{H}$ by the formula

$$\mathscr{W} := \left\{ \sum_{j=1}^{n} a_j(\mathscr{A} - \zeta_j I)^{-1}v_- + \sum_{s=1}^{m} b_s(\mathscr{A} - z_s I)^{-1}v_+, \quad v_\pm \in D_\pm \right\}, \qquad (2.13)$$

where $\zeta_j \in \mathbb{C}_-$, $z_s \in \mathbb{C}_+$, $a_j, b_s \in \mathbb{C}$, $j = 1, 2, \ldots, n < \infty$, $s = 1, 2, \ldots, m < \infty$.

Proposition 2.5. *Set* $\mathscr{W}$ *is dense in the dilation space* $\mathscr{H}$.

This proposition is equivalent to the completeness of incoming and outgoing waves of the Lax-Phillips theory [26], or completeness of incoming and outgoing eigenfunctions of continuous spectra of the dilation [38].

Proof. Since $\mathrm{s} - \lim_{t\to\infty} \pm it(\mathscr{A} \pm itI)^{-1} = I_{\mathscr{H}}$, the inclusion $D_+ \oplus D_- \subset \overline{\mathscr{W}}$ is obvious. Hence, $\mathscr{W}^\perp \subset H$. Further, $(\mathscr{A} - zI)^{-1}\mathscr{W} \subset \mathscr{W}$ and $\mathscr{A}$ is selfadjoint. It follows that $\overline{\mathscr{W}}$ and $\mathscr{W}^\perp$ are invariant subspaces of $\mathscr{A}$. Noticing that A_{B_+} is simple and $\mathscr{A}|_{\mathscr{W}^\perp} = A_{B_+}|_{\mathscr{W}^\perp}$ since $\mathscr{A}$ is the dilation of A_{B_+}, we conclude that $\mathscr{W}^\perp = \{0\}$. $\qquad\square$

Introduce a linear set W as projection of $\mathscr{W}$ onto H. According to Theorem 2.2,

$$W = \left\{ \sum_{j=1}^{n} a_j\gamma(\zeta_j)(B_+ - M(\zeta_j))^{-1}\alpha\psi_j + \sum_{s=1}^{m} b_s\gamma(z_s)(B_+^* - M(z_s))^{-1}\alpha\phi_s \right\},$$

where $\zeta_j \in \mathbb{C}_-$, $z_s \in \mathbb{C}_+$, $\psi_j, \phi_s \in E$, $a_j, b_s \in \mathbb{C}$, $j = 1, 2, \ldots, n < \infty$, $s = 1, 2, \ldots, m < \infty$.

Corollary 2.6. *The set* W *is dense in* H.

Following the example of [28], we define the spectral mapping $\Phi : \mathscr{H} \to \mathbf{H}$ initially on the dense set (D_-, W, D_+) in $\mathscr{H}$. Let $\mathscr{V} := (v_-, v_0, v_+) \in (D_-, W, D_+)$, where

$$v_0 := \sum_j a_j \gamma(\zeta_j)(B_+ - M(\zeta_j))^{-1}\alpha\psi_j + \sum_s b_s \gamma(z_s)(B_+^* - M(z_s))^{-1}\alpha\phi_s \quad (2.14)$$

in the notation introduced earlier. Let us define the mapping Φ as follows

$$\Phi : \begin{pmatrix} v_- \\ v_0 \\ v_+ \end{pmatrix} \mapsto \begin{pmatrix} \widehat{v}_+ + \frac{i}{\sqrt{2\pi}}\left[\sum_j \frac{a_j}{k-\zeta_j}S^*(\bar{\zeta}_j)\psi_j + \sum_s \frac{b_s}{k-z_s}\phi_s\right] \\ \widehat{v}_- - \frac{i}{\sqrt{2\pi}}\left[\sum_j \frac{a_j}{k-\zeta_j}\psi_j + \sum_s \frac{b_s}{k-z_s}S(z_s)\phi_s\right] \end{pmatrix}. \quad (2.15)$$

Here $\widehat{v}_\pm$ are Fourier transforms of functions $v_\pm \in L_2(\mathbb{R}_\pm, E)$. Our task is to prove that the so-defined map Φ possesses all the properties stated in Theorem 2.3.

First of all, observe that the mapping satisfying conditions (1) and (2) is unique. It follows directly from the definition of the norm in $\mathbf{H}$. (See (2.7).) Secondly, equalities $\Phi D_\pm = \mathfrak{D}_\pm$ for mapping (2.15) hold true by virtue of the Paley-Wiener theorem. Moreover, since the Fourier transform $v_\pm \mapsto \widehat{v}_\pm$ is isometric, restrictions $\Phi|_{D_\pm}$ are isometries onto $\mathfrak{D}_\pm$.

Proposition 2.7. *In notation of Corollary* 2.6

$$\Phi(0, W, 0) \subset \mathfrak{H}.$$

Proof. We need to show that vectors on the right-hand side of (2.15) where $v_\pm = 0$ are orthogonal to $\mathfrak{D}_\pm$. Due to linearity and linear independence, it is sufficient to show that for each $j = 1, 2, \ldots, n$ and $s = 1, 2, \ldots, m$ the vectors

$$\frac{1}{k - \zeta_j}\begin{pmatrix} S^*(\bar{\zeta}_j)\psi_j \\ -\psi_j \end{pmatrix}, \quad \frac{1}{k - z_s}\begin{pmatrix} \phi_s \\ -S(z_s)\phi_j \end{pmatrix}$$

are orthogonal to $(H_2^+(E), H_2^-(E))$ in $\mathbf{H}$. Let $h_\pm \in H_2^\pm(E)$ be two vector functions, so that $(h_+, h_-) \in (H_2^+(E), H_2^-(E))$. Then omitting index j, we have for $\zeta \in \mathbb{C}_-$

$$\left(\frac{1}{k-\zeta}\begin{pmatrix} S^*(\bar{\zeta})\psi \\ -\psi \end{pmatrix}, \begin{pmatrix} h_+ \\ h_- \end{pmatrix}\right)_{\mathbf{H}}$$
$$= \left((k-\zeta)^{-1}S^*(\bar{\zeta})\psi, (h_+ + S^*h_-)\right)_{L^2(E)} - \left((k-\zeta)^{-1}\psi, (Sh_+ + h_-)\right)_{L^2(E)}$$
$$= \left((k-\zeta)^{-1}S^*(\bar{\zeta})\psi, h_+\right)_{L^2(E)} - \left((k-\zeta)^{-1}\psi, Sh_+\right)_{L^2(E)}$$
$$= -\left(\frac{S^*(k) - S^*(\bar{\zeta})}{k-\zeta}\psi, h_+\right)_{L^2(E)} = 0.$$

Similarly, for $z \in \mathbb{C}_+$

$$\left(\frac{1}{k-z}\begin{pmatrix} \phi \\ -S(z)\phi \end{pmatrix}, \begin{pmatrix} h_+ \\ h_- \end{pmatrix}\right)_{\mathbf{H}} = \left(\frac{S(k) - S(z)}{k-z}\phi, h_-\right)_{L^2(E)} = 0.$$

Here we used the inclusions $(k - \zeta)^{-1} \in H_2^+$, $(k - z)^{-1} \in H_2^-$ and analytical continuation of bounded operator functions S and S^* to the upper and lower half-planes correspondingly. The proof is complete. $\qquad\square$

Later it will be shown that Φ maps the space H on the whole $\mathfrak{H}$ isometrically, therefore $\Phi(0, \mathcal{W}, 0)$ is dense in $\mathfrak{H}$.

Proposition 2.8. *Almost everywhere on the real axis*

$$\begin{pmatrix} I & S^* \\ S & I \end{pmatrix} \Phi \mathscr{V} = \begin{pmatrix} \mathscr{F}_+ \mathscr{V} \\ \mathscr{F}_- \mathscr{V} \end{pmatrix}$$

where $\mathscr{V} = (v_-, v_0, v_+) \in (D_-, \mathcal{W}, D_+)$.

Proof. The statement is obviously true if $\mathscr{V}$ belongs to the set $D_- \oplus 0 \oplus D_+$. We only need to consider the case $\mathscr{V} = (0, v_0, 0)$ with $v_0 \in \mathcal{W}$, see (2.14). Arguments of linearity and independence of terms in (2.14) show that it is sufficient to verify the statement only when each sum consists of just one element. Using definitions of $\mathscr{F}_\pm$ we reduce the claim to the following equalities where indices are omitted for convenience:

$$i\alpha\Gamma_0(A_{B_+} - k + i0)^{-1}\big[a\gamma(\zeta)(B_+ - M(\zeta))^{-1}\alpha\psi + b\gamma(z)(B_+^* - M(z))^{-1}\alpha\phi\big]$$

$$= \frac{a}{k - \zeta}[S^*(\bar\zeta) - S^*(k)]\psi + \frac{b}{k - z}[I - S^*(k)S(z)]\phi$$

$$i\alpha\Gamma_0(A_{B_+}^* - k - i0)^{-1}\big[a\gamma(\zeta)(B_+ - M(\zeta))^{-1}\alpha\psi + b\gamma(z)(B_+^* - M(z))^{-1}\alpha\phi\big]$$

$$= -\frac{a}{k - \zeta}[I - S(k)S^*(\bar\zeta)]\psi + \frac{b}{k - z}[S(k) - S(z)]\phi.$$

Regrouping terms we come to four relations to be proven for almost all $k \in \mathbb{R}$:

$$-\frac{S^*(k) - S^*(\bar\zeta)}{k - \zeta}\psi = i\alpha\Gamma_0(A_{B_+} - k + i0)^{-1}\gamma(\zeta)(B_+ - M(\zeta))^{-1}\alpha\psi$$

$$\frac{I - S^*(k)S(z)}{k - z}\phi = i\alpha\Gamma_0(A_{B_+} - k + i0)^{-1}\gamma(z)(B_+^* - M(z))^{-1}\alpha\phi$$

$$-\frac{I - S(k)S^*(\bar\zeta)}{k - \zeta}\psi = i\alpha\Gamma_0(A_{B_+}^* - k - i0)^{-1}\gamma(\zeta)(B_+ - M(\zeta))^{-1}\alpha\psi \qquad (2.16)$$

$$\frac{S(k) - S(z)}{k - z}\phi = i\alpha\Gamma_0(A_{B_+}^* - k - i0)^{-1}\gamma(z)(B_+^* - M(z))^{-1}\alpha\phi.$$

Let $\lambda = k - i\varepsilon$, $k \in \mathbb{R}$, $\varepsilon > 0$. Then, since $S^*(\bar\lambda) = I - i\alpha(B_+ - M(\lambda))^{-1}\alpha$ and $M(\lambda) - M(\zeta) = (\lambda - \zeta)\gamma^*(\bar\lambda)\gamma(\zeta)$ (see (1.16) and Theorem 1.2):

$$S^*(\bar\lambda) - S^*(\bar\zeta) = -i\alpha(B_+ - M(\lambda))^{-1}\alpha + i\alpha(B_+ - M(\zeta))^{-1}\alpha$$

$$= i\alpha(B_+ - M(\lambda))^{-1}\big[-(B_+ - M(\zeta)) + (B_+ - M(\lambda))\big](B_+ - M(\zeta))^{-1}\alpha$$

$$= -i\alpha(B_+ - M(\lambda))^{-1}\big[M(\lambda) - M(\zeta)\big](B_+ - M(\zeta))^{-1}\alpha$$

$$= -i(\lambda - \zeta)\alpha(B_+ - M(\lambda))^{-1}\gamma^*(\bar\lambda)\gamma(\zeta)(B_+ - M(\zeta))^{-1}\alpha.$$

Now the first relation of (2.12) yields:

$$-\frac{S^*(\bar\lambda) - S^*(\bar\zeta)}{\lambda - \zeta}\psi = i\alpha\Gamma_0(A_{B_+} - \lambda I)^{-1}\gamma(\zeta)(B_+ - M(\zeta))^{-1}\alpha\psi.$$

In accordance with the limiting procedure (2.8), we obtain the first formula in (2.16) as $\varepsilon \downarrow 0$.

Similarly,

$$
\begin{aligned}
I - S^*(\bar\lambda)S(z) &= i\alpha(B_+ - M(\lambda))^{-1}\alpha - i\alpha(B_+^* - M(z))^{-1}\alpha \\
&\quad + i^2\alpha(B_+ - M(\lambda))^{-1}\alpha^2(B_+^* - M(z))^{-1}\alpha \\
&= i\alpha(B_+ - M(\lambda))^{-1}\big[(B_+^* - M(z)) - (B_+ - M(\lambda)) + i\alpha^2\big](B_+^* - M(z))^{-1}\alpha \\
&= i\alpha(B_+ - M(\lambda))^{-1}\big[M(\lambda) - M(z)\big](B_+^* - M(z))^{-1}\alpha.
\end{aligned}
$$

The last expression was calculated at the previous step. The same line of reasoning applied to this case proves correctness of the second formula in (2.16) for almost all $k \in \mathbb{R}$.

Two last relations in (2.16) are verified analogously. Finally, the statement of the proposition is valid on the whole space $\mathscr{H}$ due to the uniqueness of the mapping satisfying conditions (1), (2) of Theorem 2.3. The proof is complete. $\square$

Proposition 2.9. *Operator Φ defined in (2.15) is an isometry from the dilation space $\mathscr{H}$ to the model space $\mathbf{H}$.*

Due to this proposition the mapping (2.15) is uniquely extended to the isometry from the whole space $\mathscr{H}$ into $\mathbf{H}$. In what follows we will use the same symbol Φ for this extension.

Proof. It is sufficient to prove that restriction of Φ to the space H is an isometry. To that end compute the norm of the vector $\Phi(0, v_0, 0)$ in $\mathbf{H}$. Denote $\mathscr{V} = (0, v_0, 0)$, where v_0 is defined in (2.14). Then, slightly abusing the notation, we have

$$
\begin{aligned}
\|\Phi\mathscr{V}\|_{\mathbf{H}}^2 &= \left(\begin{pmatrix} I & S^* \\ S & I \end{pmatrix}\Phi\mathscr{V}, \Phi\mathscr{V}\right)_{L_2 \oplus L_2} = \left(\begin{pmatrix} \mathscr{F}_+\mathscr{V} \\ \mathscr{F}_-\mathscr{V} \end{pmatrix}, \Phi\mathscr{V}\right)_{L_2 \oplus L_2} \\
&= \left(\begin{pmatrix} \mathscr{F}_+ v_0 \\ \mathscr{F}_- v_0 \end{pmatrix}, \begin{pmatrix} \frac{i}{\sqrt{2\pi}}\big[\sum_j \frac{a_j}{k-\zeta_j}S^*(\bar\zeta_j)\psi_j + \sum_s \frac{b_s}{k-z_s}\phi_s\big] \\ -\frac{i}{\sqrt{2\pi}}\big[\sum_j \frac{a_j}{k-\zeta_j}\psi_j + \sum_s \frac{b_s}{k-z_s}S(z_s)\phi_s\big] \end{pmatrix}\right)_{L_2 \oplus L_2}.
\end{aligned}
$$

Since $\mathscr{F}_+ v_0 \in H_2^-(E)$, $\mathscr{F}_- v_0 \in H_2^+(E)$, $(k - \zeta_j)^{-1} \in H_2^+$, and $(k - z_s)^{-1} \in H_2^-$, we obtain by the residue method that

$$
\begin{aligned}
\|\Phi\mathscr{V}\|_{\mathbf{H}}^2 &= \frac{i}{\sqrt{2\pi}}\Big[\sum_s \bar{b}_s\big(\mathscr{F}_+ v_0, (k-z_s)^{-1}\phi_s\big)_{L_2(E)} - \sum_j \bar{a}_j\big(\mathscr{F}_- v_0, (k-\zeta_j)^{-1}\psi_j\big)_{L_2(E)}\Big] \\
&= \frac{i}{\sqrt{2\pi}}\Big[2\pi i \sum_s \bar{b}_s\big((\mathscr{F}_+ v_0)(\bar{z}_s), \phi_s\big)_E + 2\pi i \sum_j \bar{a}_j\big((\mathscr{F}_- v_0)(\bar\zeta_j), \psi_j\big)_E\Big] \\
&= \sum_s \bar{b}_s\big(\alpha\Gamma_0(A_{B_+} - \bar{z}_s)^{-1}v_0, \phi_s\big) + \sum_j \bar{a}_j\big(\alpha\Gamma_0(A_{B_+}^* - \bar\zeta_j)^{-1}v_0, \psi_j\big).
\end{aligned}
$$

It follows from (2.12) that

$$\|\Phi\mathscr{V}\|_{\mathbf{H}}^2 = \sum_s \bar{b}_s(\alpha(B_+ - M(\bar{z}_s))^{-1}\gamma^*(z_s)v_0, \phi_s)$$

$$+ \sum_j \bar{a}_j(\alpha(B_+^* - M(\bar{\zeta}_j))^{-1}\gamma^*(\zeta_j)v_0, \psi_j)$$

$$= \left(v_0, \sum_j a_j\gamma(\zeta_j)(B_+ - M(\zeta_j))^{-1}\alpha\psi_j + \sum_s b_s\gamma(z_s)(B_+^* - M(z_s))^{-1}\alpha\phi_s\right)$$

$$= \|v_0\|^2.$$

Thus, Φ is an isometry from $\mathscr{H}$ to $\mathbf{H}$. $\qquad\square$

Proposition 2.10.

$$\Phi \circ (\mathscr{A} - zI)^{-1} = (k - z)^{-1} \circ \Phi, \quad z \notin \mathbb{R}.$$

Proof. The statement is a consequence of Proposition 2.9, property (2) of Theorem 2.3, which is proven in Proposition 2.8, and equalities

$$\mathscr{F}_\pm \circ (\mathscr{A} - zI)^{-1} = (k - z)^{-1} \circ \mathscr{F}_\pm, \quad z \notin \mathbb{R}$$

to be established. For $(h_-, h_0, h_+) \in \mathscr{H}$ and $z \in \mathbb{C}_+$ denote as (h'_-, h'_0, h'_+) the vector $(\mathscr{A} - zI)^{-1}(h_-, h_0, h_+)$. Since

$$h_\pm = \left(i\frac{d}{d\xi} - z\right)h'_\pm,$$

by exercising integration by parts, we obtain for the Fourier transforms $\widehat{h}'_\pm$, $\widehat{h}_\pm$:

$$(k - z)\widehat{h}'_\pm = \widehat{h}_\pm \pm \frac{i}{\sqrt{2\pi}}h'_\pm(0).$$

Then, according to the definition of $\mathscr{F}_-$ and Theorem 2.2,

$$\mathscr{F}_-(h'_-, h'_0, h'_+)$$

$$= -\frac{1}{\sqrt{2\pi}}\alpha\Gamma_0(A_{B_+}^* - k - i0)^{-1}h'_0 + \widehat{h}'_-(k) + S(k)\widehat{h}'_+(k)$$

$$= -\frac{1}{\sqrt{2\pi}}\alpha\Gamma_0(A_{B_+}^* - k - i0)^{-1}[(A_{B_+}^* - z)^{-1}h_0 - \gamma(z)(B_+^* - M(z))^{-1}\alpha h'_+(0)]$$

$$+ \frac{1}{k - z}[(\widehat{h}_- + S\widehat{h}_+) + \frac{i}{\sqrt{2\pi}}(Sh'_+(0) - h'_-(0))]$$

$$= \frac{1}{k - z}\mathscr{F}_-(h_-, h_0, h_+)$$

$$+ \frac{1}{k - z}\frac{1}{\sqrt{2\pi}}\alpha\Gamma_0(A_{B_+}^* - z)^{-1}h_0$$

$$+ \frac{1}{\sqrt{2\pi}}\alpha\Gamma_0(A_{B_+}^* - k - i0)^{-1}\gamma(z)(B_+^* - M(z))^{-1}\alpha h'_+(0)$$

$$+ \frac{1}{k - z}\frac{i}{\sqrt{2\pi}}[h'_+(0) - h'_-(0) + i\alpha(B_+^* - M(k + i0))^{-1}\alpha h'_+(0)].$$

We need to show that the sum of last three terms is equal to zero. To that end we consider the sum of the first and the third summands at the non-real point $\lambda = k + i\varepsilon$, $k \in \mathbb{R}$, $\varepsilon > 0$, $\lambda \neq z$. Substitute $h'_+(0) - h'_-(0) = i\alpha\Gamma_0 h'_0$ and $h'_0 = (A^*_{B_+} - zI)^{-1}h_0 - \gamma(z)(B^*_+ - M(z))^{-1}\alpha h'_+(0)$ and conduct necessary computations.

$$\frac{1}{\lambda - z}\frac{1}{\sqrt{2\pi}}\left[\alpha\Gamma_0(A^*_{B_+} - z)^{-1}h_0 - \alpha\Gamma_0 h'_0 - \alpha(B^*_+ - M(\lambda))^{-1}\alpha h'_+(0)\right]$$

$$= \frac{1}{\lambda - z}\frac{1}{\sqrt{2\pi}}\left[\alpha\Gamma_0\gamma(z)(B^*_+ - M(z))^{-1}\alpha h'_+(0) - \alpha(B^*_+ - M(\lambda))^{-1}\alpha h'_+(0)\right]$$

$$= -\frac{1}{\lambda - z}\frac{1}{\sqrt{2\pi}}\alpha(B^*_+ - M(\lambda))^{-1}\left[M(\lambda) - M(z)\right](B^*_+ - M(z))^{-1}\alpha h'_+(0)$$

$$= -\frac{1}{\sqrt{2\pi}}\alpha(B^*_+ - M(\lambda))^{-1}\gamma^*(\bar{\lambda})\gamma(z)(B^*_+ - M(z))^{-1}\alpha h'_+(0)$$

$$= -\frac{1}{\sqrt{2\pi}}\alpha\Gamma_0(A^*_{B_+} - \lambda I)^{-1}\gamma(z)(B^*_+ - M(z))^{-1}\alpha h'_+(0),$$

where at the last step we employed relation (2.12). According to Lemma 2.4, this vector function is analytic in the upper half-plane $\lambda \in \mathbb{C}_+$. More precisely, it belongs to the Hardy class $H_2^+(E)$. The only thing left is to observe that its boundary values as $\varepsilon \downarrow 0$ annihilate the second term in the expression for $\mathscr{F}_-(h'_-, h'_0, h'_+)$ above.

Now we turn to the lengthier computation of $\mathscr{F}_+(\mathscr{A} - zI)^{-1}(h_-, h_0, h_+)$.

$$\mathscr{F}_+(\mathscr{A} - zI)^{-1}(h_-, h_0, h_+) = \mathscr{F}_+(h'_-, h'_0, h'_+)$$

$$= -\frac{1}{\sqrt{2\pi}}\alpha\Gamma_0(A_{B_+} - k + i0)^{-1}h'_0 + S^*(k)\widehat{h'_-}(k) + \widehat{h'_+}(k)$$

$$= -\frac{1}{\sqrt{2\pi}}\alpha\Gamma_0(A_{B_+} - k + i0)^{-1}[(A^*_{B_+} - z)^{-1}h_0 - \gamma(z)(B^*_+ - M(z))^{-1}\alpha h'_+(0)]$$

$$+ \frac{1}{k - z}\left[(S^*\widehat{h_-} + \widehat{h_+}) + \frac{i}{\sqrt{2\pi}}(h'_+(0) - S^*h'_-(0))\right].$$

Let $\lambda = k - i\varepsilon$, $k \in \mathbb{R}$, $\varepsilon > 0$ be a number in the lower half-plane. Let us compute vectors $h'_+(0) - S^*(\bar{\lambda})h'_-(0)$ and $\alpha\Gamma_0(A_{B_+} - \lambda)^{-1}(A^*_{B_+} - z)^{-1}h_0$. Using Theorem 2.2, we have

$$h'_+(0) - S^*(\bar{\lambda})h'_-(0)$$

$$= h'_+(0) - S^*(\bar{\lambda})\left[-i\alpha\Gamma_0(A^*_{B_+} - z)^{-1}h_0 + S(z)h'_+(0)\right]$$

$$= \left(I - S^*(\bar{\lambda})S(z)\right)h'_+(0) + iS^*(\bar{\lambda})\alpha\Gamma_0(A^*_{B_+} - z)^{-1}h_0$$

$$= i(\lambda - z)\alpha\Gamma_0(A_{B_+} - \lambda)^{-1}\gamma(z)(B^*_+ - M(z))^{-1}\alpha h'_+(0)$$

$$+ iS^*(\bar{\lambda})\alpha\Gamma_0(A^*_{B_+} - z)^{-1}h_0$$

where at the last step we make use of computations for $I - S^*(\bar{\lambda})S(z)$ conducted in the proof of Proposition 2.8. Note that almost everywhere on the real axis there exist boundary values of both sides of this formula as $\varepsilon \downarrow 0$.

With the help of Theorem 1.2 and relations (2.12) we obtain

$$\alpha\Gamma_0(A_{B_+} - \lambda)^{-1}(A_{B_+}^* - z)^{-1}h_0$$
$$= \alpha(B_+ - M(\lambda))^{-1}\Gamma_1(A_\infty - \lambda)^{-1}$$
$$\times \left[(A_\infty - z)^{-1} + \gamma(z)(B_+^* - M(z))^{-1}\gamma^*(\bar{z})\right]h_0$$
$$= (\lambda - z)^{-1}\alpha(B_+ - M(\lambda))^{-1}\Gamma_1\left[(A_\infty - \lambda)^{-1} - (A_\infty - z)^{-1}\right]h_0$$
$$\qquad + \alpha(B_+ - M(\lambda))^{-1}\gamma^*(\bar{\lambda})\gamma(z)(B_+^* - M(z))^{-1}\gamma^*(\bar{z})h_0$$
$$= (\lambda - z)^{-1}\alpha\Gamma_0(A_{B_+} - \lambda)^{-1}h_0$$
$$\qquad + (\lambda - z)^{-1}\alpha(B_+ - M(\lambda))^{-1}$$
$$\times \left[-(B_+^* - M(z)) + (M(\lambda) - M(z))\right](B_+^* - M(z))^{-1}\gamma^*(\bar{z})h_0$$
$$= (\lambda - z)^{-1}\alpha\Gamma_0(A_{B_+} - \lambda)^{-1}h_0$$
$$\qquad - (\lambda - z)^{-1}\alpha(B_+ - M(\lambda))^{-1}\left[B_+ - M(\lambda) - i\alpha^2\right](B_+^* - M(z))^{-1}\gamma^*(\bar{z})h_0$$
$$= (\lambda - z)^{-1}\alpha\Gamma_0(A_{B_+} - \lambda)^{-1}h_0$$
$$\qquad - (\lambda - z)^{-1}\left[I - i\alpha(B_+ - M(\lambda))^{-1}\alpha\right]\alpha(B_+^* - M(z))^{-1}\gamma^*(\bar{z})h_0$$
$$= (\lambda - z)^{-1}\alpha\Gamma_0(A_{B_+} - \lambda)^{-1}h_0 - (\lambda - z)^{-1}S^*(\bar{\lambda})\alpha\Gamma_0(A_{B_+}^* - z)^{-1}h_0.$$

Again, both sides of this relation have boundary values almost everywhere on the real axis, since they both belong to the Hardy class $H_2^-(E)$. Passing $\varepsilon \downarrow 0$, substitute obtained results to the calculations of $\mathscr{F}_+(h_-', h_0', h_+')$ started above.

$$\mathscr{F}_+(\mathscr{A} - zI)^{-1}(h_-, h_0, h_+)$$
$$= -\frac{1}{\sqrt{2\pi}}\alpha\Gamma_0(A_{B_+} - k + i0)^{-1}(A_{B_+}^* - z)^{-1}h_0$$
$$\qquad + \frac{1}{\sqrt{2\pi}}\alpha\Gamma_0(A_{B_+} - k + i0)^{-1}\gamma(z)(B_+^* - M(z))^{-1}\alpha h_+'(0)$$
$$\qquad + \frac{1}{k - z}(S^*\widehat{h}_- + \widehat{h}_+) + \frac{1}{k - z}\frac{i}{\sqrt{2\pi}}$$
$$\times \left[i(k - z)\alpha\Gamma_0(A_{B_+} - k + i0)^{-1}\gamma(z)(B_+^* - M(z))^{-1}\alpha h_+'(0)\right.$$
$$\left. + iS^*(k)\alpha\Gamma_0(A_{B_+}^* - z)^{-1}h_0\right]$$
$$= \frac{1}{k - z}\left[-\frac{1}{\sqrt{2\pi}}\alpha\Gamma_0(A_{B_+} - k + i0)^{-1}h_0 + S^*\widehat{h}_- + \widehat{h}_+\right]$$
$$\qquad + \frac{1}{k - z}\frac{1}{\sqrt{2\pi}}\left[S^*(k)\alpha\Gamma_0(A_{B_+}^* - z)^{-1}h_0 - S^*(k)\alpha\Gamma_0(A_{B_+}^* - z)^{-1}h_0\right]$$
$$= (k - z)^{-1}\mathscr{F}_+(h_-, h_0, h_+).$$

The desired equality is established. Finally, the case $z \in \mathbb{C}_-$ can be considered analogously. The proof is complete. $\square$

Proposition 2.11. *The isometrical operator Φ maps $\mathscr{H}$ onto $\mathbf{H}$.*

Proof. As specified above, we use the same symbol Φ for the closure of the mapping defined in (2.15). We only need to show that $\mathcal{R}(\Phi)$ coincides with the whole space $\mathbf{H}$. It is already known that Φ maps $D_- \oplus 0 \oplus D_+$ onto $\mathfrak{D}_+ \oplus \mathfrak{D}_-$ isometrically and the linear set $\vee_{\lambda \notin \mathbb{R}} (\mathscr{A} - \lambda I)^{-1}(D_- \oplus 0 \oplus D_+)$ is dense in $\mathscr{H}$. Owing to Proposition 2.10, this set is mapped by the isometry Φ into the set $\vee_{\lambda \notin \mathbb{R}} (k - \lambda)^{-1}(\mathfrak{D}_+ \oplus \mathfrak{D}_-)$, which is dense in $(L_2(E), L_2(E))$. By the definition of norm in $\mathbf{H}$, this set is dense in $\mathbf{H}$. The range of an isometry is a closed subspace, and that observation completes the proof. $\qquad\square$

2.2. Model of non-selfadjoint non-dissipative operator

In the paper [28] S. Naboko proposed a solution to the problem of the functional model construction for a non-selfadjoint non-dissipative operator. His method was revisited later in the work [27] where it was taken as a foundation for the functional model of an arbitrary bounded operator. The key idea of this approach is to use the Sz.-Nagy-Foiaş model of a dissipative operator that is "close" in a certain sense to the initial operator and to describe the properties of the latter in this model space. It turned out that such dissipative operator can be pointed to in a very natural, but not obvious, way. Namely, one arrives at that operator by replacing the imaginary part of the initial non-dissipative operator with its absolute value. In other words, the "close" dissipative operator for $A + iV$, where $A = A^*$ and $V = V^*$ is A-bounded operator with the relative bound lesser than 1, is the operator $A + i|V|$. Similar results are obtained in [27] for a bounded operator considered as an additive perturbation of a unitary one.

The theory developed in [28] becomes inapplicable in the general situation of an unbounded non-dissipative operator, since it could not be represented as a sum of its real and imaginary parts with the imaginary part relatively bounded. The Makarov-Vasyunin schema [27] still holds its value in this case and could be employed for the model construction, provided that one works with the Cayley transform of the initial unbounded operator. However, in applications to problems arising in physics, the computational complexity and inherited inconvenience of Cayley transforms makes this method less attractive than the direct approach of [28].

Almost solvable extensions of a symmetric operator are an example when the functional model can be constructed by the method of paper [28] without resorting to the Cayley transform. In this section we will use notation introduced earlier and explain how to obtain the formulae for the resolvent $(A_B - zI)^{-1}$ acting on the Sz.-Nagy-Foiaş model space of the "close" dissipative operator A_{B_+}. Essentially, all the computations are based on some relationships between the resolvents $(A_B - zI)^{-1}$ and $(A_{B_+} - zI)^{-1}$, quite similar to the identities between the resolvents of operators A_∞ and A_{B_+} obtained in the previous section.

Let $\zeta \in \rho(A_B) \cap \mathbb{C}_-$, $\phi \in \mathcal{D}(A_B)$ and $(A_B - \zeta I)\phi = h$. We will represent ϕ as a sum of two vectors $\phi = f + g$, where $f \in \ker(A^* - \zeta I)$ and $g = (A_{B_+} - \zeta I)^{-1}h$.

Noting that $\Gamma_1\phi = B\Gamma_0\phi$ and $\Gamma_1 g = B_+\Gamma_0 g$ we obtain:

$$0 = \Gamma_1\phi - B\Gamma_0\phi = (\Gamma_1 - B\Gamma_0)(f + g)$$
$$= \Gamma_1 f - B\Gamma_0 f + \Gamma_1 g - B\Gamma_0 g = M(\zeta)\Gamma_0 f - B\Gamma_0 f + B_+\Gamma_0 g - B\Gamma_0 g$$
$$= (M(\zeta) - B)\,\Gamma_0 f + (B_+ - B)\Gamma_0 g.$$

Therefore, $\Gamma_0 f = (B - M(\zeta))^{-1}(B_+ - B)\Gamma_0 g$, so that for $\Gamma_0\phi = \Gamma_0 f + \Gamma_0 g$ we have

$$\Gamma_0\phi = \left[I + (B - M(\zeta))^{-1}(B_+ - B)\right]\Gamma_0 g.$$

Now we apply the operator α to both sides of this equation and recall that

$$\phi = (A_B - \zeta I)^{-1}h, \quad g = (A_{B_+} - \zeta I)^{-1}h, \quad B_+ - B = i\alpha X^-\alpha$$

where $X^- = (I_E - J)/2$. Thus for each $h \in H$:

$$\alpha\Gamma_0(A_B - \zeta I)^{-1}h = \left[I + i\alpha\,(B - M(\zeta))^{-1}\alpha X^-\right]\alpha\Gamma_0(A_{B_+} - \zeta I)^{-1}h.$$

Similar computations with the operators B and B_+ interchanged yield equality

$$\alpha\Gamma_0(A_{B_+} - \zeta I)^{-1}h = \left[I - i\alpha\,(B_+ - M(\zeta))^{-1}\alpha X^-\right]\alpha\Gamma_0(A_B - \zeta I)^{-1}h.$$

Introduce an analytic operator-function $\Theta_-(\zeta)$, $\zeta \in \mathbb{C}_-$

$$\left.\begin{aligned}
\Theta_-(\zeta) &:= I - i\alpha\,(B_+ - M(\zeta))^{-1}\alpha X^- \\
&= X^+ + S^*(\bar\zeta)X^-, \qquad \zeta \in \mathbb{C}_-
\end{aligned}\right\} \tag{2.17}$$

where $X^+ = (I_E + J)/2$ and $S(\cdot)$ is the characteristic function of the operator A_{B_+} as defined in the Theorem 2.1. The second equality (2.17) can be easily verified with the help of representation (1.16) for the characteristic function of an a.s. extension and the identity $X^+ + X^- = I_E$. Indeed, from (1.16) we obtain

$$X^+ + S^*(\bar\zeta)X^- = X^+ + [I_E - i\alpha(B_+ - M(\zeta))^{-1}\alpha]X^-$$
$$= X^+ + X^- - i\alpha(B_+ - M(\zeta))^{-1}\alpha X^- = \Theta_-(\zeta).$$

The preceding formulae now can be rewritten in the form of operator equalities:

$$\left.\begin{aligned}
\alpha\Gamma_0(A_{B_+} - \zeta I)^{-1} &= \Theta_-(\zeta)\,\alpha\Gamma_0(A_B - \zeta I)^{-1}, \quad \zeta \in \mathbb{C}_- \\
\alpha\Gamma_0(A_B - \zeta I)^{-1} &= \Theta_-^{-1}(\zeta)\,\alpha\Gamma_0(A_{B_+} - \zeta I)^{-1}, \quad \zeta \in \mathbb{C}_- \cap \rho(A_B)
\end{aligned}\right\}. \tag{2.18}$$

The inverse function $\Theta_-^{-1}(\cdot) = [\Theta_-(\cdot)]^{-1}$ has the form similar to (2.17):

$$\begin{aligned}
\Theta_-^{-1}(\zeta) &= I + i\alpha\,(B - M(\zeta))^{-1}\alpha X^- \\
&= X^+ + \Theta^*(\bar\zeta)X^-, \qquad \zeta \in \mathbb{C}_- \cap \rho(A_B)
\end{aligned} \tag{2.19}$$

where Θ is the characteristic function of A_B.

Now we turn to the similar, but lengthier, computations for the resolvents of the operators A_B and A_{B_+} in the upper half-plane. For $z \in \mathbb{C}_+ \cap \rho(A_B^*)$ and $h \in H$ we represent the vector $\phi \in \mathcal{D}(A_{B_+}^*)$ such that $(A_{B_+}^* - zI)\phi = h$ in the form $\phi = f + g$, where $f \in \ker(A^* - zI)$ and $g = (A_B^* - zI)^{-1}h$. Then

$$0 = (\Gamma_1 - B_+^*\Gamma_0)\phi = \left(M(z) - B_+^*\right)\Gamma_0 f + (B^* - B_+^*)\Gamma_0 g.$$

Therefore,

$$\Gamma_0 f = \left(B_+^* - M(z)\right)^{-1} (B^* - B_+^*)\Gamma_0 g = i\left(B_+^* - M(z)\right)^{-1} \alpha X^- \alpha \Gamma_0 g$$

and

$$\Gamma_0 \phi = \Gamma_0 f + \Gamma_0 g = \left[I + i\left(B_+^* - M(z)\right)^{-1} \alpha X^- \alpha\right]\Gamma_0 g.$$

After substitution of $\phi = (A_{B_+}^* - zI)^{-1}h$ and $g = (A_B^* - zI)^{-1}h$ we obtain

$$\alpha\Gamma_0(A_{B_+}^* - zI)^{-1}h = \left[I + i\alpha\left(B_+^* - M(z)\right)^{-1} \alpha X^-\right]\alpha\Gamma_0(A_B^* - zI)^{-1}h.$$

Since this identity is valid for each $h \in H$, in particular, for $h \in \mathcal{R}(A_B - zI)$ it follows that on the domain $\mathcal{D}(A_B)$

$$\alpha\Gamma_0(A_{B_+}^* - zI)^{-1}(A_B - zI)$$
$$= \left[I + i\alpha\left(B_+^* - M(z)\right)^{-1} \alpha X^-\right] \times J \cdot J\alpha\Gamma_0(A_B^* - zI)^{-1}(A_B - zI).$$

Noting that $J\alpha\Gamma_0(A_B^* - zI)^{-1}(A_B - zI)f = \Theta(z)J\alpha\Gamma_0 f$ for any $f \in \mathcal{D}(A_B)$ by the definition of the characteristic function (see calculations preceding (1.16)), we arrive at the formulae

$$\alpha\Gamma_0(A_{B_+}^* - zI)^{-1}(A_B - zI) = \left[I + i\alpha\left(B_+^* - M(z)\right)^{-1} \alpha X^-\right] \times J\Theta(z)J\alpha\Gamma_0$$

and, if $z \in \mathbb{C}_+ \cap \rho(A_B)$

$$\alpha\Gamma_0(A_{B_+}^* - zI)^{-1} = \left[I + i\alpha\left(B_+^* - M(z)\right)^{-1} \alpha X^-\right]J\Theta(z)J\alpha\Gamma_0(A_B - zI)^{-1}.$$

Denote Θ_+ the operator function from the right-hand side and compute it.

$$\begin{aligned}
\Theta_+(z) &= \left[I + i\alpha\left(B_+^* - M(z)\right)^{-1} \alpha X^-\right]J\Theta(z)J \\
&= \left[I + i\alpha\left(B_+^* - M(z)\right)^{-1} \alpha X^-\right]J\left[I + iJ\alpha\left(B^* - M(z)\right)^{-1} \alpha\right]J \\
&= I + i\alpha\left(B_+^* - M(z)\right)^{-1} \alpha X^- + i\alpha\left(B^* - M(z)\right)^{-1} \alpha J \\
&\quad + (i)^2\alpha\left(B_+^* - M(z)\right)^{-1} \alpha X^-\alpha\left(B^* - M(z)\right)^{-1} \alpha J \\
&= I + i\alpha\left(B_+^* - M(z)\right)^{-1} \\
&\quad \times \left[-X^-\left(B^* - M(z)\right) + \left(B_+^* - M(z)\right) + i\alpha X^-\alpha\right]\left(B^* - M(z)\right)^{-1} \alpha J \\
&= I + 2i\alpha\left(B_+^* - M(z)\right)^{-1} \\
&\quad \times \left[-X^-\left(B^* - M(z)\right) + B^* - M(z)\right]\left(B^* - M(z)\right)^{-1} \alpha J \\
&= I + i\alpha\left(B_+^* - M(z)\right)^{-1} X^+\alpha J = X^- + \left[I + i\alpha\left(B_+^* - M(z)\right)^{-1} \alpha\right]X^+ \\
&= X^- + S(z)X^+.
\end{aligned}$$

Therefore,

$$\left.\begin{aligned}
\alpha\Gamma_0(A_{B_+}^* - zI)^{-1}(A_B - zI) &= \Theta_+(z)\alpha\Gamma_0, \qquad z \in \mathbb{C}_+ \\
\text{where } \Theta_+(z) &= I + i\alpha\left(B_+^* - M(z)\right)^{-1}\alpha X^+ = X^- + S(z)X^+
\end{aligned}\right\}. \tag{2.20}$$

Values of the operator-function $\Theta_+(z)$ are invertible operators if $z \in \mathbb{C}_+ \cap \rho(A_B)$; simple computations show that

$$\Theta_+^{-1}(z) = I - i\alpha\,(B - M(z))^{-1}\,\alpha X^+ = X^- + \Theta^*(\bar{z})X^+. \qquad (2.21)$$

Finally we obtain the counterpart for (2.18):

$$\left.\begin{aligned}
\alpha\Gamma_0(A_{B_+}^* - zI)^{-1} &= \Theta_+(z)\alpha\Gamma_0(A_B - zI)^{-1}, \quad z \in \mathbb{C}_+ \\
\alpha\Gamma_0(A_B - zI)^{-1} &= \Theta_+^{-1}(z)\alpha\Gamma_0(A_{B_+}^* - zI)^{-1}, \quad z \in \mathbb{C}_+ \cap \rho(A_B)
\end{aligned}\right\}. \qquad (2.22)$$

Now we can compute how the spectral mappings $\mathscr{F}_\pm$ translate the resolvent of the operator A_B into the "model" terms. For $\lambda_0 \in \mathbb{C}_- \cap \rho(A_B)$, $\zeta \in \mathbb{C}_-$ and $h \in H$ with the assistance of (2.18) we have

$$\begin{aligned}
&\alpha\Gamma_0(A_{B_+} - \zeta I)^{-1}(A_B - \lambda_0 I)^{-1}h \\
&= \Theta_-(\zeta)\alpha\Gamma_0(A_B - \zeta I)^{-1}(A_B - \lambda_0 I)^{-1}h \\
&= (\zeta - \lambda_0)^{-1}\Theta_-(\zeta)\alpha\Gamma_0[(A_B - \zeta I)^{-1} - (A_B - \lambda_0 I)^{-1}]h \\
&= (\zeta - \lambda_0)^{-1}\big[\alpha\Gamma_0(A_{B_+} - \zeta I)^{-1} - \Theta_-(\zeta)\alpha\Gamma_0(A_B - \lambda_0 I)^{-1}\big]h \\
&= (\zeta - \lambda_0)^{-1}\big[\alpha\Gamma_0(A_{B_+} - \zeta I)^{-1} - \Theta_-(\zeta)\Theta_-^{-1}(\lambda_0)\alpha\Gamma_0(A_{B_+} - \lambda_0 I)^{-1}\big]h.
\end{aligned}$$

Assume $\zeta = k - i\varepsilon$, $k \in \mathbb{R}$, $\varepsilon > 0$. We obtain the expression for $\mathscr{F}_+(A_B - \lambda_0 I)^{-1}h$ when $\varepsilon \to 0$. (See definitions of $\mathscr{F}_\pm$ after Lemma 2.4.) Taking into account assertion (2) of Theorem 2.3 and noting that boundary values $\Theta_-(k - i0)$ of the bounded analytic operator-function Θ_- exist in the strong operator topology almost everywhere on the real axis (see (2.17)), we deduce from the formula above that for $(\widetilde{g}, g) = \Phi h$, $k \in \mathbb{R}$:

$$\begin{aligned}
&\big[\mathscr{F}_+(A_B - \lambda_0 I)^{-1}h\big](k) \\
&= (k - \lambda_0)^{-1}\big[(\widetilde{g} + S^*g)(k - i0) - \Theta_-(k - i0)\Theta_-^{-1}(\lambda_0)(\widetilde{g} + S^*g)(\lambda_0)\big].
\end{aligned}$$

The model representations of functions $\mathscr{F}_-(A_B - \lambda_0 I)^{-1}h$ and $\mathscr{F}_\pm(A_B - \mu_0 I)^{-1}h$, where $\mu_0 \in \mathbb{C}_+ \cap \rho(A_B)$ are computed quite similarly and below we sum up all these formulae:

$$\mathscr{F}_+(A_B - \lambda_0 I)^{-1}h = \frac{1}{k - \lambda_0}\big[(\widetilde{g} + S^*g)(k) - \Theta_-(k)\Theta_-^{-1}(\lambda_0)(\widetilde{g} + S^*g)(\lambda_0)\big]$$

$$\mathscr{F}_-(A_B - \lambda_0 I)^{-1}h = \frac{1}{k - \lambda_0}\big[(S\widetilde{g} + g)(k) - \Theta_+(k)\Theta_-^{-1}(\lambda_0)(\widetilde{g} + S^*g)(\lambda_0)\big]$$

$$\mathscr{F}_+(A_B - \mu_0 I)^{-1}h = \frac{1}{k - \mu_0}\big[(\widetilde{g} + S^*g)(k) - \Theta_-(k)\Theta_+^{-1}(\mu_0)(S\widetilde{g} + g)(\mu_0)\big]$$

$$\mathscr{F}_-(A_B - \mu_0 I)^{-1}h = \frac{1}{k - \mu_0}\big[(S\widetilde{g} + g)(k) - \Theta_+(k)\Theta_+^{-1}(\mu_0)(S\widetilde{g} + g)(\mu_0)\big]$$

where $h \in H$, $(\widetilde{g}, g) = \Phi h$, $\lambda_0 \in \mathbb{C}_- \cap \rho(A_B)$, $\mu_0 \in \mathbb{C}_+ \cap \rho(A_B)$, and for almost all $k \in \mathbb{R}$ there exist strong limits $\Theta_\pm(k) := s - \lim_{\varepsilon \downarrow 0} \Theta_\pm(k \pm i\varepsilon)$.

150 V. Ryzhov

The main theorem describes the action of operator A_B in the model space $\mathbf{H} = L_2\begin{pmatrix} I & S^* \\ S & I \end{pmatrix}$ of dissipative operator A_{B_+}. As before, for the notational convenience we use the same symbols for objects whose unitary equivalence is established by the isometry Φ.

Theorem 2.5. *For $\lambda_0 \in \mathbb{C}_- \cap \rho(A_B)$, $\mu_0 \in \mathbb{C}_+ \cap \rho(A_B)$, $(\widetilde{g}, g) \in H$*

$$(A_B - \lambda_0 I)^{-1}\begin{pmatrix} \widetilde{g} \\ g \end{pmatrix} = P_H(k - \lambda_0)^{-1}\begin{pmatrix} \widetilde{g} \\ g - X^-\Theta_-^{-1}(\lambda_0)(\widetilde{g} + S^*g)(\lambda_0) \end{pmatrix}$$

$$(A_B - \mu_0 I)^{-1}\begin{pmatrix} \widetilde{g} \\ g \end{pmatrix} = P_H(k - \mu_0)^{-1}\begin{pmatrix} \widetilde{g} - X^+\Theta_+^{-1}(\mu_0)(S\widetilde{g} + g)(\mu_0) \\ g \end{pmatrix}.$$

Here P_H is the orthogonal projection from $\mathbf{H}$ onto H.

Proof. The proof is identical to the proof of the corresponding result of [28]. For the most part it is based on the identities for $\mathscr{F}_\pm(A_B - \lambda_0 I)^{-1}$, $\mathscr{F}_\pm(A_B - \mu_0 I)^{-1}$ obtained earlier.

Let us verify the theorem's assertion for $\lambda_0 \in \mathbb{C}_- \cap \rho(A_B)$. The case of the resolvent in the upper half-plane is considered analogously. According to Theorem 2.3 we only need to show that functions $(S\widetilde{g}' + g')$ and $(\widetilde{g}' + S^*g')$ where $(\widetilde{g}', g')$ is the vector on the right-hand side of the corresponding formula satisfy the following conditions

$$\mathscr{F}_+(A_B - \lambda_0 I)^{-1}h = (\widetilde{g}' + S^*g')$$
$$\mathscr{F}_-(A_B - \lambda_0 I)^{-1}h = (S\widetilde{g}' + g')$$

with $\Phi h = (\widetilde{g}, g)$. Since

$$\begin{pmatrix} \widetilde{g}' \\ g' \end{pmatrix} = P_H(k - \lambda_0)^{-1}\begin{pmatrix} \widetilde{g} \\ g - X^-\Theta_-^{-1}(\lambda_0)(\widetilde{g} + S^*g)(\lambda_0) \end{pmatrix}$$

$$= \begin{pmatrix} \frac{\widetilde{g}}{k - \lambda_0} - P_+\frac{1}{k-\lambda_0}\left[\widetilde{g} + S^*g - S^*X^-\Theta_-^{-1}(\lambda_0)(\widetilde{g} + S^*g)(\lambda_0)\right] \\ \frac{g - X^-\Theta_-^{-1}(\lambda_0)(\widetilde{g} + S^*g)(\lambda_0)}{k - \lambda_0} - P_-\frac{1}{k-\lambda_0}\left[S\widetilde{g} + g - X^-\Theta_-^{-1}(\lambda_0)(\widetilde{g} + S^*g)(\lambda_0)\right] \end{pmatrix}$$

$$= \frac{1}{k - \lambda_0}\begin{pmatrix} \widetilde{g} - (\widetilde{g} + S^*g)(\lambda_0) + S^*(\lambda_0)X^-\Theta_-^{-1}(\lambda_0)(\widetilde{g} + S^*g)(\lambda_0) \\ g - X^-\Theta_-^{-1}(\lambda_0)(\widetilde{g} + S^*g)(\lambda_0) \end{pmatrix},$$

we have with the help of (2.17) and (2.20)

$$\widetilde{g}' + S^*g'$$

$$= \frac{1}{k - \lambda_0}\left[(\widetilde{g} + S^*g) - (\widetilde{g} + S^*g)(\lambda_0) + (S^*(\bar{\lambda}_0) - S^*)X^-\Theta_-^{-1}(\lambda_0)(\widetilde{g} + S^*g)(\lambda_0)\right]$$

$$= \frac{1}{k - \lambda_0}\left[(\widetilde{g} + S^*g) - (\Theta_-(\lambda_0) - (S^*(\bar{\lambda}_0) - S^*)X^-)\Theta_-^{-1}(\lambda_0)(\widetilde{g} + S^*g)(\lambda_0)\right]$$

$$= \frac{1}{k - \lambda_0}\left[(\widetilde{g} + S^*g) - \Theta_-(k)\Theta_-^{-1}(\lambda_0)(\widetilde{g} + S^*g)(\lambda_0)\right] = \mathscr{F}_+(A_B - \lambda_0 I)^{-1}u$$

and

$$\widetilde{S}g' + g'$$

$$= \frac{1}{k - \lambda_0}\left[(S\widetilde{g} + g) - S(\widetilde{g} + S^*g)(\lambda_0) - (I - SS^*(\bar{\lambda}_0))X^-\Theta_-^{-1}(\lambda_0)(\widetilde{g} + S^*g)(\lambda_0)\right]$$

$$= \frac{1}{k - \lambda_0}\left[(S\widetilde{g} + g) - \left(S\Theta_-(\lambda_0) + X^- - SS^*(\bar{\lambda}_0)X^-\right)\Theta_-^{-1}(\lambda_0)(\widetilde{g} + S^*g)(\lambda_0)\right]$$

$$= \frac{1}{k - \lambda_0}\left[(S\widetilde{g} + g) - \left(SX^+ + X^-\right)\Theta_-^{-1}(\lambda_0)(\widetilde{g} + S^*g)(\lambda_0)\right]$$

$$= \frac{1}{k - \lambda_0}\left[(S\widetilde{g} + g) - \Theta_+(k)\Theta_-^{-1}(\lambda_0)(\widetilde{g} + S^*g)(\lambda_0)\right] = \mathscr{F}_+(A_B - \lambda_0 I)^{-1}u.$$

The proof is complete. $\qquad\square$

Remark 2.12. Operators $X^-\Theta_-^{-1}(\lambda_0)$, $X^+\Theta_+^{-1}(\mu_-)$ in Theorem 2.5 can be replaced with $X^-\Theta^*(\bar{\lambda}_0)X^-$ and $X^+\Theta^*(\bar{\mu}_0)X^+$, respectively. For the proof see (2.19), (2.21) and identities $X^-X^+ = X^+X^- = 0$.

Remark 2.13. All assertions of Theorem 2.5 remain valid if the operator J is formally substituted by $-J$ or $\pm I_E$. Compare with [28] for details. The following theorem is a consequence of this observation obtained from Theorem 2.5 by the substitution $J \to -J$. Note that its claim can be verified independently by passing on to adjoint operators in the formulae of Theorem 2.5.

Theorem 2.6. *For* $\lambda_0 \in \mathbb{C}_- \cap \rho(A_B^*)$, $\mu_0 \in \mathbb{C}_+ \cap \rho(A_B^*)$, $(\widetilde{g}, g) \in H$

$$(A_B^* - \lambda_0 I)^{-1}\begin{pmatrix}\widetilde{g}\\ g\end{pmatrix} = P_H(k - \lambda_0)^{-1}\begin{pmatrix}\widetilde{g}\\ g - X^+\Theta(\lambda_0)X^+(\widetilde{g} + S^*g)(\lambda_0)\end{pmatrix}$$

$$(A_B^* - \mu_0 I)^{-1}\begin{pmatrix}\widetilde{g}\\ g\end{pmatrix} = P_H(k - \mu_0)^{-1}\begin{pmatrix}\widetilde{g} - X^-\Theta(\mu_0)X^-(S\widetilde{g} + g)(\mu_0)\\ g\end{pmatrix}.$$

Assuming $J = I_E$ in the statement of Theorem 2.5, we arrive at the Sz.-Nagy-Foiaş model of dissipative operator A_{B_+}, see (2.6) and Theorem 2.4.

Remark 2.14. It is unknown whether the operator A_∞ can be efficiently represented in the model space $\mathbf{H} = L_2\begin{pmatrix}I & S^*\\ S & I\end{pmatrix}$. The computations, analogous to the carried out above, fail to yield "resolvent identities" that could be used for the desired model representation of the operator A_∞.

At this point we close our discussion of the functional model of the operator A_B and turn to the illustrations of the developed theory.

3. Examples

In this section we offer two examples of calculation of the Weyl function.

The first example is a symmetric operator that models the finite set of δ-interactions of quantum mechanics ([14]). A recently published preprint [13] offers a description of the boundary triple of this operator in the case of a single δ-interaction. It does not touch upon more general situation; however, a generalization to the case considered below is quite evident. The paper [13] is not concerned with any questions related to the functional model of non-selfadjoint extensions.

The second example is the symmetric operator generated by the differential expression $l[y] = -y'' + q(z)y$ in $L^2(0, \infty)$ with a real-valued potential $q(x)$ such that the Weyl limit circle case at infinity is observed. Explicit construction of the selfadjoint dilation of a dissipative extension of this operator and subsequent spectral analysis in terms of its characteristic function are carried out in the paper [3] in complete accordance with B. Pavlov's schema.

In this section we content ourselves with the description of convenient boundary triples and the computation of the corresponding Weyl functions. The construction of the functional models is not given here, since the model perspective on any a.s. non-selfadjoint extension of these operators can be easily derived from the exposition of Section 2.

3.1. Point interactions in $\mathbb{R}^3$

Let $\{x_s\}_{s=1}^n$ $(n < \infty)$ be the finite set of distinct points in $\mathbb{R}^3$. We define a symmetric operator A as a closure of the restriction of the Laplace operator $-\Delta$ on $H = L_2(\mathbb{R}^3)$ to the set of smooth functions vanishing in the neighborhood of $\cup_s x_s$. It is known ([14], [36]), that

$$\mathcal{D}(A) = \left\{ u \in W_2^2(\mathbb{R}^3) \ : \ u(x_s) = 0, \quad s = 1, 2, \ldots, n \right\}.$$

The deficiency indices $n_\pm(A)$ are equal to (n, n). The domain of conjugate operator A^* is described in the following theorem borrowed from [36].

Theorem 3.1. *The domain $\mathcal{D}(A^*)$ of conjugate operator A^* consists of the functions $u \in L_2(\mathbb{R}^3) \cap W_2^2(\mathbb{R}^3 \setminus \cup_s x_s)$ with the following asymptotic expansion in the neighborhood of $\{x_s\}_{s=1}^n$*

$$u(x) \sim u_-^s/|x - x_s| + u_0^s + O(|x - x_s|^{1/2}), \quad x \to x_s, \quad s = 1, 2, \ldots, n.$$

For given vectors $u, v \in \mathcal{D}(A^*)$ the analogue of the second Green formula holds:

$$(A^*u, v)_H - (u, A^*v)_H = \sum_{s=1}^n \left(u_0^s \bar{v}_-^s - u_-^s \bar{v}_0^s \right).$$

It is easy to show that the boundary triple $\{\mathcal{H}, \Gamma_0, \Gamma_1\}$ for the operator A^* can be chosen in the form $(u \in \mathcal{D}(A^*))$:

$$\mathcal{H} = \mathbb{C}^n, \quad \Gamma_0 u = (u_-^1, u_-^2, \ldots, u_-^n)^T, \quad \Gamma_1 u = (u_0^1, u_0^2, \ldots, u_0^n)^T.$$

In order to compute the Weyl function corresponding to this boundary triple let us fix a complex number $z \in \mathbb{C}_- \cup \mathbb{C}_+$ and let y_z be a vector from $\ker(A^* - zI)$, so that $y_z \in \mathcal{D}(A^*)$ and $-\Delta y_z = zy_z$. Note that vector y_z is uniquely represented in the form of a linear combination

$$y_z(x) = \sum_{s=1}^n C_s \frac{\exp(ik|x - x_s|)}{|x - x_s|},$$

where $k = \sqrt{z}$, $\operatorname{Im} z > 0$, and $\{C_s\}_{s=1}^n$ are some constants. Noting that in the neighborhoods of the points $\{x_s\}_{s=1}^n$ asymptotically

$$\frac{\exp(ik|x - x_s|)}{|x - x_s|} \sim \frac{1}{|x - x_s|} + ik + O(|x - x_s|), \qquad \text{as } x \to x_s$$

and obviously

$$\lim_{x \to x_j} \frac{\exp(ik|x - x_s|)}{|x - x_s|} = \frac{\exp(ik|x_j - x_s|)}{|x_j - x_s|}, \quad j \neq s,$$

we easily compute both vectors $\Gamma_0 y_z$, $\Gamma_1 y_z$.

$$\Gamma_0 y_z = (C_1, C_2, \ldots, C_n)^T$$

$$\Gamma_1 y_z = \left(ik \cdot C_1 + \sum_{s=2}^n C_s \frac{\exp(ik|x_1 - x_s|)}{|x_1 - x_s|}, \; \ldots \right.$$

$$\ldots \; ik \cdot C_j + \sum_{s \neq j}^n C_s \frac{\exp(ik|x_j - x_s|)}{|x_j - x_s|}, \; \ldots$$

$$\left. \ldots \; ik \cdot C_n + \sum_{s=1}^{n-1} C_s \frac{\exp(ik|x_{n-1} - x_s|)}{|x_{n-1} - x_s|} \right)^T.$$

Comparison of these formulae with the definition $\Gamma_1 y_z = M(z)\Gamma_0 y_z$ of the Weyl function yields its explicit form. It is a $(n \times n)$-matrix function $M(z) = \|M_{sj}(z)\|_1^n$ with elements

$$M_{sj}(z) = \begin{cases} ik, & s = j \\ \langle s, j \rangle, & s \neq j \end{cases}$$

where $k = \sqrt{z}$, $k \in \mathbb{C}_+$ and

$$\langle s, j \rangle := \frac{\exp(ik|x_s - x_j|)}{|x_s - x_j|}, \quad s \neq j, \quad s, j = 1, 2, \ldots, n.$$

Note that the selfadjoint operator A_∞ defined as a restriction of A^* to the set $\{y \in \mathcal{D}(A^*) : \Gamma_0 y = 0\}$ is the Laplace operator $-\Delta$ in $L_2(\mathbb{R}^3)$ with the domain $\mathcal{D}(A_\infty) = W_2^2(\mathbb{R}^3)$. At the same time it is the Friedrichs extension of the symmetric operator A. The special role of extension A_∞ with regard to the functional model construction was pointed out in Section 2.

3.2. Schrödinger operator in the Weyl limit circle case

The second example is the symmetric operator A defined as a closure in the Hilbert space $H = L_2(\mathbb{R}_+)$ of the minimal operator generated by the differential expression

$$l[y] = -y'' + q(x)y \tag{3.1}$$

on the domain $C_0^\infty(\mathbb{R}_+)$. We assume the potential $q(x)$ to be a real-valued continuous function such that for the expression (3.1) the Weyl limit circle case at infinity is observed. The deficiency indices of A are equal to $(2, 2)$ and both solutions of equation $l[y] = \lambda y$ are functions from $L_2(\mathbb{R}_+)$ for any $\lambda \in \mathbb{C}$, see [46], [17]. The conjugate operator A^* is generated by the same differential expression (3.1) on the class of absolutely continuous functions y from $L_2(\mathbb{R}_+)$ whose derivatives are locally absolutely continuous and $l[y]$ is square integrable.

Let $v_1(x)$, $v_2(x)$, $x \in \mathbb{R}$ be two linearly independent solutions of the equation $l[y] = 0$ satisfying the conditions at $x = 0$:

$$v_1(0) = 1, \quad v_1'(0) = 0, \quad v_2(0) = 0, \quad v_2'(0) = 1.$$

For our purposes we will use the boundary triple $\{\mathcal{H}, \Gamma_0, \Gamma_1\}$ for the operator A^* described in [3]. The space $\mathcal{H}$ is two-dimensional: $\mathcal{H} = \mathbb{C}^2$, and the mappings $\Gamma_0, \Gamma_1 :$ $\mathcal{D}(A^*) \to \mathbb{C}^2$ are defined as

$$\Gamma_0 y = \begin{pmatrix} y'(0) \\ \mathcal{W}[y, v_2]\big|_\infty \end{pmatrix}, \quad \Gamma_1 y = \begin{pmatrix} -y(0) \\ \mathcal{W}[y, v_1]\big|_\infty \end{pmatrix}, \quad y \in \mathcal{D}(A^*) \tag{3.2}$$

where $\mathcal{W}[f, g] := fg' - f'g$ is the Wronsky determinant of two functions f, g from $\mathcal{D}(A^*)$.

In order to compute the corresponding Weyl function $M(\cdot)$ let us fix a complex number $\lambda \in \mathbb{C}_+$ and let ψ_λ, $\phi_\lambda(x)$ be the solutions of the equation $l[y] = \lambda y$ satisfying

$$\begin{aligned} \psi_\lambda(0) &= 1, \quad \psi_\lambda'(0) = 0, \\ \phi_\lambda(0) &= 0, \quad \phi_\lambda'(0) = 1. \end{aligned} \tag{3.3}$$

Both functions ϕ_λ, ψ_λ are square integrable on the real half-axis $\mathbb{R}_+$, their Wronsky determinant is independent on $x \in \mathbb{R}_+$ and is equal to one: $\mathcal{W}[\psi_\lambda, \phi_\lambda] = 1$. The functions ψ_λ, ϕ_λ are linearly independent vectors in $L_2(\mathbb{R}_+)$ and any solution y_λ of the equation $(A^* - \lambda I)y_\lambda = 0$ is their linear combination $y_\lambda = C_1 \psi_\lambda + C_2 \phi_\lambda$ with some constants $C_1, C_2 \in \mathbb{C}$. According to (3.2),

$$\begin{aligned} \Gamma_0 y_\lambda &= \begin{pmatrix} y_\lambda'(0) \\ \mathcal{W}[y_\lambda, v_2]\big|_\infty \end{pmatrix} = \begin{pmatrix} C_2 \\ C_1 \cdot \mathcal{W}[\psi_\lambda, v_2]\big|_\infty + C_2 \cdot \mathcal{W}[\phi_\lambda, v_2]\big|_\infty \end{pmatrix} \\ \Gamma_1 y_\lambda &= \begin{pmatrix} -y_\lambda(0) \\ \mathcal{W}[y_\lambda, v_1]\big|_\infty \end{pmatrix} = \begin{pmatrix} -C_1 \\ C_1 \cdot \mathcal{W}[\psi_\lambda, v_1]\big|_\infty + C_2 \cdot \mathcal{W}[\phi_\lambda, v_1]\big|_\infty \end{pmatrix}. \end{aligned}$$

Let $M(\lambda) = \|m_{ij}(\lambda)\| = \begin{pmatrix} m_{11}(\lambda) & m_{12}(\lambda) \\ m_{21}(\lambda) & m_{22}(\lambda) \end{pmatrix}$ be the Weyl function being sought. Since $\Gamma_1 y_\lambda = M(\lambda)\Gamma_0 y_\lambda$ by the definition, the equalities

$$-C_1 = m_{11}(\lambda) \cdot C_2 + m_{12}(\lambda) \cdot \left\{ C_1 \cdot \mathcal{W}[\psi_\lambda, v_2]\big|_\infty + C_2 \cdot \mathcal{W}[\phi_\lambda, v_2]\big|_\infty \right\}$$

$$C_1 \cdot \mathcal{W}[\psi_\lambda, v_1]\big|_\infty + C_2 \cdot \mathcal{W}[\phi_\lambda, v_1]\big|_\infty$$

$$= m_{21}(\lambda) \cdot C_2 + m_{22}(\lambda) \cdot \left\{ C_1 \cdot \mathcal{W}[\psi_\lambda, v_2]\big|_\infty + C_2 \cdot \mathcal{W}[\phi_\lambda, v_2]\big|_\infty \right\}$$

should be valid for any $C_1, C_2 \in \mathbb{C}$. The solution of this linear system is easy to compute:

$$m_{11}(\lambda) = \left(\mathcal{W}[\phi_\lambda, v_2]/\mathcal{W}[\psi_\lambda, v_2]\right)\big|_\infty$$
$$m_{12}(\lambda) = (-1)/\mathcal{W}[\psi_\lambda, v_2]\big|_\infty$$
$$m_{21}(\lambda) = \mathcal{W}[\psi_\lambda, v_1]\big|_\infty - \left(\mathcal{W}[\phi_\lambda, v_1]/\mathcal{W}[\phi_\lambda, v_2]\right)\big|_\infty \cdot \mathcal{W}[\psi_\lambda, v_2]\big|_\infty$$
$$m_{22}(\lambda) = \left(\mathcal{W}[\phi_\lambda, v_1]/\mathcal{W}[\phi_\lambda, v_2]\right)\big|_\infty.$$

The expression for $m_{21}(\lambda)$ above can be further simplified

$$\begin{aligned}
m_{21}(\lambda) &= \mathcal{W}[\phi_\lambda, v_2]^{-1} \cdot \left(\mathcal{W}[\phi_\lambda, v_1] \cdot \mathcal{W}[\psi_\lambda, v_2] - \mathcal{W}[\psi_\lambda, v_1] \cdot \mathcal{W}[\phi_\lambda, v_2]\right)\big|_\infty \\
&= \left(\mathcal{W}[\phi_\lambda, v_2]|_\infty\right)^{-1} \cdot \lim_{b\to\infty} \big((\phi_\lambda v_1' - \phi_\lambda' v_1)(\psi_\lambda v_2' - \psi_\lambda' v_2) \\
&\qquad\qquad\qquad - (\psi_\lambda v_1' - \psi_\lambda' v_1)(\phi_\lambda v_2' - \phi_\lambda' v_2)\big)\big|_b \\
&= \left(\mathcal{W}[\phi_\lambda, v_2]|_\infty\right)^{-1} \cdot \lim_{b\to\infty} \left(\phi_\lambda \psi_\lambda'(v_1 v_2' - v_1' v_2) - \phi_\lambda' \psi_\lambda(v_1 v_2' - v_1' v_2)\right)\big|_b \\
&= \left(\mathcal{W}[\phi_\lambda, v_2]|_\infty\right)^{-1} \cdot \lim_{b\to\infty} \mathcal{W}[\phi_\lambda, \psi_\lambda]|_b \cdot \mathcal{W}[v_1, v_2]|_b \\
&= -\left(\mathcal{W}[\phi_\lambda, v_2]|_\infty\right)^{-1}.
\end{aligned}$$

Finally, for the Weyl function we obtain the formula

$$M(\lambda) = \left\{\mathcal{W}[\psi_\lambda, v_2]|_\infty\right\}^{-1} \begin{pmatrix} \mathcal{W}[\phi_\lambda, v_2]|_\infty & -1 \\ -1 & \mathcal{W}[\psi_\lambda, v_1]|_\infty \end{pmatrix}, \quad \lambda \in \mathbb{C}_+. \qquad (3.4)$$

There exists another representation of the Weyl function (3.4) derived from the work of M.G. Krein [23]. Introduce the following functions

$$\left.\begin{aligned}
D_0(x, \lambda) &= -\lambda \int_0^x \phi_\lambda(s)v_2(s)ds & D_1(x, \lambda) &= 1 + \lambda \int_0^x \phi_\lambda(s)v_1(s)ds \\
E_0(x, \lambda) &= 1 - \lambda \int_0^x \psi_\lambda(s)v_2(s)ds & E_1(x, \lambda) &= \lambda \int_0^x \psi_\lambda(s)v_1(s)ds\,.
\end{aligned}\right\} \qquad (3.5)$$

Noticing that the Cauchy function of the differential operator $-\frac{d^2}{dx^2}+q(x)$ coincides with $v_1(x)v_2(s) - v_1(s)v_2(x)$, after a short computation we conclude that

$$\begin{aligned}
\mathcal{W}[\psi_\lambda, v_2] &= E_0(x, \lambda) & \mathcal{W}[\phi_\lambda, v_2] &= D_0(x, \lambda) \\
\mathcal{W}[\psi_\lambda, v_1] &= -E_1(x, \lambda) & \mathcal{W}[\phi_\lambda, v_1] &= -D_1(x, \lambda).
\end{aligned}$$

Consequently, the Weyl function (3.4) can be rewritten in the form

$$M(\lambda) = \left(E_0(\lambda)\right)^{-1} \begin{pmatrix} D_0(\lambda) & -1 \\ -1 & -E_1(\lambda) \end{pmatrix}, \quad \lambda \in \mathbb{C}_+$$

where

$$D_0(\lambda) := \lim_{x\to+\infty} D_0(x, \lambda), \qquad E_j(\lambda) := \lim_{x\to+\infty} E_j(x, \lambda), \quad j = 0, 1.$$

These limits exist due to the square integrability of the functions ψ_λ, ϕ_λ, v_1, v_2, when $\lambda \in \mathbb{C}_+$, see (3.5). Moreover, these limits are entire functions of the variable $\lambda \in \mathbb{C}$.

The selfadjoint operator A_∞ is generated by the expression (3.1) and the boundary condition $\Gamma_0 y = \left(\begin{smallmatrix} y'(0) \\ \mathcal{W}[y,v_2]|_\infty \end{smallmatrix} \right) = 0$. It is well known that the spectrum of the operator A_∞ consists of pure eigenvalues with the multiplicity equal to one. By the definition (3.3) the solution ψ_λ satisfies $\Gamma_0 \psi_\lambda = 0$ if the Wronsky determinant $\mathcal{W}[\psi_\lambda, v_2] = E_0(x, \lambda)$ tends to zero as $x \to \infty$. It means that the zeroes of the entire function $E_0(\lambda)$ in the "denominator" of the Weyl function are the eigenvalues of the operator A_∞ with the corresponding eigenvectors ψ_λ.

References

[1] R.A. Adams, J.F. Fournier, *Sobolev Spaces*, Elsevier Science, Oxford, UK, 2003.

[2] V.M. Adamyan, B.S. Pavlov, *Trace formula for dissipative operators* (Russian) Vestnik Leningrad. Univ. Mat. Mekh. Astronom., **2** (1979), 5–9.

[3] B.P. Allahverdiev, *On dilation theory and spectral analysis of dissipative Schrödinger operators in Weyl's limit-circle case* (Russian), Izv. Akad. Nauk. SSSR Ser. Mat. **54** (1990), 242–257; English transl.: Math. USSR Izv. **36** (1991), 247–262.

[4] B.P. Allahverdiev, *Spectral analysis of nonselfadjoint Schrödinger operators with a matrix potential*, J. Math. Anal. Appl. **303** (2005), 208–219.

[5] B.P. Allahverdiev, *A nonself-adjoint singular Sturm-Liouville problem with a spectral parameter in the boundary condition*, Math. Nachr. **278, 7-8** (2005), 743–755.

[6] B.P. Allahverdiev, *Spectral analysis of dissipative Dirac operators with general boundary conditions*, J. Math. Anal. Appl. **283** (2003), 287–303.

[7] B.P. Allahverdiev, *Dilation and Functional Model of Dissipative Operator Generated by an Infinite Jacobi Matrix*, Math. and Comp. Modelling **38, 3** (2003), 989–1001.

[8] D.Z. Arov, *Passive linear stationary dynamic systems.* (Russian) Sibirsk. Mat. Zh. **20, 2** (1979), 211–228.; English Transl.: Siberian Mathematical Journal, 20 (2), 149–162.

[9] T.Ya. Azizov, I.S. Iokhvidov, *Linear Operators in Spaces With an Indefinite Metric*, John Wiley & Sons Inc., 1989.

[10] M. Baro, H. Neidhardt *Dissipative Schrödinger-type operators as a model for generation and recombination*, J. Math. Phys. **44, 6** (2003), 2373–2401.

[11] M. Baro, H.-Chr. Kaiser, H. Neidhardt, J. Rehberg, *Dissipative Schrödinger–Poisson systems*, J. Math. Phys. **45, 1** (2004), 21–43.

[12] M. Baro, H.-Chr. Kaiser, H. Neidhardt, J. Rehberg, *A Quantum Transmitting Schrödinger–Poisson System*, Rev. Math. Phys. **16, 3** (2004), 281–330.

[13] J. Behrndt, M. Malamud, H. Neidhardt, *Scattering matrices and Weyl functions*, Preprint (2006).

[14] F.A. Berezin, L.D. Faddeev, *Remark on the Schrödinger equation with singular potential.* (Russian) Dokl. Akad. Nauk SSSR **137** (1961), 1011–1014.

[15] H.-J. Bremermann, *Distributions, Complex Variables and Fourier Transforms*, Addison-Wesley, Reading, Massachusetts, 1965.

[16] V.M. Bruk, *A certain class of boundary value problems with a spectral parameter in the boundary condition.* (Russian) Mat. Sb. (N.S.) **100 (142)**, **2** (1976), 210–216.

[17] E.A. Coddington, N. Levinson, *Theory of ordinary differential equations*, New York-Toronto-London: McGraw-Hill, 1955.

[18] V.A. Derkach and M.M. Malamud, *Generalized resolvents and the boundary value problems for Hermitian operators with gaps*, J. Funct. Anal. **95** (1991), 1–95.

[19] V.I. Gorbachuk and M.L. Gorbachuk, *Boundary value problems for operator differential equations.* Translated and revised from the 1984 Russian original. Mathematics and its Applications (Soviet Series), 48. Kluwer Academic Publishers Group, Dordrecht, 1991.

[20] H.-Ch. Kaiser, H. Neidhardt, J. Rehberg, On 1-*dimensional dissipative Schrödinger-type operators, their dilations, and eigenfunction expansions.* Math. Nachr. **252** (2003), 51–69.

[21] A.N. Kočhubeĭ, *On extension of symmetric operators and symmetric binary relations* (Russian) Mat. Zametki **17** (1975), 41–48.

[22] A.N. Kočhubeĭ, *Characteristic functions of symmetric operators and their extensions.* (Russian) Izv. Akad. Nauk Armyan. SSR Ser. Mat. **15, 3** (1980), 219–232.

[23] M.G. Kreĭn, *On the indeterminate case of the Sturm-Liouville boundary problem in the interval* $(0, \infty)$. (Russian) Izvestiya Akad. Nauk SSSR. Ser. Mat. **16** (1952). 293–324.

[24] Yu. Kudryashov, *Symmetric and selfadjoint dilations of dissipative operators.* (Russian) Teor. Funktsiĭ, Funktsional. Anal. i Prilozhen. **37** (1982), 51–54.

[25] Yu.A. Kuperin, S.N. Naboko, R.V. Romanov, *Spectral analysis of transport operator: Functional model approach*, Indiana Univ. Math. J. **51, 6** (2002), 1389-1426.

[26] P. Lax, R. Phillips, *Scattering theory.* Pure and Applied Mathematics, Vol. 26 Academic Press, New York-London, 1967.

[27] N.G. Makarov, V.I. Vasyunin, *A model for noncontractions and stability of the continuous spectrum.* Lect.Notes in Math., **864** (1981), 365–412.

[28] S.N. Naboko, *Functional model of perturbation theory and its applications to scattering theory.* (Russian) Boundary value problems of mathematical physics, 10. Trudy Mat. Inst. Steklov. **147** (1980), 86–114.

[29] S.N. Naboko, *Absolutely continuous spectrum of a nondissipative operator, and a functional model. I.* (Russian) Investigations on linear operators and the theory of functions, VII. Zap. Naučn. Sem. Leningrad. Otdel Mat. Inst. Steklov. (LOMI) **65** (1976), 90–102. English transl.: J. Sov. Math. **16, 3** (1981)

[30] S.N. Naboko, *Absolutely continuous spectrum of a nondissipative operator, and a functional model. II.* (Russian) Investigations on linear operators and the theory of functions, VIII. Zap. Nauchn. Sem. Leningrad. Otdel. Mat. Inst. Steklov. (LOMI) **73** (1977). English transl.: J. Math. Sci. **34, 6** (1986), 2090–2101.

[31] S.N. Naboko, R.V. Romanov, *Spectral singularities, Szökefalvi-Nagy-Foias functional model and the spectral analysis of the Boltzmann operator.* Recent advances in operator theory and related topics (Szeged, 1999), 473–490, Oper. Theory Adv. Appl., **127**, Birkhäuser, Basel, 2001.

[32] B. Szökefalvi-Nagy, C. Foiaş, *Harmonic Analysis of Operators on Hilbert Space*, North Holland, New York, 1970.

[33] N.K. Nikolskiĭ, *Treatise on the Shift Operator*, Springer-Verlag, Berlin etc., 1986.

[34] N.K. Nikolskiĭ, *Operators, Functions, and Systems: An Easy reading, v. 2: Model Operators and Systems*, AMS, 2002.

[35] B.S. Pavlov, *On a non-selfadjoint Schrödinger operator.* (Russian) Probl. Math. Phys., No. 1, Spectral Theory and Wave Processes (1966) pp. 102–132 Izdat. Leningrad. Univ., Leningrad.

[36] B.S. Pavlov, *A model of zero-radius potential with internal structure.* (Russian) Teoret. Mat. Fiz. **59, 3** (1984), 345–353. English transl.: Theoretical and Mathematical Physics, **59, 3**, pp. 544–550, Springer, New York.

[37] B.S. Pavlov, *On conditions for separation of the spectral components of a dissipative operator*, Izv. Akad. Nauk SSSR, Ser. Matem., **39** (1975), 123–148 (Russian); English transl.: Math USSR Izvestiya, **9** (1976), 113–137.

[38] B.S. Pavlov, *Selfadjoint dilation of a dissipative Schrödinger operator, and expansion in its eigenfunction.* (Russian) Mat. Sb. (N.S.) **102 (144), 4** (1977), 511–536.

[39] B.S. Pavlov, *Dilation theory and spectral analysis of nonselfadjoint differential operators*, Proc. 7th Winter School, Drobobych 1974, 3-69 (1976) (Russian); English transl: Transl. II. Ser., Am. Math. Soc. **115** (1981), 103–142.

[40] B.S. Pavlov, *A Remark on Spectral Meaning of the Symmetric Functional Model*, in Operator Theory: Advances and Applications, Vol. **154**, 163–177, Birkhäuser Verlag, Basel/Switzerland, 2004.

[41] A.M. Petrov, *Spectral projectors of some classes of nonselfadjoint operators.* (Russian) Dynamical systems, No. **7** (Russian), 109–114, 160, "Vishcha Shkola", Kiev, 1988.; English transl. in J. Math. Sci. 65 (1993), no. 2, 1575–1578.

[42] M. Rosenblum, J. Rovnyak, *Hardy classes and operator theory*, Oxford University Press, 1985.

[43] B.S. Solomyak, *A functional model for dissipative operators. A coordinate-free approach.* (Russian) Zap. Nauchn. Sem. Leningrad. Otdel. Mat. Inst. Steklov. (LOMI) **178** (1989), Issled. Linein. Oper. Teorii Funktsii. 18, 57–91, 184–185; Translation in J. Soviet Math. **61** (1992), no. **2**, 1981–2002.

[44] A.V. Štraus, *Characteristic functions of linear operators.* (Russian) Izv. Akad. Nauk. SSSR, Ser. Mat., **24, 1** (1960), 43–74; English transl. in AMS Transl. **(2) 40** (1964), 1–37.

[45] A.V. Štraus, *On Extensions and characteristic function of a symmetric operator.* (Russian), Izv. Akad. Nauk SSSR Ser. Mat. **32** (1968), 186–207; English transl. in Math. USSR - Izvestija, **2** (1968), 181–204.

[46] E.S. Titchmarsh, *Eigenfunction Expansions Associated with Second-Order Differential Equations. Part I.* Second Edition Clarendon Press, Oxford 1962.

Vladimir Ryzhov
Suite 200 – 1628 Dickson Ave.
Kelowna, BC V1Y 9X1, Canada
e-mail: `vryzhov@bridges.com`

Operator Theory:
Advances and Applications, Vol. 174, 159–172

Lyapunov Exponents at Anomalies of $\mathrm{SL}(2, \mathbb{R})$-actions

Hermann Schulz-Baldes

Abstract. Anomalies are known to appear in the perturbation theory for the one-dimensional Anderson model. A systematic approach to anomalies at critical points of products of random matrices is developed, classifying and analysing their possible types. The associated invariant measure is calculated formally. For an anomaly of so-called second degree, it is given by the ground-state of a certain Fokker-Planck equation on the unit circle. The Lyapunov exponent is calculated to lowest order in perturbation theory with rigorous control of the error terms.

1. Introduction

Anomalies in the perturbative calculation of the Lyapunov exponent and the density of states were first found and analysed by Kappus and Wegner [6] when they studied the center of the band in a one-dimensional Anderson model. Further anomalies, albeit in higher order perturbation theory, were then treated by Derrida and Gardner [4] as well as Bovier and Klein [2]. More recently, anomalies also appeared in the study of random polymer models [5]. Quite some effort has been made to understand anomalies in the particular case of the Anderson model also from a more mathematical point of view [3, 9]. However, Campanino and Klein [3] need to suppose decay estimates on the characteristic function of the random potential, and Shubin, Vakilian and Wolff [9] appeal to rather complicated techniques from harmonic analysis (allowing only to give the correct scaling of the Lyapunov exponent, but not a precise perturbative formula for it).

It is the purpose of this work to present a more conceptual approach to anomalies of products of random matrices. In fact, various types may appear and only those of second degree (in the sense of the definition below) seem to have been studied previously. Indeed, this is the most difficult and interesting case to analyse, and the main insight of the present work is to exhibit an associated Fokker-Planck operator, the spectral gap of which is ultimately responsible for the positivity of the Lyapunov exponent. In the special case of the Anderson model, a related operator already appeared in [2]. Here it is, however, possible to circumvent the spectral analysis of the Fokker-Planck operator and prove the asymptotics of the Lyapunov exponent more directly (*cf.* Section 5.4). The other cases of various first degree anomalies are more elementary to analyse. Examples for different types of anomalies are given in Section 6.

2. Definition of anomalies

Let us consider families $(T_{\lambda,\sigma})_{\lambda\in\mathbb{R},\sigma\in\Sigma}$ of matrices in $\mathrm{SL}(2,\mathbb{R})$ depending on a random variable σ in some probability space $(\Sigma,\mathbf{p})$ as well as a real coupling parameter λ. In order to avoid technicalities, we suppose that $\mathbf{p}$ has compact support. The dependence on λ is supposed to be smooth. The expectation value w.r.t. $\mathbf{p}$ will be denoted by $\mathbf{E}$.

Definition 2.1. The value $\lambda = 0$ is anomaly of first order of the family $(T_{\lambda,\sigma})_{\lambda\in\mathbb{R},\sigma\in\Sigma}$ if for all $\sigma \in \Sigma$:

$$T_{0,\sigma} \;=\; \pm\mathbf{1}\,, \tag{2.1}$$

with a sign that may depend on $\sigma \in \Sigma$. In order to further classify the anomalies and for later use, let us introduce $P_\sigma, Q_\sigma \in \mathrm{sl}(2,\mathbb{R})$ by

$$MT_{\lambda,\sigma}M^{-1} \;=\; \pm\exp\left(\lambda\,P_\sigma + \lambda^2\,Q_\sigma + \mathcal{O}(\lambda^3)\right)\,, \tag{2.2}$$

where $M \in \mathrm{SL}(2,\mathbb{R})$ is a λ- and σ-independent basis change to be chosen later. An anomaly is said to be of first degree if $\mathbf{E}(P_\sigma)$ is non-vanishing, and then it is called elliptic if $\det(\mathbf{E}(P_\sigma)) > 0$, hyperbolic if $\det(\mathbf{E}(P_\sigma)) < 0$ and parabolic if $\det(\mathbf{E}(P_\sigma)) = 0$. Note that all these notions are independent of the choice of M.

If $\mathbf{E}(P_\sigma) = 0$, but the variance of P_σ is non-vanishing, then an anomaly is said to be of second degree.

Furthermore, for $k \in \mathbb{N}$, set $\hat\sigma = (\sigma(k),\ldots,\sigma(1)) \in \hat\Sigma = \Sigma^{\times k}$, as well as $\hat{\mathbf{p}} = \mathbf{p}^{\times k}$ and $T_{\lambda,\hat\sigma} = T_{\lambda,\sigma(k)}\cdots T_{\lambda,\sigma(1)}$. Then $\lambda = 0$ is anomaly of kth order of the family $(T_{\lambda,\sigma})_{\lambda\in\mathbb{R},\sigma\in\Sigma}$ if the family $(T_{\lambda,\hat\sigma})_{\lambda\in\mathbb{R},\hat\sigma\in\hat\Sigma}$ has an anomaly of first order at $\lambda = 0$ in the above sense. The definitions of degree and nature transpose to kth order anomalies.

As is suggested in the definition and will be further explained below, we may (and will) restrict ourselves to the analysis of anomalies of first order. In the examples, however, anomalies of higher order do appear and can then be studied by the present techniques (*cf.* Section 6). Furthermore, by a change of variables in λ, anomalies of degree higher than 2 can be analysed like an anomaly of second degree.

Anomalies are particular cases of so-called critical points studied in [5, 8], namely $\lambda = 0$ is by definition a critical point of the family $(T_{\lambda,\sigma})_{\lambda\in\mathbb{R},\sigma\in\Sigma}$ if for all $\sigma,\sigma' \in \Sigma$:

$$[T_{0,\sigma}, T_{0,\sigma'}] \;=\; 0\,, \qquad \text{and} \qquad |\mathrm{Tr}(T_{0,\sigma})| \;<\; 2 \ \text{ or } \ T_{0,\sigma} = \pm\mathbf{1}\,. \tag{2.3}$$

Critical points appear in many applications like the Anderson model and the random polymer model. In these situations anomalies appear for special values of the parameters, such as the energy or the coupling constant, *cf.* Section 6.

3. Phase shift dynamics

The bijective action $\mathcal{S}_T$ of a matrix $T \in \mathrm{SL}(2,\mathbb{R})$ on $S^1 = [0, 2\pi)$ is given by

$$e_{\mathcal{S}_T(\theta)} = \frac{T e_\theta}{\|T e_\theta\|}, \qquad e_\theta = \begin{pmatrix} \cos(\theta) \\ \sin(\theta) \end{pmatrix}, \quad \theta \in [0, 2\pi) . \tag{3.1}$$

This defines a group action, namely $\mathcal{S}_{TT'} = \mathcal{S}_T \mathcal{S}_{T'}$. In particular the map $\mathcal{S}_T$ is invertible and $\mathcal{S}_T^{-1} = \mathcal{S}_{T^{-1}}$. Note that this is actually an action on $\mathbb{RP}(1)$, and S^1 appears as a double cover here. In order to shorten notations, we write

$$\mathcal{S}_{\lambda,\sigma} = \mathcal{S}_{M T_{\lambda,\sigma} M^{-1}} .$$

Next we need to iterate this dynamics. Associated to a given semi-infinite code $\omega = (\sigma_n)_{n \geq 1}$ with $\sigma_n \in \Sigma$ is a sequence of matrices $(T_{\lambda,\sigma_n})_{n \geq 1}$. Codes are random and chosen independently according to the product law $\mathbf{p}^{\otimes \mathbb{N}}$. Averaging w.r.t. $\mathbf{p}^{\otimes \mathbb{N}}$ is also denoted by $\mathbf{E}$. Then one defines iteratively for $N \in \mathbb{N}$

$$\mathcal{S}_{\lambda,\omega}^N(\theta) = \mathcal{S}_{\lambda,\sigma_N}\big(\mathcal{S}_{\lambda,\omega}^{N-1}(\theta)\big) , \qquad \mathcal{S}_{\lambda,\omega}^0(\theta) = \theta . \tag{3.2}$$

This is a discrete time random dynamical system on S^1. Let us note that at an anomaly of first order, one has $\mathcal{S}_{\lambda,\sigma}(\theta) = \theta + \mathcal{O}(\lambda)$ or $\mathcal{S}_{\lambda,\sigma}(\theta) = \theta + \pi + \mathcal{O}(\lambda)$ depending on the sign in (2.1). As all the functions appearing below will be π-periodic we can neglect the summand π, meaning that we may suppose that there is a sign $+$ in (2.1) for all σ (this reflects that the action is actually on projective space).

In order to do perturbation theory in λ, we need some notations. Introducing the unit vector $v = \frac{1}{\sqrt{2}} \begin{pmatrix} 1 \\ -\imath \end{pmatrix}$, we define the first order polynomials in $e^{2\imath\theta}$

$$p_\sigma(\theta) = \Im m \left(\frac{\langle v | P_\sigma | e_\theta \rangle}{\langle v | e_\theta \rangle} \right) , \qquad q_\sigma(\theta) = \Im m \left(\frac{\langle v | Q_\sigma | e_\theta \rangle}{\langle v | e_\theta \rangle} \right) ,$$

as well as

$$\alpha_\sigma = \langle v | P_\sigma | v \rangle , \qquad \beta_\sigma = \langle \overline{v} | P_\sigma | v \rangle .$$

Hence $p_\sigma(\theta) = \Im m(\alpha_\sigma - \beta_\sigma e^{2\imath\theta})$. Now starting from the identity

$$e^{2\imath \mathcal{S}_{\lambda,\sigma}(\theta)} = \frac{\langle v | M T_\sigma M^{-1} | e_\theta \rangle}{\langle \overline{v} | M T_\sigma M^{-1} | e_\theta \rangle} ,$$

the definition (2.2) and the identity $\langle v | e_\theta \rangle = \frac{1}{\sqrt{2}} e^{\imath\theta}$, one can verify that

$$\mathcal{S}_{\lambda,\sigma}(\theta) = \theta + \Im m \left(\lambda \frac{\langle v | P_\sigma | e_\theta \rangle}{\langle v | e_\theta \rangle} + \frac{\lambda^2}{2} \frac{\langle v | 2 Q_\sigma + P_\sigma^2 | e_\theta \rangle}{\langle v | e_\theta \rangle} - \frac{\lambda^2}{2} \frac{\langle v | P_\sigma | e_\theta \rangle^2}{\langle v | e_\theta \rangle^2} \right) ,$$

162   H. Schulz-Baldes

with error term of order $\mathcal{O}(\lambda^3)$. As one readily verifies that

$$P_\sigma^2 \;=\; -\det(P_\sigma)\,\mathbf{1}\,, \qquad \Im m\left(\frac{\langle v|P_\sigma|e_\theta\rangle^2}{\langle v|e_\theta\rangle^2}\right) \;=\; -\,p_\sigma(\theta)\,\partial_\theta\,p_\sigma(\theta)\,,$$

it follows that

$$\mathcal{S}_{\lambda,\sigma}(\theta) \;=\; \theta + \lambda\,p_\sigma(\theta) + \lambda^2\,q_\sigma(\theta) + \frac{1}{2}\,\lambda^2\,p_\sigma(\theta)\,\partial_\theta\,p_\sigma(\theta) + \mathcal{O}(\lambda^3)\,. \qquad (3.3)$$

Finally let us note that

$$\mathcal{S}_{\lambda,\sigma}^{-1}(\theta) \;=\; \theta - \lambda\,p_\sigma(\theta) - \lambda^2\,q_\sigma(\theta) + \frac{1}{2}\,\lambda^2\,p_\sigma(\theta)\,\partial_\theta p_\sigma(\theta) + \mathcal{O}(\lambda^3)\,, \qquad (3.4)$$

as one verifies immediately because $\mathcal{S}_{\lambda,\sigma}(\mathcal{S}_{\lambda,\sigma}^{-1}(\theta)) = \theta + \mathcal{O}(\lambda^3)$, or can deduce directly just as above from the identity

$$\exp\left(\lambda\,P_\sigma + \lambda^2\,Q_\sigma + \mathcal{O}(\lambda^3)\right)^{-1}\exp\left(-\lambda\,P_\sigma - \lambda^2\,Q_\sigma + \mathcal{O}(\lambda^3)\right)\,.$$

4. Formal perturbative formula for the invariant measure

For each λ, the family $(MT_{\lambda,\sigma}M^{-1})_{\sigma\in\Sigma}$ and the probability $\mathbf{p}$ define an invariant probability measure ν_λ on S^1 by the equation

$$\int d\nu_\lambda(\theta)\,f(\theta) \;=\; \mathbf{E}\int d\nu_\lambda(\theta)\,f(\mathcal{S}_{\lambda,\sigma}(\theta))\,, \qquad f\in C(S^1)\,. \qquad (4.1)$$

Furstenberg proved that this invariant measure is unique whenever the Lyapunov exponent of the associated product of random matrices (discussed below) is positive (*e.g.*, [1]) and in this situation ν_λ is also known to be Hölder continuous, so, in particular, it does not contain a point component. For the study of the invariant measure at an anomaly of order k, it is convenient to iterate (4.1):

$$\int d\nu_\lambda(\theta)\,f(\theta) \;=\; \mathbf{E}\int d\nu_\lambda(\theta)\,f(\mathcal{S}_{\lambda,\omega}^k(\theta))\,, \qquad f\in C(S^1)\,.$$

Replacing $(\Sigma,\mathbf{p})$ by $(\hat{\Sigma},\hat{\mathbf{p}})$ therefore shows that the families $(T_{\lambda,\sigma})_{\lambda\in\mathbb{R},\sigma\in\Sigma}$ and $(T_{\lambda,\hat{\sigma}})_{\lambda\in\mathbb{R},\hat{\sigma}\in\hat{\Sigma}}$ have the same invariant measure. Hence it is sufficient to study anomalies of first order.

The aim of this section is to present a formal perturbative expansion of the invariant measure under the hypothesis that it is absolutely continuous, that is $d\nu_\lambda(\theta) = \rho_\lambda(\theta)\frac{d\theta}{2\pi}$ with $\rho_\lambda = \rho_0 + \lambda\rho_1 + \mathcal{O}(\lambda^2)$. Then (4.1) leads to

$$\mathbf{E}\left(\partial_\theta\mathcal{S}_{\lambda,\omega}^{-1}(\theta)\rho_\lambda(\mathcal{S}_{\lambda,\omega}^{-1}(\theta))\right) \;=\; \rho_\lambda(\theta)\,, \qquad (4.2)$$

which with equation (3.4) gives, with error terms of order $\mathcal{O}(\lambda^3)$,

$$\rho_\lambda - \lambda\,\partial_\theta\left(\mathbf{E}(p_\sigma)\rho_\lambda\right) + \lambda^2\,\frac{1}{2}\,\partial_\theta\left(\mathbf{E}(p_\sigma^2)\,\partial_\theta\rho_\lambda + \mathbf{E}(p_\sigma\partial_\theta p_\sigma)\,\rho_\lambda - 2\,\mathbf{E}(q_\sigma)\,\rho_\lambda\right) \;=\; \rho_\lambda\,.$$

We first consider an anomaly of first degree. As $\mathbf{E}(P_\sigma) \neq 0$, it follows that $\mathbf{E}(p_\sigma)$ is not vanishing identically. Therefore the above perturbative equation is non-trivial to first order in λ, hence $\mathbf{E}(p_\sigma)\rho_0$ should be constant. If now the first degree anomaly is elliptic, then $\det \mathbf{E}(P_\sigma) > 0$ which can easily be seen to be equivalent to $|\mathbf{E}(\alpha_\sigma)| > |\mathbf{E}(\beta_\sigma)|$, which in turn is equivalent to the fact that $\mathbf{E}(p_\sigma(\theta))$ does not vanish for any $\theta \in S^1$. For an elliptic anomaly of first degree, the lowest order of the invariant measure is therefore

$$\rho_0 \;=\; \frac{c}{\mathbf{E}(p_\sigma)} \;,$$

with an adequate normalization constant $c \in \mathbb{R}$. If on the other hand, the anomaly is hyperbolic (resp. parabolic), then $\mathbf{E}(p_\sigma)$ has four (resp. two) zeros on S^1. In this situation, the only possible (formal) solution is that ρ_0 is given by Dirac peaks on these zeros (which is, of course, only formal because the invariant measure is known to be Hölder continuous). In Section 5.2, we shall see that for the calculation of certain expectation values w.r.t. the invariant measure, it looks as if it were given by a sum of two Dirac peaks, concentrated on the stable fixed points of the averaged phase shift dynamics. These fixed points are two of the zeros of $\mathbf{E}(p_\sigma)$.

Next we consider an anomaly of second degree. As then $\mathbf{E}(p_\sigma) = 0$, it follows that the equation for the lowest order of the invariant measure is

$$\frac{1}{2}\,\partial_\theta \left(\mathbf{E}(p_\sigma^2)\,\partial_\theta \rho_0 \;+\; \mathbf{E}(p_\sigma \partial_\theta p_\sigma)\,\rho_0 \;-\; 2\,\mathbf{E}(q_\sigma)\,\rho_0 \right) \;=\; 0\,.$$

Now $\mathbf{E}(p_\sigma^2) > 0$ unless $\mathbf{p}$-almost all p_σ vanish simultaneously for some θ, a (rare) situation which is excluded throughout the present work. Then this is an analytic Fokker-Planck equation on the unit circle and it can be written as $\mathcal{L}\,\rho_0 = 0$ where $\mathcal{L}$ is by definition the Fokker-Planck operator. Its spectrum contains the simple eigenvalue 0 with eigenvector given by the (lowest order of the) invariant measure ρ_0 calculated next. Indeed,

$$\frac{1}{2}\,\mathbf{E}(p_\sigma^2)\,\partial_\theta \rho_0 \;+\; \frac{1}{2}\,\mathbf{E}(p_\sigma \partial_\theta p_\sigma)\,\rho_0 \;-\; \mathbf{E}(q_\sigma)\,\rho_0 \;=\; C\,,$$

where the real constant C has to be chosen such that the equation admits a positive, 2π-periodic and normalized solution ρ_0. It is a routine calculation to determine the solution using the method of variation of the constants. Setting

$$\kappa(\theta) \;=\; \int_0^\theta d\theta'\,\frac{2\,\mathbf{E}(q_\sigma(\theta'))}{\mathbf{E}(p_\sigma^2(\theta'))}\,, \qquad K(\theta) \;=\; \int_0^\theta d\theta'\,2\,\mathbf{E}(p_\sigma^2(\theta'))^{-\frac{1}{2}}\,e^{-\kappa(\theta')}\,,$$

and

$$C \;=\; \frac{e^{-\kappa(2\pi)} - 1}{K(2\pi)}\,,$$

it is given by

$$\rho_0(\theta) \;=\; \frac{c\,e^{\kappa(\theta)}}{\mathbf{E}(p_\sigma^2(\theta))^{\frac{1}{2}}}\,(C\,K(\theta) + 1)\,, \tag{4.3}$$

where c is a normalization constant. It is important to note at this point that $\rho_0(\theta)$ is an analytic function of θ. Furthermore, let us remark that the normalization constant eliminates the arbitrariness of the splitting of λ and P_σ in (2.2).

The rest of the spectrum of $\mathcal{L}$ is discrete ($\mathcal{L}$ has a compact resolvent), at most twice degenerate and has a strictly negative real part, all facts that can be proven as indicated in [7]. As already stated in the introduction, we do not need to use this spectral information directly.

5. The Lyapunov exponent

The asymptotic behavior of the products of the random sequence of matrices $(T_{\lambda,\sigma_n})_{n\geq 1}$ is characterized by the Lyapunov exponent [1, A.III.3.4]

$$\gamma(\lambda) \;=\; \lim_{N\to\infty} \frac{1}{N}\, \mathbf{E}\, \log\left(\left\|\prod_{n=1}^{N} T_{\lambda,\sigma_n}\, e_\theta\right\|\right)\,, \tag{5.1}$$

where θ is an arbitrary initial condition. One may also average over θ w.r.t. an arbitrary continuous measure before taking the limit [5, Lemma 3]. A result of Furstenberg states a criterion for having a positive Lyapunov exponent [1]. A quantitative control of the Lyapunov exponent in the vicinity of a critical point is given in [8, Proposition 1], however, only in the case where the critical point is not an anomaly of first or second order. The latter two cases are dealt with in the present work.

Let us first suppose that the anomaly is of first order. Because the boundary terms vanish in the limit, it is possible to use the matrices $MT_{\lambda,\sigma_n}M^{-1}$ instead of T_{λ,σ_n} in (5.1). Furthermore, the random dynamical system (3.2) allows to expand (5.1) into a telescopic sum:

$$\gamma(\lambda) \;=\; \lim_{N\to\infty} \frac{1}{N}\sum_{n=0}^{N-1} \mathbf{E}\, \log\left(\left\|MT_{\lambda,\sigma_{n+1}}M^{-1}\, e_{\mathcal{S}^n_{\lambda,\omega}(\theta)}\right\|\right)\,. \tag{5.2}$$

Up to terms of order $\mathcal{O}(\lambda^3)$, we can expand each contribution of this Birkhoff sum:

$$\log\left(\left\|MT_{\lambda,\sigma}M^{-1}\, e_\theta\right\|\right)$$
$$= \lambda\,\langle e_\theta|P_\sigma|e_\theta\rangle \;+\; \frac{1}{2}\,\lambda^2\Big(\langle e_\theta|(|P_\sigma|^2 + Q_\sigma + P_\sigma^2)|e_\theta\rangle \;-\; 2\,\langle e_\theta|P_\sigma|e_\theta\rangle^2\Big)$$
$$= \frac{1}{2}\,\Re e\left[2\,\lambda\,\beta_\sigma\, e^{2\imath\theta} \;+\; \lambda^2\Big(|\beta_\sigma|^2 \;+\; \langle\overline{v}|(|P_\sigma|^2 + 2\,Q_\sigma)|v\rangle e^{2\imath\theta} \;-\; \beta_\sigma^2\, e^{4\imath\theta}\Big)\right],$$

where we used the identity

$$\langle e_\theta|T|e_\theta\rangle \;=\; \frac{1}{2}\,\mathrm{Tr}(T) \;+\; \Re e\left(\langle\overline{v}|T|v\rangle\, e^{2\imath\theta}\right)\,,$$

holding for any real matrix T, as well as $\mathrm{Tr}(P_\sigma) = \mathrm{Tr}(Q_\sigma) = 0$ and $\mathrm{Tr}(|P_\sigma|^2 + P_\sigma^2) = 4\,|\beta_\sigma|^2$. Let us set

$$I_j(N) \;=\; \frac{1}{N}\sum_{n=0}^{N-1} \mathbf{E}\big(e^{2\imath j \mathcal{S}^n_{\lambda,\omega}(\theta)}\big)\,, \qquad j = 1, 2\,, \tag{5.3}$$

and introduce I_j by $I_j(N) = I_j + \mathcal{O}(\lambda)$ for N sufficiently large. We therefore obtain for an anomaly of first order with errors of order $\mathcal{O}(\lambda^3)$,

$$\gamma(\lambda) \;=\; \frac{1}{2}\,\mathbf{E}\,\Re e\left(2\,\lambda\,\beta_\sigma\,I_1 \;+\; \lambda^2\left(|\beta_\sigma|^2 \;+\; \langle\overline{v}|(|P_\sigma|^2 + 2\,Q_\sigma)|v\rangle\,I_1 \;-\; \beta_\sigma^2\,I_2\right)\right). \tag{5.4}$$

For an anomaly of second order, one regroups the contributions pairwise as in Definition 2.1, namely works with the family $(T_{\lambda,\hat\sigma})_{\lambda\in\mathbb{R},\hat\sigma\in\Sigma^2}$ where $T_{\lambda,\hat\sigma} = T_{\lambda,\sigma(2)}T_{\lambda,\sigma(1)}$ for $\hat\sigma = (\sigma(2),\sigma(1))$, furnished with the probability measure $\hat{\mathbf{p}} = \mathbf{p}^{\times 2}$. This family has an anomaly of first order, and its Lyapunov exponent is exactly twice that of the initial family $(T_{\lambda,\sigma})_{\lambda\in\mathbb{R},\sigma\in\Sigma}$. It is hence sufficient to study anomalies of first order.

5.1. Elliptic first degree anomaly

Let us consider the matrix $\mathbf{E}(\partial_\lambda T_{\lambda,\sigma}|_{\lambda=0}) \in \mathrm{sl}(2,\mathbb{R})$. For an elliptic anomaly, the determinant of this matrix is positive. Its eigenvalues are therefore a complex conjugate pair $\pm\imath\frac{\eta}{2}$, so that there exists a basis change $M \in \mathrm{SL}(2,\mathbb{R})$ such that

$$\mathbf{E}\big(M\partial_\lambda T_{\lambda,\sigma}|_{\lambda=0}M^{-1}\big) \;=\; \mathbf{E}(P_\sigma) \;=\; \frac{1}{2}\begin{pmatrix} 0 & -\eta \\ \eta & 0 \end{pmatrix}. \tag{5.5}$$

It follows that $\mathbf{E}(\alpha_\sigma) = \imath\frac{\eta}{2}$ and $\mathbf{E}(\beta_\sigma) = 0$, implying $\gamma(\lambda)\mathcal{O}(\lambda^2)$. Furthermore, from (3.3),

$$e^{2\imath j\mathcal{S}_{\lambda,\sigma}(\theta)} \;=\; \big(1 \,+\, 2\imath j\,\lambda\,\Im m(\alpha_\sigma \,-\, \beta_\sigma\,e^{2\imath\theta})\big)\,e^{2\imath j\theta} \;+\; \mathcal{O}(\lambda^2)\,.$$

Hence

$$I_j(N) \;=\; \frac{1}{N}\sum_{n=0}^{N-1}(1 + \imath j\lambda\eta)\,\mathbf{E}\big(e^{2\imath j\mathcal{S}^{n-1}_{\lambda,\omega}(\theta)}\big) \;+\; \mathcal{O}(\lambda^2)$$

$$=\; (1 + \imath j\lambda\eta)\,I_j(N) \;+\; \mathcal{O}\big(N^{-1},\lambda^2\big)\,,$$

so that $I_j(N) = \mathcal{O}(\lambda,(N\lambda)^{-1})$ and $I_j = 0$. Replacing into (5.4), this leads to:

Proposition 5.1. *If $\lambda = 0$ is an elliptic anomaly of first order and first degree, then*

$$\gamma(\lambda) \;=\; \frac{1}{2}\,\lambda^2\,\mathbf{E}(|\beta_\sigma|^2) \;+\; \mathcal{O}(\lambda^3)\,.$$

In order to calculate β_σ in an application, one first has to determine the basis change (5.5), then P_σ before deducing β_σ and $\mathbf{E}(|\beta_\sigma|^2)$, see Section 6 for two examples. Let us note that after the basis change, (5.5) implies $\mathbf{E}(p_\sigma(\theta)) = \eta$. Hence the invariant measure is to lowest order given by the Lebesgue measure after the basis change.

The term 'elliptic' indicates that the mean dynamics at the anomaly is to lowest order a rotation. For a hyperbolic anomaly it is an expansion in a given direction and a contraction into another one. These directions (*i.e.*, angles) can be chosen to our convenience through the basis change M, as will be done next.

5.2. Hyperbolic first degree anomaly

For a hyperbolic anomaly of first degree, the eigenvalues of $\mathbf{E}(\partial_\lambda T_{\lambda,\sigma}|_{\lambda=0})$ are $\pm\frac{\mu}{2}$ and there exists $M \in \mathrm{SL}(2,\mathbb{R})$ such that

$$\mathbf{E}\big(M\partial_\lambda T_{\lambda,\sigma}|_{\lambda=0}M^{-1}\big) \;=\; \mathbf{E}(P_\sigma) \;=\; \frac{1}{2}\left(\begin{array}{cc} \mu & 0 \\ 0 & -\mu \end{array}\right). \tag{5.6}$$

It follows that $\mathbf{E}(\alpha_\sigma) = 0$ and $\mathbf{E}(\beta_\sigma) = \frac{\mu}{2}$, so that (5.4) leads to

$$\gamma(\lambda) \;=\; \frac{1}{2}\,\lambda\mu\,\Re e(I_1) \;+\; \mathcal{O}(\lambda^2)\,.$$

We now need to evaluate I_1. Introducing the reference dynamics $\tilde{\mathcal{S}}_\lambda(\theta) = \theta - \frac{1}{2}\lambda\mu\sin(2\theta)$ as well as the centered perturbation $r_\sigma(\theta) = \Im m(\alpha_\sigma - (\beta_\sigma - \frac{\mu}{2})e^{2\imath\theta})$, it follows from (3.3) that the phase shift dynamics is

$$\mathcal{S}_{\lambda,\sigma}(\theta) \;=\; \tilde{\mathcal{S}}_\lambda(\theta) \;+\; \lambda\,r_\sigma(\theta) \;+\; \mathcal{O}(\lambda^2)\,. \tag{5.7}$$

The non-random dynamics $\tilde{\mathcal{S}}_\lambda$ has four fixed points, $\theta = 0, \frac{\pi}{2}, \pi, \frac{3\pi}{2}$. If $\lambda\mu > 0$, $\theta = 0$ and π are stable, while $\theta = \frac{\pi}{2}$ and $\frac{3\pi}{2}$ are unstable. For $\lambda\mu < 0$, the roles are exchanged, and this case will not be considered here. Unless the initial condition is the unstable fixed point, one has $\tilde{\mathcal{S}}_\lambda^n(\theta) \to 0$ or π as $n \to \infty$. Furthermore, if θ is not within a $\mathcal{O}(\lambda^{\frac{1}{2}})$-neighborhood of an unstable fixed point, it takes $n = \mathcal{O}(\lambda^{-\frac{3}{2}})$ iterations of $\tilde{\mathcal{S}}_\lambda$ in order to attain a $\mathcal{O}(\lambda^{\frac{1}{2}})$-neighborhood of 0. We also need to expand iterations of $\hat{\mathcal{S}}_\lambda$:

$$\tilde{\mathcal{S}}_\lambda^k\big(\theta + \lambda\,r_\sigma(\theta') + \mathcal{O}(\lambda^2)\big) \;=\; \tilde{\mathcal{S}}_\lambda^k(\theta) \;+\; \lambda\,\partial_\theta\tilde{\mathcal{S}}_\lambda^k(\theta)\,r_\sigma(\theta') \;+\; \mathcal{O}(\lambda^2)\,,$$

where the corrective term $\mathcal{O}(\lambda^2)$ on the r.h.s. is bounded uniformly in k as one readily realizes when thinking of the dynamics induced by $\tilde{\mathcal{S}}_\lambda$. Furthermore, $\partial_\theta\tilde{\mathcal{S}}_\lambda^k(\theta) = \mathcal{O}(1)$ uniformly in k. Iteration thus shows:

$$\mathcal{S}_{\lambda,\omega}^n(\theta) \;=\; \tilde{\mathcal{S}}_\lambda^n(\theta) \;+\; \lambda\sum_{k=1}^{n}\partial_\theta\tilde{\mathcal{S}}_\lambda^{n-k}\big(\tilde{\mathcal{S}}_\lambda(\mathcal{S}_{\lambda,\omega}^{k-1}(\theta))\big)\,r_{\sigma_k}(\mathcal{S}_{\lambda,\omega}^{k-1}(\theta)) \;+\; \mathcal{O}(n\lambda^2)\,.$$

Let us denote the coefficient in the sum over k by s_k. Then s_k is a random variable that depends only on σ_l for $l \leq k$. Moreover, s_k is centered when a conditional expectation over σ_k is taken. Taking successively conditional expectations thus shows

$$\mathbf{E}\big(e^{2\imath\mathcal{S}_{\lambda,\omega}^n(\theta)}\big) \;=\; e^{2\imath\tilde{\mathcal{S}}_\lambda^n(\theta)}\left(\mathbf{E}\big(e^{2\imath\lambda\sum_{k=1}^{n-1}s_k + \mathcal{O}(n\lambda^2)}\big) + \mathcal{O}(\lambda^2)\right)$$

$$\phantom{\mathbf{E}\big(e^{2\imath\mathcal{S}_{\lambda,\omega}^n(\theta)}\big)} \;=\; e^{2\imath\tilde{\mathcal{S}}_\lambda^n(\theta)} \;+\; \mathcal{O}(n\lambda^2)\,. \tag{5.8}$$

Choosing $n = \lambda^{-\frac{3}{2}}$ gives according to the above

$$\mathbf{E}\big(e^{2\imath\mathcal{S}_{\lambda,\omega}^n(\theta)}\big) \;=\; 1 \;+\; \mathcal{O}(\lambda^{\frac{1}{2}})\,, \tag{5.9}$$

unless θ is within a $\mathcal{O}(\lambda^{\frac{1}{2}})$-neighborhood of $\frac{\pi}{2}$ or $\frac{3\pi}{2}$. In the latter cases, an elementary argument based on the central limit theorem shows that it takes of order $\mathbf{E}(r_\sigma(\frac{\pi}{2})^2)\lambda^{-\frac{3}{2}}$ iterations to diffuse out of these regions left out before. Supposing that one does not have $r_\sigma(\frac{\pi}{2}) = 0$ for $\mathbf{p}$-almost all σ, one can conclude that (5.9) holds for all initial conditions θ. Consequently $I_1 = 1 + \mathcal{O}(\lambda^{\frac{1}{2}})$ so that:

Proposition 5.2. *If $\lambda = 0$ is an hyperbolic anomaly of first order and first degree, and $r_\sigma(\frac{\pi}{2})$ does not vanish for $\mathbf{p}$-almost all σ, one has*

$$\gamma(\lambda) \;=\; \frac{1}{2}\,|\lambda\mu| \;+\; \mathcal{O}(\lambda^{\frac{3}{2}})\,.$$

The argument above shows that the random phase dynamics is such that the angles $\mathcal{S}^n_{\lambda,\omega}(\theta)$ are for most n and ω in a neighborhood of size $\lambda^{\frac{1}{2}}$ of the stable fixed points $\theta = 0, \pi$. This does not mean that for some n and ω, the angles are elsewhere; in particular, the rotation number of the dynamics does not vanish. However, this leads to corrections which do not enter into the lowest order term for the Lyapunov exponent.

5.3. Parabolic first degree anomaly

This may seem like a pathological and exceptional case. It turns out to be the mathematically most interesting anomaly of first degree, though, and its analysis is similar to that of the Lyapunov exponent at the band edge (this will be discussed elsewhere). First of all, at a parabolic anomaly of first degree, there exists $M \in \mathrm{SL}(2, \mathbb{R})$ allowing to attain the Jordan normal form:

$$\mathbf{E}\big(M\partial_\lambda T_{\lambda,\sigma}|_{\lambda=0}M^{-1}\big) \;=\; \mathbf{E}(P_\sigma) \;=\; \begin{pmatrix} 0 & 1 \\ 0 & 0 \end{pmatrix}\,. \tag{5.10}$$

Thus $\mathbf{E}(\alpha_\sigma) = \mathbf{E}(\beta_\sigma) = -\imath\,\frac{1}{2}$ so that $\gamma(\lambda)\frac{1}{2}\,\lambda\,\Im m(I_1) + \mathcal{O}(\lambda^2)$. Introducing the reference dynamics $\hat{\mathcal{S}}_\lambda(\theta) = \theta + \frac{\lambda}{2}(\cos(2\theta) - 1)$ as well as the centered perturbation $r_\sigma(\theta) = \Im m(\alpha_\sigma + \frac{\imath}{2} - (\beta_\sigma + \frac{\imath}{2})\,e^{2\imath\theta})$, the dynamics can then be decomposed as in (5.7). Moreover, the argument leading to (5.8) directly transposes to the present case. However, this does not allow to calculate the leading order contribution, but shows that $\gamma(\lambda) = \mathcal{O}(\lambda^{\frac{3}{2}})$.

5.4. Second degree anomaly

At a second degree anomaly one has $\mathbf{E}(\beta_\sigma) = 0$, so in order to calculate the lowest order of the Lyapunov exponent one needs to, according to (5.4), evaluate I_1 and I_2. For this purpose let us introduce an analytic change of variables $Z : S^1 \to S^1$ using the density ρ_0 given in (4.3):

$$\hat{\theta} \;=\; Z(\theta) \;=\; \int_0^\theta d\theta'\, \rho_0(\theta')\,.$$

According to Section 4, one expects that the distribution of $\hat{\theta}$ is the Lebesgue measure. We will only need to prove that this holds perturbatively in a weak sense when integrating analytic functions.

We need to study the transformed dynamics $\hat{\mathcal{S}}_{\lambda,\sigma} = Z \circ \mathcal{S}_{\lambda,\sigma} \circ Z^{-1}$ and write it again in the form:

$$\hat{\mathcal{S}}_{\lambda,\sigma}(\hat{\theta}) \;=\; \hat{\theta} \,+\, \lambda \hat{p}_\sigma(\hat{\theta}) \,+\, \lambda^2 \hat{q}_\sigma(\hat{\theta}) \,+\, \frac{1}{2}\lambda^2\, \hat{p}_\sigma(\hat{\theta})\, \partial_{\hat{\theta}}\,\hat{p}_\sigma(\hat{\theta}) \,+\, \mathcal{O}(\lambda^3)\,. \qquad (5.11)$$

As, $Z \circ \mathcal{S}_{\lambda,\sigma}(\theta)$ is up to order $\mathcal{O}(\lambda^3)$ equal to

$$Z(\theta) \,+\, \left(\lambda\, p_\sigma(\theta) + \frac{1}{2}\lambda^2\, p_\sigma(\theta)\, \partial_\theta p_\sigma(\theta) + \lambda^2\, q_\sigma(\theta)\right) \partial_\theta Z(\theta) \,+\, \frac{1}{2}\lambda^2\, p_\sigma^2\, \partial_\theta^2 Z(\theta)\,,$$

one deduces from

$$\partial_\theta Z \;=\; \rho_0\,, \qquad \partial_\theta^2 Z \;=\; \frac{2}{\mathbf{E}(p_\sigma^2)}\left(C \,-\, \frac{1}{2}\,\mathbf{E}(p_\sigma \partial_\theta p_\sigma)\,\rho_0 \,+\, \mathbf{E}(q_\sigma)\,\rho_0\right)\,,$$

that, with $\theta = Z^{-1}(\hat{\theta})$,

$$\hat{p}_\sigma(\hat{\theta}) \;=\; p_\sigma(\theta)\,\rho_0(\theta)\,, \qquad \hat{q}_\sigma(\hat{\theta}) \;=\; q_\sigma(\theta)\,\rho_0(\theta)\,.$$

A short calculation shows that the expectation values satisfy

$$\mathbf{E}(\hat{p}_\sigma(\hat{\theta})) \;=\; 0\,, \qquad \mathbf{E}(\hat{q}_\sigma(\hat{\theta})) \;=\; \frac{1}{2}\,\mathbf{E}\left(\hat{p}_\sigma(\hat{\theta})\,\partial_{\hat{\theta}}\,\hat{p}_\sigma(\hat{\theta})\right) - C\,, \qquad (5.12)$$

where C is as in (4.3).

Given any analytic function $\hat{f}$ on S^1, let us introduce its Fourier coefficients

$$\hat{f}(\hat{\theta}) \;=\; \sum_{m\in\mathbb{Z}} \hat{f}_m\, e^{\imath m\hat{\theta}}\,, \qquad \hat{f}_m \;=\; \int_0^{2\pi} \frac{d\hat{\theta}}{2\pi}\,\hat{f}(\hat{\theta})\, e^{-\imath m\hat{\theta}}\,.$$

There exist $a, \xi > 0$ such that $\hat{f}_m \le a\,e^{-\xi|m|}$. We are interested in

$$\hat{I}_{\hat{f}}(N) \;=\; \frac{1}{N}\,\mathbf{E}\sum_{n=0}^{N-1} \hat{f}(\hat{\theta}_n)\,,$$

where, for sake of notational simplicity, we introduced $\hat{\theta}_n = \hat{\mathcal{S}}_{\lambda,\omega}^n(\hat{\theta})$ for iterations defined just as in (3.2).

Lemma 5.3. *Suppose* $\mathbf{E}(p_\sigma^2) > 0$ *and that* $\hat{f}$ *is analytic. Then*

$$\hat{I}_{\hat{f}}(N) \;=\; \hat{f}_0 \,+\, \mathcal{O}(\lambda, (\lambda^2 N)^{-1})\,,$$

with an error that depends on $\hat{f}$.

Proof. Set $\hat{r} = \frac{1}{2}\mathbf{E}(\hat{p}_\sigma^2)$. This is an analytic function which is strictly positive on S^1. Furthermore let $\hat{F}$ be an auxiliary analytic function with Fourier coefficients $\hat{F}_m$. Then, using (5.11) and the identities (5.12), we deduce that $\hat{I}_{\hat{F}}(N)$ is equal to

$$\frac{1}{N}\,\mathbf{E}\sum_{n=0}^{N-1}\sum_{m\in\mathbb{Z}} \hat{F}_m\, e^{\imath m\hat{\theta}_n}\left(1 - \imath m C\lambda^2 + m\lambda^2(\imath\partial_{\hat{\theta}} - m)\hat{r}(\hat{\theta}_n) + \mathcal{O}(m^3\lambda^3, N^{-1})\right)$$

As the term of order $\mathcal{O}(1)$ gives back $\hat{I}_{\hat{F}}(N)$, we conclude after summing up the error terms:

$$\frac{\lambda^2}{N}\,\mathbf{E}\,\sum_{n=0}^{N-1}\sum_{m\in\mathbb{Z}}\hat{F}_m\,e^{\imath m\hat{\theta}_n}\left(-\imath m C\lambda^2 + m\lambda^2(\imath\partial_{\hat{\theta}} - m)\hat{r}(\hat{\theta}_n)\right) = \mathcal{O}(\lambda^3, N^{-1}).$$

If a prime denotes a derivative, we therefore deduce for any analytic function $\hat{F}$

$$\hat{I}_{-C\hat{F}'+(\hat{r}\hat{F}')'}(N) = \mathcal{O}(\lambda, (\lambda^2 N)^{-1})\,. \tag{5.13}$$

Now, extracting the constant term, it is clearly sufficient to show $\hat{I}_{\hat{f}}(N) = \mathcal{O}(\lambda, (\lambda^2 N)^{-1})$ for an analytic function $\hat{f}$ with $\hat{f}_0 = 0$. But for such an $\hat{f}$ one can solve the equation

$$\hat{f} = -C\hat{F}' + (\hat{r}\hat{F}')' \tag{5.14}$$

for an analytic and periodic function $\hat{F}$ and then conclude due to (5.13). Indeed, by the method of variation of constants one can always solve (5.14) for an analytic $\hat{F}'$. This then has an antiderivative $\hat{F}$ as long as $\hat{F}'$ does not have a constant term, $i.e.$, the zeroth order Fourier coefficient of the solution of (5.14) vanishes. Integrating (5.14) w.r.t. $\hat{\theta}$, one sees that this is precisely the case when $\hat{f}_0 = 0$ as long as $C \neq 0$. If on the other hand $C = 0$, then (5.14) can be integrated once, and the antiderivative $\int \hat{f}$ of $\hat{f}$ chosen such that $\hat{r}^{-1}\int \hat{f}$ does not have a constant term. Then a second antiderivative can be taken, giving the desired function $\hat{F}$ in this case. $\qquad\square$

In order to use this result for the evaluation of $I_j(N)$ defined in (5.3), let us note that

$$I_j(N) = \frac{1}{N}\,\mathbf{E}\,\sum_{n=0}^{N-1} e^{2\imath j Z^{-1}(\hat{\theta}_n)}\,.$$

Hence up to corrections the result is given by the zeroth order Fourier coefficient of the analytic function $\hat{f}(\hat{\theta}) = e^{2\imath j Z^{-1}(\hat{\theta})}$, so that after a change of variables one gets:

$$I_j(N) = \int_0^{2\pi}\frac{d\theta}{2\pi}\,\rho_0(\theta)\,e^{2\imath j\theta} + \mathcal{O}(\lambda, (\lambda^2 N)^{-1})\,.$$

Proposition 5.4. *If* $\lambda = 0$ *is an anomaly of first order and second degree and* $\mathbf{E}(p_\sigma^2) > 0$, *then one has, with* ρ_0 *given by (4.3) and up to errors of order* $\mathcal{O}(\lambda^3)$:

$$\gamma(\lambda) = \frac{\lambda^2}{2}\,\Re e\int_0^{2\pi}\frac{d\theta}{2\pi}\,\rho_0(\theta)\left[\mathbf{E}(|\beta_\sigma|^2) + \mathbf{E}(\langle\bar{v}|(|P_\sigma|^2 + 2\,Q_\sigma)|v\rangle)\,e^{2\imath\theta} - \mathbf{E}(\beta_\sigma^2)\,e^{4\imath\theta}\right]\,.$$

In the above, anomalies of first degree were classified into elliptic, hyperbolic and parabolic. Second degree anomalies should be called 'diffusive' (strictly if $\mathbf{E}(p_\sigma^2) > 0$). The random dynamics of the phases is diffusive on S^1, with a varying diffusion coefficient and furthermore submitted to a mean drift, also varying with the position. It does not seem possible to transform this complex situation into a simple normal form by an adequate basis change M.

Let us note that the Lyapunov exponent at an anomaly does depend on the higher order term Q_σ in the expansion (2.2), while away from an anomaly it does not depend on Q_σ [8]. Of course, the coefficient of λ^2 in Proposition 5.4 cannot be negative. Up to now, no general argument could be found showing this directly (a problem that was solved in [8, Proposition 1] away from anomalies). For this purpose, it might be of help to choose an adequate basis change M.

6. Examples

6.1. Center of band of the Anderson model

The transfer matrices of the Anderson model are given by

$$T_{\lambda,\sigma} \;=\; \begin{pmatrix} \lambda\,v_\sigma - E & -1 \\ 1 & 0 \end{pmatrix}, \tag{6.1}$$

where v_σ is a real random variable and $E \in \mathbb{R}$ is the energy. The band center is given at $E = 0$. In order to study the behavior of the Lyapunov exponent at its vicinity, we set $E = \epsilon\lambda^2$ for some fixed $\epsilon \in \mathbb{R}$. Then the associated family of i.i.d. random matrices has an anomaly of second order because

$$\begin{aligned}
T_{\lambda,\hat\sigma} \;&=\; T_{\lambda,\sigma_2} T_{\lambda,\sigma_1} \\
&=\; -\,\exp\left(\lambda \begin{pmatrix} 0 & v_{\sigma_2} \\ -v_{\sigma_1} & 0 \end{pmatrix} + \lambda^2 \begin{pmatrix} -\frac{1}{2} v_{\sigma_1} v_{\sigma_2} & -\epsilon \\ \epsilon & \frac{1}{2} v_{\sigma_1} v_{\sigma_2} \end{pmatrix} \right),
\end{aligned}$$

where $\hat\sigma = (\sigma_2,\sigma_1)$ and errors of order $\mathcal{O}(\lambda^3)$ were neglected. It follows that $\alpha_{\hat\sigma} = \frac{1}{2i}(v_{\sigma_2} + v_{\sigma_1})$ and $\beta_{\hat\sigma} = \frac{1}{2i}(v_{\sigma_2} - v_{\sigma_1})$. If now v_σ is not centered, then one has an elliptic anomaly of first degree and Proposition 5.1, combined with the factor $1/2$ due to the order of the anomaly, implies directly (no basis change needed here) that

$$\gamma(\lambda) \;=\; \frac{1}{8} \lambda^2 \left(\mathbf{E}(v_\sigma^2) - \mathbf{E}(v_\sigma)^2 \right) + \mathcal{O}(\lambda^3) \,.$$

If, on the other hand, v_σ is centered, one has a second degree anomaly and can apply Proposition 5.4. One readily verifies that $\mathbf{E}(|\beta_{\hat\sigma}|^2) = \frac{1}{2}\mathbf{E}(v_\sigma^2)$ and $\mathbf{E}(\beta_{\hat\sigma}^2) = -\frac{1}{2}\mathbf{E}(v_\sigma^2)$, and furthermore that the second term in Proposition 5.4 always vanishes, so that

$$\gamma(\lambda) \;=\; \frac{1}{8} \lambda^2\, \mathbf{E}(v_\sigma^2) \int_0^{2\pi} \frac{d\theta}{2\pi}\, \rho_0(\theta)\, (1 + \cos(4\theta)) + \mathcal{O}(\lambda^3) \,,$$

which is strictly positive unless v_σ vanishes identically (it was already supposed to be centered). If one wants to go further, $\mathbf{E}(|p_{\hat\sigma}(\theta)|^2) = \frac{1}{2}\mathbf{E}(v_\sigma^2)(1 + \cos^2(2\theta))$. In the case $\epsilon = 0$, one then has $\mathbf{E}(Q_{\hat\sigma}) = 0$ so that $\rho_0(\theta) = c(1 + \cos^2(2\theta))^{-\frac{1}{2}}$ with a normalization constant c that can be calculated by a contour integration. This proves the formula given in [4].

6.2. A particular random dimer model

In the random dimer model, the transfer matrix is given by the square of (6.1), with a potential that can only take two values $\lambda v_\sigma = \sigma v$ where $v \in \mathbb{R}$ and $\sigma \in \{-1, 1\}$ (*e.g.*, [5]). Hence for a given energy $E \in \mathbb{R}$ it is

$$T_\sigma^E = \begin{pmatrix} \sigma\,v - E & -1 \\ 1 & 0 \end{pmatrix}^2 .$$

Now the energy is chosen to be $E = v + \lambda$. Then $T_{\lambda,\sigma} = T_\sigma^{v+\lambda}$ has a critical point if $|v| < 2$. For the particular value $v = \frac{1}{\sqrt{2}}$ fixing hence a special type of dimer model, one has:

$$
\begin{aligned}
T_{\lambda,\sigma} &= \begin{pmatrix} \frac{1}{\sqrt{2}}\,(\sigma - 1) - \lambda & -1 \\ 1 & 0 \end{pmatrix}^2 \\
&= \begin{pmatrix} -\sigma - \sqrt{2}\,(\sigma - 1)\,\lambda & \frac{1}{\sqrt{2}}\,(1 - \sigma) + \lambda \\ \frac{1}{\sqrt{2}}\,(\sigma - 1) - \lambda & -1 \end{pmatrix} + \mathcal{O}(\lambda^2) .
\end{aligned}
$$

This family has now an anomaly of second order and first degree because

$$
(T_{\lambda,\sigma})^2 = \sigma \, \exp\left(\lambda \begin{pmatrix} \sqrt{2}\,(\sigma - 1) & -3 + \sigma \\ 3 - \sigma & -\sqrt{2}\,(\sigma - 1) \end{pmatrix} + \mathcal{O}(\lambda^2) \right) .
$$

One readily verifies that the determinant of $\mathbf{E}(M^{-1} P_\sigma M)$ is equal to $7 - 2\mathbf{E}(\sigma) - \mathbf{E}(\sigma)^2$ and hence positive so that the anomaly is elliptic. Therefore the general result of Section 5.1 can be applied. Let us set $e = \mathbf{E}(\sigma)$. The adequate basis change (without normalization of the determinant) is

$$
M = \begin{pmatrix} \sqrt{7 - 2e - e^2} & 0 \\ \sqrt{2}(1 - e) & 3 - e \end{pmatrix} .
$$

A calculation then gives

$$
P_\sigma = \frac{1}{3 - e} \begin{pmatrix} 2\sqrt{2}(\sigma - e) & (\sigma - 3)\sqrt{7 - 2e - e^2} \\ \frac{4(1-e)(\sigma-e)-(e-3)(7-\sigma-e-e^2)}{\sqrt{7-2e-e^2}} & -2\sqrt{2}(\sigma - e) \end{pmatrix} ,
$$

allowing to extract β_σ and then $\mathbf{E}(|\beta_\sigma|^2)$, leading to (this contains a factor $1/2$ because the anomaly is of second order)

$$
\gamma(\lambda) = \lambda^2 \, \frac{2(1 - e^2)}{(3 - e)^2} \left(1 + \frac{2(e - 1)^2}{7 - 2e - e^2} \right) + \mathcal{O}(\lambda^3) .
$$

Note that if $e = 1$ or $e = -1$ so that there is no randomness, the coefficient vanishes. This special case was left out in [5] (the condition $\mathbf{E}(e^{4i\eta}) \neq 1$ in the theorems of [5] is violated). Within the wide class of polymer models discussed in [5], models with all types of anomalies can be constructed, and then be analysed by the techniques of the present work.

Acknowledgement

First of all, I would like to thank the organizers for the invitation to the wonderful and inspiring conference in the convivial atmosphere of Bedlewo, allowing for quite unexpected encounters; second of all, many thanks to my new colleges Andreas Knauf and Hajo Leschke for a few patient discussions on the matters of the present work.

References

[1] P. Bougerol, J. Lacroix, *Products of Random Matrices with Applications to Schrö-dinger Operators*, (Birkhäuser, Boston, 1985).

[2] A. Bovier, A. Klein, *Weak disorder expansion of the invariant measure for the one-dimensional Anderson model*, J. Stat. Phys. **51**, 501–517 (1988).

[3] M. Campanino, A. Klein, *Anomalies in the one-dimensional Anderson model at weak disorder*, Commun. Math. Phys. **130**, 441–456 (1990).

[4] B. Derrida, E.J. Gardner, *Lyapunov exponent of the one dimensional Anderson model: weak disorder expansion*, J. Physique **45**, 1283 (1984).

[5] S. Jitomirskaya, H. Schulz-Baldes, G. Stolz, *Delocalization in random polymer chains*, Commun. Math. Phys. **233**, 27–48 (2003).

[6] M. Kappus, F. Wegner, *Anomaly in the band centre of the one-dimensional Anderson model*, Z. Phys. **B 45**, 15–21 (1981).

[7] H. Risken, *The Fokker-Planck equation*, Second Edition, (Springer, Berlin, 1988).

[8] R. Schrader, H. Schulz-Baldes, A. Sedrakyan, *Perturbative test of single parameter scaling for 1D random media*, Ann. H. Poincaré **5**, 1159–1180 (2004).

[9] C. Shubin, R. Vakilian, T. Wolff, *Some harmonic analysis questions suggested by Anderson-Bernoulli models*, Geom. Funct. Anal. **8**, 932–964 (1998).

Hermann Schulz-Baldes
Mathematisches Institut
Universität Erlangen-Nürnberg
Bismarkstr. $1\frac{1}{2}$
D-91054 Erlangen, Germany
e-mail: `schuba@mi.uni-erlangen.de`

Operator Theory:
Advances and Applications, Vol. 174, 173–186
© 2007 Birkhäuser Verlag Basel/Switzerland

Uniform and Smooth Benzaid-Lutz Type Theorems and Applications to Jacobi Matrices

Luis O. Silva

Abstract. Uniform and smooth asymptotics for the solutions of a parametric system of difference equations are obtained. These results are the uniform and smooth generalizations of the Benzaid-Lutz theorem (a Levinson type theorem for discrete linear systems) and are used to develop a technique for proving absence of accumulation points in the pure point spectrum of Jacobi matrices. The technique is illustrated by proving discreteness of the spectrum for a class of unbounded Jacobi operators.

Mathematics Subject Classification (2000). 39A11, 39A12, 47B36, 47A10.

Keywords. Difference equations, Asymptotics, Jacobi matrices, Discrete Spectrum.

1. Introduction

The asymptotic behavior of solutions of discrete linear systems can be obtained by means of discrete Levinson type theorems [4, 6]. Here we are mainly concerned with asymptotically diagonal linear systems to which the Benzaid-Lutz theorem can be applied.

Consider the system

$$\vec{x}_{n+1} = (\Lambda_n + R_n)\vec{x}_n, \qquad n \geq n_0, \tag{1}$$

where $\vec{x}_n$ is a d-dimensional vector, $\Lambda_n + R_n$ is an invertible $d \times d$ matrix, and $\Lambda_n = \mathrm{diag}\{\nu_n^{(k)}\}_{k=1}^d$. The Benzaid-Lutz theorem [2, 4, 6] asserts that, when the sequence $\{\Lambda_n\}_{n \geq n_0}$ satisfies the Levinson condition for $k = 1, \dots, d$ (see below Def. 2.1) and

$$\sum_{n=n_0}^{\infty} \frac{\|R_n\|}{|\nu_n^{(k)}|} < \infty, \qquad k = 1, \dots, d,$$

Partially supported by project PAPIIT IN 101902, DGAPA-UNAM.

then, there is a basis $\{\vec{x}_n^{(k)}\}_{n \geq n_0}$ $(k = 1, \ldots, d)$ in the space of solutions of (1) such that

$$\left\| \frac{\vec{x}_n^{(k)}}{\prod_{i=n_0}^{n-1} \nu_i^{(k)}} - \vec{e}_k \right\| \to 0 \,, \quad \text{as } n \to \infty \,, \quad \text{for } k = 1, \ldots, d \,,$$

where $\{\vec{e}_k\}_{k=1}^d$ is the canonical basis in $\mathbb{C}^d$. This result has its counterpart for linear systems of ordinary differential equations [3]. Loosely speaking, if the conditions of the Benzaid-Lutz theorem hold, the solutions $\{\vec{x}_n^{(k)}\}_{n \geq n_0}$ of (1) asymptotically behave as the solutions of the unperturbed system

$$\vec{x}_{n+1} = \Lambda_n \vec{x}_n \,, \qquad n \geq n_0 \,.$$

Let us now consider the second order difference equation for the sequence $\{x_n\}_{n=1}^\infty$,

$$b_{n-1} x_{n-1} + q_n x_n + b_n x_{n+1} = \lambda x_n \,, \quad \lambda \in \mathbb{R} \,, \quad n \geq 2 \,, \tag{2}$$

where $\{q_n\}_{n=1}^\infty$ and $\{b_n\}_{n=1}^\infty$ are real sequences and $b_n \neq 0$ for any $n \in \mathbb{N}$. This equation can be written as follows

$$\vec{x}_{n+1} = B_n(\lambda) \vec{x}_n \,, \quad \lambda \in \mathbb{R} \,, \quad n \geq 2 \,. \tag{3}$$

where, $\vec{x}_n := \begin{pmatrix} x_{n-1} \\ x_n \end{pmatrix}$ and $B_n(\lambda) := \begin{pmatrix} 0 & 1 \\ -\frac{b_{n-1}}{b_n} & \frac{\lambda - q_n}{b_n} \end{pmatrix}$. In general, difference equations of order d can be reduced to similar systems with $d \times d$ matrices.

It is well known that the spectral analysis of Jacobi operators having the matrix representation

$$\begin{pmatrix} q_1 & b_1 & 0 & 0 & \cdots \\ b_1 & q_2 & b_2 & 0 & \cdots \\ 0 & b_2 & q_3 & b_3 & \\ 0 & 0 & b_3 & q_4 & \ddots \\ \vdots & \vdots & & \ddots & \ddots \end{pmatrix} \,, \qquad \begin{array}{l} 0 \neq b_n \in \mathbb{R} \,, \ \forall n \in \mathbb{N} \,, \\[2mm] q_n \in \mathbb{R} \,, \ \forall n \in \mathbb{N} \,, \end{array}$$

with respect to the canonical basis in $l_2(\mathbb{N})$, can be carried out on the basis of the asymptotic behavior of the solutions of (2), for example using Subordinacy Theory [5, 8]. In its turn, in certain cases (see Sec. 5), the asymptotics of solutions of (3) (and therefore of (2)) can be obtained by the Benzaid-Lutz theorem applied point-wise with respect to $\lambda \in \mathbb{R}$.

In this paper we obtain sufficient conditions for a parametric Benzaid-Lutz system of the form

$$\vec{x}_{n+1}(\lambda) = (\Lambda_n(\lambda) + R_n(\lambda))\vec{x}_n(\lambda) \,, \qquad n \geq n_0 \,,$$

to have solutions with certain smooth behavior with respect to λ (see Sec. 4). This result, together with a uniform (also with respect to λ) estimate of the asymptotic remainder of solutions of (2) obtained in [12], is used to develop a technique for excluding accumulation points in the pure point spectrum of difference operators. The technique is illustrated in a simple example.

2. Preliminaries

Throughout this work, unless otherwise stated, $\mathfrak{I}$ denotes some real interval. Besides, we shall refer in multiple occasions to the sequence of matrices Λ defined as follows

$$\Lambda := \{\Lambda_n(\lambda)\}_{n=n_0}^{\infty}\,, \quad \text{where } \Lambda_n := \operatorname{diag}\{\nu_n^{(k)}(\lambda)\}_{k=1}^{d}\,, \quad \lambda \in \mathfrak{I}\,. \tag{4}$$

Definition 2.1. The sequence Λ, given by (4), is said to satisfy the Levinson condition for k (denoted $\Lambda \in \mathcal{L}(k)$) if there exist an $N \geq n_0$ such that $\nu_n^{(k)}(\lambda) \neq 0$, for any $n \geq N$ and $\lambda \in \mathfrak{I}$, and if for some constant number $M > 1$, with k being fixed, each j $(1 \leq j \leq d)$ falls into one and only one of the two classes I_1 or I_2, where

(a) $j \in I_1$ if $\forall \lambda \in \mathfrak{I}$

$$\frac{\left|\prod_{i=N}^{n} \nu_i^{(k)}(\lambda)\right|}{\left|\prod_{i=N}^{n} \nu_i^{(j)}(\lambda)\right|} \to \infty \quad \text{as } n \to \infty\,, \quad \text{and}$$

$$\frac{\left|\prod_{i=n}^{n'} \nu_i^{(k)}(\lambda)\right|}{\left|\prod_{i=n}^{n'} \nu_i^{(j)}(\lambda)\right|} > \frac{1}{M}\,, \quad \forall\, n'\,, n \text{ such that } n' \geq n \geq N\,.$$

(b) $j \in I_2$ if $\forall \lambda \in \mathfrak{I}$

$$\frac{\left|\prod_{i=n}^{n'} \nu_i^{(k)}(\lambda)\right|}{\left|\prod_{i=n}^{n'} \nu_i^{(j)}(\lambda)\right|} < M\,, \quad \forall\, n'\,, n \text{ such that } n' \geq n \geq N\,.$$

Definition 2.2. Fix the natural numbers k $(k \leq d)$ and n_1, and assume that $\nu_n^{(k)}(\lambda) \neq 0$, $\forall n \geq n_1$ and $\forall \lambda \in \mathfrak{I}$. Let $X_k(n_1)$ denote the normed space containing all sequences $\vec{\varphi} = \{\vec{\varphi}_n(\lambda)\}_{n=n_1+1}^{\infty}$ of functions defined on $\mathfrak{I}$ and with range in $\mathbb{C}^d$, such that

$$\sup_{n > n_1} \sup_{\lambda \in \mathfrak{I}} \left\{ \|\vec{\varphi}_n(\lambda)\|_{\mathbb{C}^d} \frac{1}{|\prod_{i=n_1}^{n-1} \nu_i^{(k)}(\lambda)|} \right\} < \infty$$

and where the norm is defined by

$$\|\vec{\varphi}\|_{X_k(n_1)} = \sup_{n > n_1} \sup_{\lambda \in \mathfrak{I}} \left\{ \|\vec{\varphi}_n(\lambda)\|_{\mathbb{C}^d} \frac{1}{|\prod_{i=n_1}^{n-1} \nu_i^{(k)}(\lambda)|} \right\}\,. \tag{5}$$

Clearly, $X_k(n_1)$ is complete. It will be also considered the subspace $X_k^0(n_1)$ which contains all functions of $X_k(n_1)$ such that

$$\sup_{\lambda \in \mathfrak{I}} \left\{ \|\vec{\varphi}_n(\lambda)\|_{\mathbb{C}^d} \frac{1}{|\prod_{i=n_1}^{n-1} \nu_i^{(k)}(\lambda)|} \right\} \to 0 \text{ as } n \to \infty\,. \tag{6}$$

In $\mathbb{C}^d$ consider the canonical orthonormal basis $\{\vec{e}_k\}_{k=1}^d$. The $d \times d$ diagonal matrix $\mathrm{diag}\{\delta_{kl}\}_{l=1}^d$, where δ_{kl} $(k, l = 1, \ldots, d)$ is the Kronecker symbol, is a projector to the one dimensional space generated by $\vec{e}_k$.

Definition 2.3. Assuming $\Lambda \in \mathcal{L}(k)$ (for some $k = 1, \ldots, d$), let $\mathcal{P}_i(\Lambda, k) = \mathcal{P}_i$ be defined by

$$\mathcal{P}_i = \sum_{j \in I_i} \mathrm{diag}\{\delta_{jl}\}_{l=1}^d \qquad i = 1, 2\,, \tag{7}$$

where I_1 and I_2 are the classes of Definition 2.1.

3. Uniform asymptotics of solutions

The following result has been proven in [12].

Lemma 3.1. *Let the sequence Λ be defined as in (4). For any $n \in \mathbb{N}$, $\lambda \in \mathfrak{I}$, let $R_n(\lambda)$ be a $d \times d$ complex matrix. Fix the natural number $k \le d$ and assume that the following conditions hold:*

(i) $\Lambda \in \mathcal{L}(k)$.

(ii) $\sup_{\lambda \in \mathfrak{I}} \sum_{n=N}^{\infty} \dfrac{\|R_n(\lambda)\|}{|\nu_n^{(k)}(\lambda)|} < \infty$ *(N is given by the previous condition, see Def. 2.1).*

(iii) *for any $\epsilon > 0$ there exists an N_ϵ (which depends only on ϵ) such that $\forall \lambda \in \mathfrak{I}$ we have*

$$\sum_{n=N_\epsilon}^{\infty} \frac{\|R_n(\lambda)\|}{|\nu_n^{(k)}(\lambda)|} < \epsilon\,.$$

Then, for some $N_0 \ge N$ and any bounded continuous function, denoted by $\varphi_{N_0}(\lambda)$ $(\lambda \in \mathfrak{I})$, the operator T_k defined on any $\vec{\varphi} = \{\vec{\varphi}_n(\lambda)\}_{n=N_0+1}^{\infty}$ in $X_k(N_0)$ by

$$\begin{aligned}
(T_k\vec{\varphi})_n(\lambda) = {} & \mathcal{P}_1 \prod_{i=N_0}^{n-1} \Lambda_i(\lambda) \sum_{m=N_0}^{n-1} \left(\prod_{i=N_0}^{m} \Lambda_i(\lambda) \right)^{-1} R_m(\lambda)\vec{\varphi}_m(\lambda) \\
& - \mathcal{P}_2 \prod_{i=N_0}^{n-1} \Lambda_i(\lambda) \sum_{m=n}^{\infty} \left(\prod_{i=N_0}^{m} \Lambda_i(\lambda) \right)^{-1} R_m(\lambda)\vec{\varphi}_m(\lambda)\,, \quad n > N_0\,,
\end{aligned} \tag{8}$$

has the following properties

1. $\|T_k\| < 1$
2. $T_k X_k(N_0) \subset X_k^0(N_0)$

Assuming that Λ, defined by (4), and $\{R_n(\lambda)\}_{n=n_0}^{\infty}$ satisfy the conditions of Lemma 3.1, let the sequence $\vec{\varphi} = \{\vec{\varphi}_n^{(k)}(\lambda)\}_{n=N_0}^{\infty}$ in $X_k(N_0)$ be a solution of

$$\vec{\varphi} = \vec{\psi}^{(k)} + T_k\vec{\varphi}\,, \tag{9}$$

where $\vec{\psi}^{(k)} = \{\vec{\psi}_n^{(k)}(\lambda)\}_{n=N_0+1}^{\infty}$ is defined by

$$\vec{\psi}_n^{(k)} = \prod_{i=N_0}^{n-1} \Lambda_i(\lambda)\vec{e}_k = \prod_{i=N_0}^{n-1} \nu_i^{(k)}(\lambda)\vec{e}_k\,, \qquad n > N_0\,.$$

It is straightforwardly verifiable that one obtains an identity if substitutes (9) into

$$\vec{\varphi}_{n+1}(\lambda) = (\Lambda_n(\lambda) + R_n(\lambda))\vec{\varphi}_n(\lambda)\,, \qquad n > N_0\,, \tag{10}$$

and take into account (8). Thus, $\vec{\varphi} \in X_k(N_0)$, defined as a solution of (9), is a solution of (10) for each $k \leq d$. Notice that T_k's property 2, stated in Lemma 3.1, implies

$$\vec{\varphi} - \vec{\psi}^{(k)} \in X_k^0(N_0)\,. \tag{11}$$

The following assertion is the uniform version of the Benzaid-Lutz theorem [2, 4].

Theorem 3.2. *Let the sequences Λ, given by* (4), *and $\{R_n(\lambda)\}_{n=n_0}^{\infty}$ satisfy the conditions of Lemma* 3.1 *for all $k = 1,\ldots,d$. Then one can find an $N_0 \in \mathbb{N}$ such that there exists a basis $\{\vec{\varphi}^{(k)}(\lambda)\}_{k=1}^{d}$, $\vec{\varphi}^{(k)} = \{\vec{\varphi}_n^{(k)}(\lambda)\}_{n=N_0+1}^{\infty}$, in the space of solutions of* (10) *satisfying*

$$\sup_{\lambda \in \mathfrak{I}} \left\| \frac{\vec{\varphi}_n^{(k)}(\lambda)}{\prod_{i=N_0}^{n-1} \nu_i^{(k)}(\lambda)} - \vec{e}_k \right\| \to 0\,, \qquad as\ n \to \infty\,, \quad for\ k = 1,\ldots,d\,. \tag{12}$$

Proof. We have d solutions of (10) given by (9) for $k = 1,\ldots,d$. Equation (12) follows directly from (11). That $\{\vec{\varphi}^{(k)}(\lambda)\}_{k=1}^{d}$ is a basis is a consequence of (12). Indeed, let $\Phi(n,\lambda)$ be the $d \times d$ matrix whose columns are given by the d vectors $\vec{\varphi}_n^{(k)}(\lambda)$ ($k = 1,\ldots,d$); then (12) implies that, for sufficiently big n,

$$\forall \lambda \in \mathfrak{I}\,, \quad \det \Phi(n,\lambda) \neq 0\,. \qquad \square$$

It is worth remarking that the uniform Levinson theorem (in the continuous case, i.e., for a system of ordinary differential equations) has already been proven in [10], where this result is used in the spectral analysis of a self-adjoint fourth-order differential operator.

4. Smoothness of solutions

Here we show that if the matrices $R_n(\lambda)$ and $\Lambda_n(\lambda)$ enjoy certain smooth properties with respect to λ, then the solutions of (10) obtained through Theorem 3.2 are also smooth.

Lemma 4.1. *Let the sequences $\{R_n(\lambda)\}_{n=n_0}^{\infty}$ and Λ, defined in* (4), *satisfy the conditions of Lemma* 3.1, *and suppose that the entries of $R_n(\lambda)$ and $\Lambda_n(\lambda)$, seen as functions of λ, are continuous on $\mathfrak{I}$ for every $n \geq N_0$, where N_0 is given by Lemma* 3.1. *Then the solution $\vec{\varphi} = \{\vec{\varphi}_n(\lambda)\}_{n=N_0+1}^{\infty}$ of* (9) *is such that $\vec{\varphi}_n(\lambda)$, as a function of λ, is continuous on $\mathfrak{I}$ for each $n > N_0$.*

Proof. From the definition of T_k it follows that if the sequence $\vec{\varphi} = \{\vec{\varphi}_n(\lambda)\}_{n=N_0+1}^{\infty}$ is such that $\vec{\varphi}_n(\lambda)$ is a continuous function on $\mathfrak{I}$, $\forall n > N_0$, then $(T_k\vec{\varphi})_n(\lambda)$ is continuous on $\mathfrak{I}$, $\forall n > N_0$. Indeed, from (8) one has that $(T_k\vec{\varphi})_n(\lambda)$ is a uniform convergent series of continuous functions. The assertion of the lemma then follows from the fact that the unique solution of (9) can be found by the method of successive approximations. $\qquad\square$

Lemma 4.2. *Suppose that the sequences $\{R_n(\lambda)\}_{n=n_0}^{\infty}$ and $\{\Lambda_n(\lambda)\}_{n=n_0}^{\infty}$ satisfy the conditions of Theorem 3.2 and $\mathfrak{I}$ is a closed interval. Let $R_n(\lambda)$ and $\Lambda_n(\lambda)$ be matrices whose entries are continuous functions of λ on $\mathfrak{I}$ for every $n \geq n_0$ and such that*

$$\det(\Lambda_n(\lambda) + R_n(\lambda)) \neq 0\,, \quad \lambda \in \mathfrak{I}, \quad n_0 \leq n \leq N_0\,. \tag{13}$$

Then, the solutions $\{\vec{\varphi}^{(k)}(\lambda)\}_{k=1}^{d}$, $\vec{\varphi}^{(k)} = \{\vec{\varphi}_n^{(k)}(\lambda)\}_{n=N_0+1}^{\infty}$, of (10) given by Theorem 3.2 can be extended to solutions $\vec{\varphi}^{(k)} = \{\vec{\varphi}_n^{(k)}(\lambda)\}_{n=n_0}^{\infty}$ of the system

$$\vec{\varphi}_{n+1}(\lambda) = (\Lambda_n(\lambda) + R_n(\lambda))\vec{\varphi}_n(\lambda) \qquad n \geq n_0\,,$$

having the property that, given $n \geq n_0$ fixed, for any $\epsilon > 0$ there exists δ such that

$$\forall \lambda_1, \lambda_2 \in \mathfrak{I}, \quad |\lambda_1 - \lambda_2| < \delta \ \Rightarrow \ \left\|\vec{\varphi}_n^{(k)}(\lambda_1) - \vec{\varphi}_n^{(k)}(\lambda_2)\right\| < \epsilon, \quad k = 1,\ldots,d. \tag{14}$$

Proof. The proof is again straightforward. By Theorem 3.2 there exists an $N_0 \in \mathbb{N}$ such that the basis $\{\vec{\varphi}^{(k)}(\lambda)\}_{k=1}^{d}$ in the space of solutions of (10) satisfies (12). $\vec{\varphi}_n^{(k)}(\lambda)$ is continuous on $\mathfrak{I}$ for all $n > N_0$ as a consequence of Lemma 4.1. Since $\mathfrak{I}$ is closed, each $\vec{\varphi}_n^{(k)}(\lambda)$ is actually uniform continuous. Therefore, we have (14) for $n > N_0$. Now, for $n_0 \leq p \leq N_0$, one has

$$\vec{\varphi}_p^{(k)}(\lambda) = Q(\lambda, p, N_0)\vec{\varphi}_{N_0+1}^{(k)}(\lambda)\,,$$

where

$$Q(\lambda, p, N_0) := (\Lambda_p(\lambda) + R_p(\lambda))^{-1} \ldots (\Lambda_{N_0}(\lambda) + R_{N_0}(\lambda))^{-1}.$$

Condition (13) implies that $Q(\lambda, p, N_0)$ is always well defined, and the smooth properties of $R_n(\lambda)$ and $\Lambda_n(\lambda)$ imply that the entries of $Q(\lambda, p, N_0)$ are uniform continuous on $\mathfrak{I}$ for all p. Thus, from

$$\left\|\vec{\varphi}_p^{(k)}(\lambda_1) - \vec{\varphi}_p^{(k)}(\lambda_2)\right\| \leq \left\|(Q(\lambda_1, p, N_0) - Q(\lambda_2, p, N_0))\vec{\varphi}_{N_0}^{(k)}(\lambda_2)\right\|$$
$$+ \left\|Q(\lambda_1, p, N_0)(\vec{\varphi}_{N_0}^{(k)}(\lambda_1) - \vec{\varphi}_{N_0}^{(k)}(\lambda_2))\right\|.$$

it follows that

$$\left\|\vec{\varphi}_p^{(k)}(\lambda_1) - \vec{\varphi}_p^{(k)}(\lambda_2)\right\| \to 0 \quad \text{as } \lambda_1 \to \lambda_2 \qquad\square$$

5. An application to a class of Jacobi matrices

In the Hilbert space $l_2(\mathbb{N})$, let J be the operator whose matrix representation with respect to the canonical basis in $l_2(\mathbb{N})$ is a semi-infinite Jacobi matrix of the form

$$
\begin{pmatrix}
0 & b_1 & 0 & 0 & \cdots \\
b_1 & 0 & b_2 & 0 & \cdots \\
0 & b_2 & 0 & b_3 & \\
0 & 0 & b_3 & 0 & \ddots \\
\vdots & \vdots & & \ddots & \ddots
\end{pmatrix}. \tag{15}
$$

The elements of the sequence $\{b_n\}_{n=1}^{\infty}$ are defined as follows

$$
b_n := n^{\alpha}\left(1 + \frac{c_n}{n}\right), \quad \forall n \in \mathbb{N}, \tag{16}
$$

where $\alpha > 1$ and $c_n = c_{n+2L}$ ($L \in \mathbb{N}$). We assume that $1 + \frac{c_n}{n} \neq 0$ for all n. Clearly, the Jacobi operator J is symmetric and unbounded. J is closed by definition since the unbounded symmetric operator J is said to have the matrix representation (15) with respect to the canonical basis in $l_2(\mathbb{N})$ if it is the minimal closed operator satisfying

$$
(Je_k, e_{k+1}) = (Je_{k+1}, e_k) = b_k, \quad \forall k \in \mathbb{N},
$$

where $\{e_k\}_{k=1}^{\infty}$ is the canonical basis in $l_2(\mathbb{N})$ (see [1]). The class of Jacobi matrices given by (15) and (16) is said to have rapidly growing weights. This class is based on an example suggested by A.G. Kostyuchenko and K.A. Mirzoev in [9].

On the basis of subordinacy theory [5, 8], the spectral properties of J have been studied in [6, 11, 12]. The theory of subordinacy reduces the spectral analysis of operators to the asymptotic analysis of the corresponding generalized eigenvectors. This approach has proved to be very useful in the spectral analysis of Jacobi operators. In [6] it is proven that if

$$
\left|\sum_{k=1}^{2L}(-1)^k c_k\right| \geq L(\alpha - 1), \tag{17}
$$

then $J = J^*$ and it has pure point spectrum. However, within the framework of subordinacy theory, one cannot determine if the pure point spectrum has accumulation points in some finite interval.

Equation (2) for J takes the form

$$
b_{n-1}u_{n-1} + b_n u_{n+1} = \lambda u_n, \quad n > 1, \ \lambda \in \mathbb{R}, \tag{18}
$$

with $\{b_n\}_{n=1}^{\infty}$ given by (16). As was mentioned before, the asymptotic behavior of the solutions of (18) gives information on the spectral properties of J. If a solution $u(\lambda) = \{u_n(\lambda)\}_{n=1}^{\infty}$ of (18) satisfies the "boundary condition"

$$
b_1 u_2 = \lambda u_1 \tag{19}
$$

and turns out to be in $l_2(\mathbb{N})$, then $u(\lambda)$ is an eigenvector of J^* corresponding to the eigenvalue λ.

Using the results of Sections 3 and 4, we shall develop a technique to prove that J with weights given by (16) and (17) has discrete spectrum.

It is worth remarking that there are simpler methods for proving that the spectrum of J is purely discrete. Indeed, one can use for instance the asymptotic behavior of the solutions of (18) to show that the resolvent of J is compact. This has been done for a class of Jacobi operators in [7] and the technique developed there can in fact be used to obtain estimates for the eigenvalues.

The method we develop below may, nevertheless, be advantageous in some cases since it uses and preserves more information inherent in system (18). For simplicity, operator J has been chosen to illustrate the technique, but one can easily adapt the reasoning for other Jacobi operators. Our technique seems to be especially useful for operators having simultaneously intervals of pure point and absolutely continuous spectrum [6, Th. 2.2].

We begin by deriving from (18) a system suitable for applying our previous results, but first we introduce the following notation. Given a sequence of matrices $\{M_s(\lambda)\}_{s=1}^\infty$ ($\lambda \in \mathfrak{J}$) and a sequence $\{f_s\}_{s=1}^\infty$ of real numbers, we shall say that

$$M_s(\lambda) = \widehat{O}_{\mathfrak{J}}(f_s) \quad \text{as } s \to \infty.$$

if there exists a constant $C > 0$ and $S \in \mathbb{N}$ such that

$$\sup_{\lambda \in \mathfrak{J}} \|M_s(\lambda)\| < C\,|f_s|\,, \quad \forall s > S\,.$$

Now suppose that $\mathfrak{J}$ is a finite interval and rewrite (18), with $\lambda \in \mathfrak{J}$, in the form of (3). We have

$$B_n(\lambda) = \begin{pmatrix} 0 & 1 \\ -\frac{b_{n-1}}{b_n} & \frac{\lambda}{b_n} \end{pmatrix}, \quad n \geq 2, \quad \lambda \in \mathfrak{J}\,.$$

Define the sequence of matrices $\{A_m(\lambda)\}_{m=1}^\infty$ as follows

$$A_m(\lambda) := \prod_{s=1+L(m-1)}^{Lm} B_{2s+1}(\lambda)B_{2s}(\lambda)\,, \quad m \in \mathbb{N}. \tag{20}$$

Whenever we have products of non-diagonal matrices, as in (20), we take them in "chronological" order, that is,

$$A_m(\lambda) := B_{2Lm+1}(\lambda)B_{2Lm} \ldots B_{2L(m-1)+3}(\lambda)B_{2L(m-1)+2}\,.$$

A straightforward computation shows that

$$B_{2s+1}(\lambda)B_{2s}(\lambda) = -I + \begin{pmatrix} \frac{c_{2s}-c_{2s-1}+\alpha}{2s} & 0 \\ 0 & \frac{c_{2s+1}-c_{2s}+\alpha}{2s} \end{pmatrix} + \widehat{O}_{\mathfrak{J}}(s^{-1-\epsilon})\,, \quad \epsilon > 0\,.$$

Indeed, one can easily verify that

$$B_{2s+1}(\lambda)B_{2s}(\lambda) + I - \begin{pmatrix} \frac{c_{2s}-c_{2s-1}+\alpha}{2s} & 0 \\ 0 & \frac{c_{2s+1}-c_{2s}+\alpha}{2s} \end{pmatrix} = \begin{pmatrix} r_1(s) & r_2(s) \\ r_3(s)\lambda & r_4(s) + r_5(s)\lambda^2 \end{pmatrix}$$

where $r_l(s) = O(s^{-1-\epsilon})$ for $l = 1,\ldots,5$. Clearly, up to the same asymptotic estimate, we may also write ($\epsilon > 0$)

$$B_{2s+1}(\lambda)B_{2s}(\lambda) = \begin{pmatrix} -e^{\frac{c_{2s-1}-c_{2s}-\alpha}{2s}} & 0 \\ 0 & -e^{\frac{c_{2s}-c_{2s+1}-\alpha}{2s}} \end{pmatrix} \left[I + \widehat{O}_{\mathfrak{I}}(s^{-1-\epsilon}) \right].$$

Therefore,

$$A_m(\lambda) = \prod_{s=1+L(m-1)}^{Lm} \begin{pmatrix} -e^{\frac{c_{2s-1}-c_{2s}-\alpha}{2s}} & 0 \\ 0 & -e^{\frac{c_{2s}-c_{2s+1}-\alpha}{2s}} \end{pmatrix} \prod_{s=1+L(m-1)}^{Lm} \left[I + \widehat{O}_{\mathfrak{I}}(s^{-1-\epsilon}) \right]$$

$$= (-1)^L \begin{pmatrix} \exp \sum_{s=1+L(m-1)}^{Lm} \frac{c_{2s-1}-c_{2s}-\alpha}{2s} & 0 \\ 0 & \exp \sum_{s=1+L(m-1)}^{Lm} \frac{c_{2s}-c_{2s+1}-\alpha}{2s} \end{pmatrix} \left[I + \widehat{O}_{\mathfrak{I}}(m^{-1-\epsilon}) \right]$$

Let us define, for $m \in \mathbb{N}$, $\lambda \in \mathfrak{I}$, the matrices

$$\Lambda_m := \operatorname{diag}\{\nu_m^{(1)}, \nu_m^{(2)}\},$$

where

$$\nu_m^{(1)} := (-1)^L \exp \sum_{s=1+L(m-1)}^{Lm} \frac{c_{2s-1} - c_{2s} - \alpha}{2s}$$

$$\nu_m^{(2)} := (-1)^L \exp \sum_{s=1+L(m-1)}^{Lm} \frac{c_{2s} - c_{2s+1} - \alpha}{2s},$$

(21)

and

$$R_m(\lambda) := A_m(\lambda) - \Lambda_m$$

Observe that Λ_m does not depend on λ, and $R_m(\lambda) = \widehat{O}_{\mathfrak{I}}(m^{-1-\epsilon})$ as $m \to \infty$.

Lemma 5.1. *Let $\mathfrak{I}$ be a finite closed interval. There is a basis $\vec{x}^{(k)}(\lambda) = \{\vec{x}_n^{(k)}(\lambda)\}_{n=1}^{\infty}$ ($k = 1, 2$) in the space of solutions of the system*

$$\vec{x}_{n+1}(\lambda) = A_n(\lambda)\vec{x}_n(\lambda), \qquad n \in \mathbb{N}, \quad \lambda \in \mathfrak{I}, \tag{22}$$

with $A_n(\lambda)$ given by (20), such that

$$\sup_{\lambda \in \mathfrak{I}} \left\| \frac{\vec{x}_n^{(k)}(\lambda)}{\prod_{i=1}^{n-1} \nu_i^{(k)}} - \vec{e}_k \right\| \to 0, \qquad as \ n \to \infty, \quad for \ k = 1, 2,$$

where $\nu_i^{(k)}$ is defined in (21). Moreover, for any fixed $n \in \mathbb{N}$

$$\sup_{\substack{|\lambda'-\lambda|<\delta \\ \lambda',\lambda\in\mathfrak{I}}} \left\| \vec{x}_n^{(k)}(\lambda') - \vec{x}_n^{(k)}(\lambda) \right\| \to 0, \qquad as \ \delta \to 0, \quad k = 1, 2.$$

Proof. Write $A_n(\lambda) = \Lambda_n + R_n(\lambda)$ as was done before. We first show that the sequences $\{\Lambda_n\}_{n=1}^\infty$ and $\{R_n(\lambda)\}_{n=1}^\infty$ satisfy the conditions of Theorem 3.2. Let us prove that $\{\Lambda_n\}_{n=1}^\infty \in \mathcal{L}(k)$ for $k = 1, 2$. Define

$$\gamma := \frac{1}{2L} \sum_{s=1}^{L} c_{2s-1} - 2c_{2s} + c_{2s+1} \, .$$

It is not difficult to verify that for every $n \geq 2$ there is a constant K such that

$$\prod_{i=1}^{n} \frac{|\nu_i^{(1)}|}{|\nu_i^{(2)}|} = \exp \sum_{i=1}^{n} \sum_{s=1+L(i-1)}^{Li} \frac{c_{2s-1} - 2c_{2s} + c_{2s+1}}{2s}$$

$$< K \exp\left\{ \gamma \sum_{s=1}^{n} \frac{1}{s} \right\} .$$

Analogously for some constant $\widetilde{K}$

$$\prod_{i=1}^{n} \frac{|\nu_i^{(1)}|}{|\nu_i^{(2)}|} = \exp \sum_{i=1}^{n} \sum_{s=1+L(i-1)}^{Li} \frac{c_{2s-1} - 2c_{2s} + c_{2s+1}}{2s}$$

$$> \widetilde{K} \exp\left\{ \gamma \sum_{s=1}^{n} \frac{1}{s} \right\} .$$

Clearly, one obtains similar estimates interchanging $k = 1, 2$. Thus i holds. Conditions ii and iii follow from the fact that $\nu_n^{(k)} \to 1$ as $n \to \infty$ and $R_n = \widehat{O}_{\mathfrak{J}}(n^{-1-\epsilon})$.

Now observe that (13) holds for the system (22), and for $n \in \mathbb{N}$ the entries of $R_n(\lambda)$ and Λ_n are continuous functions of $\lambda \in \mathfrak{J}$. Therefore, the conditions of Lemma 4.2 are satisfied. $\square$

Lemma 5.2. *Let $\mathfrak{J}$ be any closed finite interval of $\mathbb{R}$. Then, there exists a solution $u(\lambda) = \{u_n(\lambda)\}_{n=1}^\infty$ of (18), with $\{b_n\}_{n=1}^\infty$ given by (16) and satisfying (17), such that*

$$\sum_{n=1}^{\infty} \sup_{\lambda \in \mathfrak{J}} |u_n(\lambda)|^2 < \infty \, .$$

Moreover, for any fixed $n \in \mathbb{N}$,

$$\sup_{\substack{|\lambda' - \lambda| < \delta \\ \lambda', \lambda \in \mathfrak{J}}} |u_n(\lambda') - u_n(\lambda)| \to 0 \, , \qquad as \ \delta \to 0 \, .$$

Proof. By (20) and (22), it is clear that

$$\vec{x}_{n+1}^{(k)}(\lambda) = \begin{pmatrix} u_{2Ln+1}^{(k)}(\lambda) \\ u_{2Ln+2}^{(k)}(\lambda) \end{pmatrix} . \tag{23}$$

Thus, Lemma 5.1 yields that for $n \in \mathbb{N}$ and some constants $C, C' > 0$

$$
\sup_{\lambda \in \mathfrak{J}} \left| u_{2Ln+2}^{(1)}(\lambda) \right| \leq \sup_{\lambda \in \mathfrak{J}} \left\| \vec{x}_{n+1}^{(1)}(\lambda) \right\|
$$

$$
\leq C \left| \prod_{i=1}^{n} \nu_i^{(1)} \right|
$$

$$
= C \exp \sum_{i=1}^{n} \sum_{s=1+L(i-1)}^{Li} \frac{c_{2s-1} - c_{2s} - \alpha}{2s}
$$

$$
\leq C' \exp \left\{ \sum_{s=1}^{L} \frac{c_{2s-1} - c_{2s} - \alpha}{2L} \sum_{s=1}^{n} \frac{1}{s} \right\},
$$

where we have use the periodicity of the sequence $\{c_k\}_{k=1}^{\infty}$. Thus for some constant C'' we have

$$
\sup_{\lambda \in \mathfrak{J}} \left| u_{2Ln+2}^{(1)}(\lambda) \right| \leq C'' n^{\beta}, \quad \beta := \frac{1}{2L} \sum_{s=1}^{2L} (-1)^{s+1} c_s - \frac{\alpha}{2}.
$$

Analogously, there is a $\widetilde{C} > 0$ such that

$$
\sup_{\lambda \in \mathfrak{J}} \left| u_{2Ln+2}^{(2)}(\lambda) \right| \leq \widetilde{C} n^{\tilde{\beta}}, \quad \tilde{\beta} := \frac{1}{2L} \sum_{s=1}^{2L} (-1)^{s} c_s - \frac{\alpha}{2}.
$$

Since $\alpha > 1$, (17) implies that either for $k = 1$ or $k = 2$

$$
\sum_{n=1}^{\infty} \sup_{\lambda \in \mathfrak{J}} \left| u_{2Ln+2}^{(k)}(\lambda) \right|^2 < \infty \tag{24}
$$

The first assertion of the lemma follows from (24) and the fact that there is a constant C such that

$$
\sup_{\lambda \in \mathfrak{J}} \left\| \prod_{j=2}^{s} B_{2Ln+j}(\lambda) \right\| < C \qquad s = 2, 3, \ldots, 2L, \qquad n \in \mathbb{N}.
$$

Now, Lemma 5.1 and (23) yield, for $n \in \mathbb{N}$ and $k = 1, 2$,

$$
\sup_{\substack{|\lambda' - \lambda| < \delta \\ \lambda', \lambda \in \mathfrak{J}}} \left| u_{2Ln+2}^{(k)}(\lambda') - u_{2Ln+2}^{(k)}(\lambda) \right| \to 0, \quad \text{as } \delta \to 0. \tag{25}
$$

Since for any $s = 2, 3, \ldots, 2L$

$$
\begin{pmatrix} u_{2Ln+s}^{(k)}(\lambda) \\ u_{2Ln+s+1}^{(k)}(\lambda) \end{pmatrix} = \prod_{j=2}^{s} B_{2Ln+j}(\lambda) \begin{pmatrix} u_{2Ln+1}^{(k)}(\lambda) \\ u_{2Ln+2}^{(k)}(\lambda) \end{pmatrix},
$$

the following inequality holds for $n \in \mathbb{N}$ and $s = 2, 3, \ldots, 2L$

$$\left| u_{2Ln+s+1}^{(k)}(\lambda') - u_{2Ln+s+1}^{(k)}(\lambda) \right| \leq \left\| \left(\prod_{j=2}^{s} B_{2Ln+j}(\lambda') - \prod_{j=2}^{s} B_{2Ln+j}(\lambda) \right) \vec{x}_{n+1}^{(k)}(\lambda) \right\|$$

$$+ \left\| \prod_{j=2}^{s} B_{2Ln+j}(\lambda') \left(\vec{x}_{n+1}^{(k)}(\lambda') - \vec{x}_{n+1}^{(k)}(\lambda) \right) \right\| .$$

Taking into account the smooth properties of the finite product $\prod_{j=2}^{s} B_{2Ln+j}(\lambda)$ and Lemma 5.1, one obtains from the last inequality and (25) the second assertion of the lemma for any $n \geq 2L + 2$. To complete the proof use the invertibility and smoothness of the matrices $B_n(\lambda)$ for $n < 2L + 2$. $\qquad\square$

Remark 1. Let $u(\lambda)$ $(\lambda \in \mathfrak{J})$ be the solution mentioned in the previous lemma. If $J = J^*$ and $\lambda_0 \in \mathfrak{J}$ is such that (19) is satisfied, then λ_0 is in the point spectrum of J and $u(\lambda_0)$ is the corresponding eigenvector.

Theorem 5.3. *Let J be the Jacobi operator defined by (15), (16) and (17). Then the spectrum of J is discrete.*

Proof. It is already known that the spectrum of J, denoted $\sigma(J)$, is pure point [6]. Suppose that $\sigma(J)$ has a point of accumulation μ in some finite closed interval $\mathfrak{J}$. Let λ and λ' $(\lambda \neq \lambda')$ be arbitrarily chosen from $\sigma(J) \cap \mathfrak{J} \cap V_{\frac{\delta}{2}}(\mu)$, where $V_{\frac{\delta}{2}}(\mu)$ is a $\frac{\delta}{2}$-neighborhood of μ. Consider

$$\left| (u(\lambda), u(\lambda'))_{l_2(\mathbb{N})} \right| = \left| \sum_{n=1}^{\infty} u_n(\lambda) \overline{u_n(\lambda')} \right|$$

$$\geq \left| \sum_{n=1}^{N_1} u_n(\lambda) \overline{u_n(\lambda')} \right| - \left| \sum_{n>N_1} u_n(\lambda) \overline{u_n(\lambda')} \right| . \tag{26}$$

As a consequence of Lemmas 5.1 and 5.2, one can choose N_1, δ and $n_0 \leq N_1$ so that

$$\left| \sum_{n>N_1} u_n(\lambda) \overline{u_n(\lambda')} \right| < \frac{1}{4} |u_{n_0}(\mu)|^2 < \frac{1}{2} |u_{n_0}(\lambda)|^2 . \tag{27}$$

Now, consider the first term in the right-hand side of (26)

$$\left| \sum_{n=1}^{N_1} u_n(\lambda) \overline{u_n(\lambda')} \right| \geq \sum_{n=1}^{N_1} |u_n(\lambda)|^2 - \left| \sum_{n=1}^{N_1} u_n(\lambda) (\overline{u_n(\lambda')} - u_n(\lambda)) \right|$$

$$\geq |u_1(\lambda)|^2 - \left| \sum_{n=1}^{N_1} u_n(\lambda) (\overline{u_n(\lambda')} - u_n(\lambda)) \right| .$$

Since $|\lambda' - \lambda| < \delta$, we have

$$\left| \sum_{n=1}^{N_1} u_n(\lambda)\overline{(u_n(\lambda') - u_n(\lambda))} \right| \leq \max_{1 \leq n \leq N_1} \omega_n(\delta) \sum_{n=1}^{N_1} |u_n(\lambda)| \,,$$

where

$$\omega_n(\delta) = \sup_{\substack{|\lambda'-\lambda|<\delta \\ \lambda',\lambda\in\mathfrak{I}}} |u_n(\lambda') - u_n(\lambda)|$$

is the modulus of continuity of $u_n(\lambda)$ on $\mathfrak{I}$. By the second assertion of Lemma 5.2, taking δ sufficiently small, one obtains

$$\left| \sum_{n=1}^{N_1} u_n(\lambda)\overline{(u_n(\lambda') - u_n(\lambda))} \right| < \frac{1}{2} |u_{n_0}(\lambda)|^2 \,. \tag{28}$$

From (26), (27), and (28)

$$(u(\lambda), u(\lambda'))_{l^2(\mathbb{N})} > |u_{n_0}(\lambda)|^2 - \frac{1}{2} |u_{n_0}(\lambda)|^2 - \frac{1}{2} |u_{n_0}(\lambda)|^2 = 0 \,.$$

But this cannot be true since $J = J^*$ and it must be that $u(\lambda) \perp u(\lambda')$.

$\square$

Acknowledgment

I express my deep gratitude to the referee for very useful remarks and a hint on the literature.

References

[1] N.I. Akhiezer and I.M. Glazman. *Theory of linear operators in Hilbert space.* Dover, New York, 1993. Two volumes bound as one.

[2] Z. Benzaid and D.A. Lutz. Asymptotic representation of solutions of perturbed systems of linear difference equations. *Stud. Appl. Math.,* **77** (1987) 195–221.

[3] E.A. Coddington and N. Levinson. *Theory of ordinary differential equations.* McGraw-Hill, New York-Toronto-London, 1955.

[4] S.N. Elaydi. *An introduction to difference equations.* Undergraduate Texts in Mathematics. Springer-Verlag, New York, second edition, 1999.

[5] D.J. Gilbert and D.B. Pearson. On subordinacy and analysis of the spectrum of one-dimensional Schrödinger operators. *J. Math. Anal. Appl.,* **128** (1987) 30–56.

[6] J. Janas and M. Moszyński. Spectral properties of Jacobi matrices by asymptotic analysis. *J. Approx. Theory,* **120** (2003) 309–336.

[7] J. Janas and S. Naboko. Spectral analysis of selfadjoint Jacobi matrices with periodically modulated entries. *J. Funct. Anal.,* **191** (2002) 318–342.

[8] S. Khan and D.B. Pearson. Subordinacy and spectral theory for infinite matrices. *Helv. Phys. Acta,* **65** (1992) 505–527.

[9] A.G. Kostyuchenko and K.A. Mirzoev. Generalized Jacobi matrices and deficiency indices of ordinary differential operators with polynomial coefficients. *Funct. Anal. Appl.*, **33** (1999) 30–45.

[10] C. Remling. Spectral analysis of higher-order differential operators. II. Fourth-order equations. *J. London Math. Soc.* (2), **59** (1999) 188–206.

[11] L.O. Silva. Spectral properties of Jacobi matrices with rapidly growing power-like weights. *Oper. Theory Adv. Appl.*, **132** (2002) 387–394.

[12] L.O. Silva. Uniform Levinson type theorems for discrete linear systems. *Oper. Theory Adv. Appl.*, **154** (2004) 203–218.

Luis O. Silva
Departamento de Métodos Matemáticos y Numéricos
IIMAS, Universidad Nacional Autónoma de México
Apdo. Postal 20-726
México, D.F. 01000
e-mail: `silva@leibniz.iimas.unam.mx`

Operator Theory:
Advances and Applications, Vol. 174, 187–203

An Example of Spectral Phase Transition Phenomenon in a Class of Jacobi Matrices with Periodically Modulated Weights

Sergey Simonov

Abstract. We consider self-adjoint unbounded Jacobi matrices with diagonal $q_n = n$ and weights $\lambda_n = c_n n$, where c_n is a 2-periodical sequence of real numbers. The parameter space is decomposed into several separate regions, where the spectrum is either purely absolutely continuous or discrete. This constitutes an example of the spectral phase transition of the first order. We study the lines where the spectral phase transition occurs, obtaining the following main result: either the interval $(-\infty; \frac{1}{2})$ or the interval $(\frac{1}{2}; +\infty)$ is covered by the absolutely continuous spectrum, the remainder of the spectrum being pure point. The proof is based on finding asymptotics of generalized eigenvectors via the Birkhoff-Adams Theorem. We also consider the degenerate case, which constitutes yet another example of the spectral phase transition.

Mathematics Subject Classification (2000). 47A10, 47B36.

Keywords. Jacobi matrices, Spectral phase transition, Absolutely continuous spectrum, Pure point spectrum, Discrete spectrum, Subordinacy theory, Asymptotics of generalized eigenvectors.

1. Introduction

In the present paper we study a class of Jacobi matrices with unbounded entries: a linearly growing diagonal and periodically modulated linearly growing weights.

We first define the operator J on the linear set of vectors $l_{\mathrm{fin}}(\mathbb{N})$ having finite number of non-zero elements:

$$(Ju)_n = \lambda_{n-1}u_{n-1} + q_n u_n + \lambda_n u_{n+1}, \ n \geq 2 \tag{1.1}$$

with the initial condition $(Ju)_1 = q_1 u_1 + \lambda_1 u_2$, where $q_n = n$, $\lambda_n = c_n n$, and c_n is a real 2-periodic sequence, generated by the parameters c_1 and c_2.

Let $\{e_n\}_{n\in\mathbb{N}}$ be the canonical basis in $l^2(\mathbb{N})$. With respect to this basis the operator J admits the following matrix representation:

$$J = \begin{pmatrix} q_1 & \lambda_1 & 0 & \cdots \\ \lambda_1 & q_2 & \lambda_2 & \cdots \\ 0 & \lambda_2 & q_3 & \cdots \\ \vdots & \vdots & \vdots & \ddots \end{pmatrix}$$

Due to the Carleman condition [2] $\sum_{n=1}^{\infty} \frac{1}{\lambda_n} = \infty$, the operator J is essentially self-adjoint. We will therefore assume throughout the paper, that J is a closed self-adjoint operator in $l^2(\mathbb{N})$, defined on its natural domain $D(J) = \{u \in l^2(\mathbb{N}) : Ju \in l^2(\mathbb{N})\}$.

We base our spectrum investigation on the subordinacy theory due to Gilbert and Pearson [6], generalized to the case of Jacobi matrices by Khan and Pearson [12]. Using this theory, we study an example of spectral phase transition of the first order. This example was first obtained by Naboko and Janas in [9] and [10]. In cited articles, the authors managed to demonstrate that the space of parameters $(c_1; c_2) \in \mathbb{R}^2$ can be naturally decomposed into a set of regions of two types. In the regions of the first type, the spectrum of the operator J is purely absolutely continuous and covers the real line $\mathbb{R}$, whereas in the regions of the second type the spectrum is discrete.

Due to [9] and [10], spectral properties of Jacobi matrices of our class are determined by the location of the point zero relative to the absolutely continuous spectrum of a certain periodic matrix J_{per}, constructed based on the modulation parameters c_1 and c_2. In our case this leads to:

$$J_{\mathrm{per}} = \begin{pmatrix} 1 & c_1 & 0 & 0 & \cdots \\ c_1 & 1 & c_2 & 0 & \cdots \\ 0 & c_2 & 1 & c_1 & \cdots \\ 0 & 0 & c_1 & 1 & \cdots \\ \vdots & \vdots & \vdots & \vdots & \ddots \end{pmatrix}$$

Considering the characteristic polynomial

$$d_{J_{\mathrm{per}}}(\lambda) = Tr\left(\begin{pmatrix} 0 & 1 \\ -\frac{c_1}{c_2} & \frac{\lambda-1}{c_2} \end{pmatrix} \begin{pmatrix} 0 & 1 \\ -\frac{c_2}{c_1} & \frac{\lambda-1}{c_1} \end{pmatrix} \right) = \frac{(\lambda-1)^2 - c_1^2 - c_2^2}{c_1 c_2},$$

the location of the absolutely continuous spectrum $\sigma_{\mathrm{ac}}(J_{\mathrm{per}})$ of J_{per} can then be determined from the following condition [2]:

$$\lambda \in \sigma_{\mathrm{ac}}(J_{\mathrm{per}}) \Leftrightarrow \left| d_{J_{\mathrm{per}}}(\lambda) \right| \leq 2. \tag{1.2}$$

This leads to the following result [10], concerning the spectral structure of the operator J.

If $\left| d_{J_{\mathrm{per}}}(0) \right| < 2$, then the spectrum of the operator J is purely absolutely continuous, covering the whole real line.

If, on the other hand, $\left| d_{J_{\mathrm{per}}}(0) \right| > 2$, then the spectrum of the operator J is discrete.

Thus, the condition $\left| \frac{1-c_1^2-c_2^2}{c_1 c_2} \right| = 2$, equivalent to $\{ \, |c_1| + |c_2| = 1 \text{ or } \left| |c_1| - |c_2| \right| = 1 \, \}$, determines the boundaries of the above-mentioned regions on the plane $(c_1; c_2)$ where one of the cases holds and the spectrum of the operator J is either purely absolutely continuous or discrete (see Figure 1 on page 195).

Note also, that Jacobi matrices with modulation parameters equal to $\pm c_1$ and $\pm c_2$ are unitarily equivalent. Thus the situation can be reduced to studying the case $c_1,\ c_2 > 0$.

In the present paper we attempt to study the spectral structure on the lines, where the spectral phase transition occurs, i.e., on the lines separating the aforementioned regions.

The paper is organized as follows.

Section 2 deals with the calculation of the asymptotics of generalized eigenvectors of the operator J. This calculation is mainly based on the Birkhoff-Adams Theorem [4]. The asymptotics are then used to characterize the spectral structure of the operator via the Khan-Pearson Theorem [12]. It turns out, that on the lines where the spectral phase transition occurs the spectrum is neither purely absolutely continuous nor pure point, but a combination of both.

In Section 3, we attempt to ascertain whether the pure point part of the spectrum is actually discrete. In doing so, we establish a criterion that guarantees that the operator J is semibounded from below, for all $(c_1; c_2) \in \mathbb{R}^2$. This semiboundedness is then used in conjunction with classical methods of operator theory to prove, that in at least one situation the discreteness of the pure point spectrum is guaranteed.

Section 4 is dedicated to the study of the degenerate case, i.e., the case when one of the modulation parameters turns to zero. In this situation, one can explicitly calculate all eigenvalues of the operator. On this route we obtain yet another "hidden" example of the spectral phase transition of the first order as the point $(c_1; c_2)$ moves along one of the critical lines in the space of parameters.

2. Generalized eigenvectors and the spectrum of the operator J

In this section, we calculate asymptotics of generalized eigenvectors of the operator J. Consider the recurrence relation [12]

$$\lambda_{n-1} u_{n-1} + (q_n - \lambda) u_n + \lambda_n u_{n+1} = 0, \ n \geq 2. \tag{2.1}$$

We reduce it to a form such that the Birkhoff-Adams Theorem is applicable. To this end, we need to have a recurrence relation of the form:

$$x_{n+2} + F_1(n) x_{n+1} + F_2(n) x_n = 0, \ n \geq 1, \tag{2.2}$$

where $F_1(n)$ and $F_2(n)$ admit the following asymptotical expansions as $n \to \infty$:

$$F_1(n) \sim \sum_{k=0}^{\infty} \frac{a_k}{n^k}, \quad F_2(n) \sim \sum_{k=0}^{\infty} \frac{b_k}{n^k} \tag{2.3}$$

with $b_0 \neq 0$. Consider the characteristic equation $\alpha^2 + a_0\alpha + b_0 = 0$ and denote its roots α_1 and α_2. Then [4]:

Theorem (Birkhoff-Adams). *There exist two linearly independent solutions $x_n^{(1)}$ and $x_n^{(2)}$ of the recurrence relation (2.2) with the following asymptotics as $n \to \infty$:*

1.
$$x_n^{(i)} = \alpha_i^n n^{\beta_i}\left(1 + O\left(\frac{1}{n}\right)\right), \quad i = 1, 2,$$

   if the roots α_1 and α_2 are different, where $\beta_i = \frac{a_1\alpha_i + b_1}{a_0\alpha_i + 2b_0}$, $i = 1, 2$.

2.
$$x_n^{(i)} = \alpha^n e^{\delta_i\sqrt{n}} n^{\beta}\left(1 + O\left(\frac{1}{\sqrt{n}}\right)\right), \quad i = 1, 2,$$

   if the roots α_1 and α_2 coincide, $\alpha := \alpha_1 = \alpha_2$, and an additional condition $a_1\alpha + b_1 \neq 0$ holds, where $\beta = \frac{1}{4} + \frac{b_1}{2b_0}$, $\delta_1 = 2\sqrt{\frac{a_0 a_1 - 2b_1}{2b_0}} = -\delta_2$.

This theorem is obviously not directly applicable in our case, due to wrong asymptotics of coefficients at infinity. In order to deal with this problem, we study a pair of recurrence relations, equivalent to (2.1), separating odd and even components of a vector u. This allows us to apply the Birkhoff-Adams Theorem to each of the recurrence relations of the pair, which yields the corresponding asymptotics. Combining the two asymptotics together, we then obtain the desired result for the solution of (2.1).

Denoting $v_k := u_{2k-1}$ and $w_k := u_{2k}$, we rewrite the recurrence relation (2.1) for the consecutive values of n: $n = 2k$ and $n = 2k + 1$.

$$\lambda_{2k-1}v_k + (q_{2k} - \lambda)w_k + \lambda_{2k}v_{k+1} = 0,$$

$$\lambda_{2k}w_k + (q_{2k+1} - \lambda)v_{k+1} + \lambda_{2k+1}w_{k+1} = 0.$$

Then we exclude w in order to obtain the recurrence relation for v:

$$w_k = -\frac{\lambda_{2k-1}v_k + \lambda_{2k}v_{k+1}}{q_{2k} - \lambda},$$

$$v_{k+2} + P_1(k)v_{k+1} + P_2(k)v_k = 0, \quad k \geq 1, \tag{2.4}$$

where

$$P_1(k)\frac{q_{2k+2} - \lambda}{q_{2k} - \lambda}\frac{\lambda_{2k}^2}{\lambda_{2k+1}\lambda_{2k+2}} - \frac{(q_{2k+1} - \lambda)(q_{2k+2} - \lambda)}{\lambda_{2k+1}\lambda_{2k+2}} + \frac{\lambda_{2k+1}}{\lambda_{2k+2}},$$

$$P_2(k) = \frac{q_{2k+2} - \lambda}{q_{2k} - \lambda}\frac{\lambda_{2k-1}\lambda_{2k}}{\lambda_{2k+1}\lambda_{2k+2}}.$$

In our case ($\lambda_n = c_n n$ and $q_n = n$) this yields the following asymptotic expansions (cf. (2.3)) for $P_1(k)$ and $P_2(k)$ as k tends to infinity:

$$P_1(k) = \sum_{j=0}^{\infty} \frac{a_j}{k^j}, \quad P_2(k) = \sum_{j=0}^{\infty} \frac{b_j}{k^j}$$

with

$$a_0 = \frac{c_1^2 + c_2^2 - 1}{c_1 c_2}, \quad a_1 - \frac{c_1^2 + c_2^2 - 2\lambda}{2 c_1 c_2} = -\frac{a_0}{2} + \frac{\lambda - \frac{1}{2}}{c_1 c_2}, \qquad (2.5)$$

$$b_0 = 1, \ b_1 = -1.$$

The remaining coefficients $\{a_j\}_{j=2}^{+\infty}$, $\{b_j\}_{j=2}^{+\infty}$ can also be calculated explicitly.

On the same route one can obtain the recurrence relation for even components w_k of the vector u:

$$w_{k+2} + R_1(k) w_{k+1} + R_2(k) w_k = 0, \ k \geq 1. \qquad (2.6)$$

Note, that if k is substituted in (2.4) by $k + \frac{1}{2}$ and v by w, the equation (2.4) turns into (2.6). Therefore,

$$R_1(k) = P_1\left(k + \frac{1}{2}\right), \quad R_2(k) = P_2\left(k + \frac{1}{2}\right),$$

and thus as $k \to \infty$,

$$R_1(k) = a_0 + \frac{a_1}{k} + O\left(\frac{1}{k^2}\right),$$

$$R_2(k) = b_0 + \frac{b_1}{k} + O\left(\frac{1}{k^2}\right),$$

with a_0, a_1, b_0, b_1 defined by (2.5).

Applying now the Birkhoff-Adams Theorem we find the asymptotics of solutions of recurrence relations (2.4) and (2.6). This leads to the following result.

Lemma 2.1. *Recurrence relations* (2.4) *and* (2.6) *have solutions* v_n^+, v_n^- *and* w_n^+, w_n^-, *respectively, with the following asymptotics as* $k \to \infty$:

1.
$$v_k^{\pm}, \ w_k^{\pm} = \alpha_{\pm}^k k^{\beta_{\pm}}\left(1 + O\left(\frac{1}{k}\right)\right), \qquad (2.7)$$

if $\left|\frac{c_1^2 + c_2^2 - 1}{c_1 c_2}\right| \neq 2$, *where* α_+ *and* α_- *are the roots of the equation* $\alpha^2 + a_0 \alpha + b_0 = 0$ *and* $\beta_{\pm} - \frac{a_1 \alpha_{\pm} + b_1}{a_0 \alpha_{\pm} + 2 b_0}$ *with* a_0, a_1, b_0, b_1 *defined by* (2.5).
Moreover, if $\left|\frac{c_1^2 + c_2^2 - 1}{c_1 c_2}\right| > 2$ *then* $\alpha_{\pm}$ *are real and* $|\alpha_-| < 1 < |\alpha_+|$, *whereas if* $\left|\frac{c_1^2 + c_2^2 - 1}{c_1 c_2}\right| < 2$ *then* $\alpha_+ = \overline{\alpha_-}$, $\beta_+ = \overline{\beta_-}$ *and the vectors* v^+, v^-, w^+, w^- *are not in* $l^2(\mathbb{N})$.

2.
$$v_k^{\pm}, \ w_k^{\pm} = \alpha^k k^{-\frac{1}{4}} e^{\delta_{\pm}\sqrt{k}}\left(1 + O\left(\frac{1}{\sqrt{k}}\right)\right), \qquad (2.8)$$

if $\left|\frac{c_1^2+c_2^2-1}{c_1 c_2}\right| = 2$ *and* $\lambda \neq \frac{1}{2}$, *where* $\alpha = \alpha_+ = \alpha_-$.

Moreover, if $\frac{c_1^2+c_2^2-1}{c_1 c_2} = 2$, *then* $\delta_+ = 2\sqrt{\frac{2\lambda-1}{2c_1 c_2}} = -\delta_-$,

whereas if $\frac{c_1^2+c_2^2-1}{c_1 c_2} = -2$, *then* $\delta_+ = 2\sqrt{\frac{1-2\lambda}{2c_1 c_2}} = -\delta_-$.

Proof. Consider recurrence relation (2.4) and let the constants a_0, a_1, b_0, b_1 be defined by (2.5). Consider the characteristic equation $\alpha^2 + a_0\alpha + b_0 = 0$. It has different roots, $\alpha_- < \alpha_+$, when the discriminant D differs from zero: $D = \left(\frac{c_1^2+c_2^2-1}{c_1 c_2}\right)^2 - 4 \neq 0$. Note that $\alpha_+\alpha_- = 1$.

Consider the case $D < 0$. A direct application of the Birkhoff-Adams Theorem yields:

$$v_k^{\pm} = \alpha_{\pm}^k k^{\beta_{\pm}} \left(1 + O\left(\frac{1}{k}\right)\right), \ k \to \infty,$$

where $\beta_{\pm} = \frac{a_1\alpha_{\pm}+b_1}{a_0\alpha_{\pm}+2b_0}$. Then $\alpha_+ = \overline{\alpha_-}$, $|\alpha_+| = |\alpha_-| = 1$ and $\beta_+ = \overline{\beta_-}$. Note also, that $v^{\pm}$ are not in l^2:

$$\mathrm{Re}\ \beta_+ = \mathrm{Re}\ \beta_- = -\frac{1}{2} + \frac{2\lambda - 1}{2c_1 c_2}\, \mathrm{Re}\left(\frac{1}{a_0 + 2\alpha_-}\right) = -\frac{1}{2}.$$

In the case $D > 0$, α_+ and α_- are real and $|\alpha_-| < 1 < |\alpha_+|$, hence v^- lies in l^2.

Ultimately, in the case $D = 0$, the roots of the characteristic equation coincide and are equal to $\alpha = -\frac{a_0}{2}$, with $|\alpha| = 1$, and the additional condition $a_0 a_1 \neq 2b_1$ is equivalent to

$$-\frac{a_0^2}{2} + \frac{a_0(\lambda - \frac{1}{2})}{c_1 c_2} \neq -2 \Leftrightarrow \lambda \neq \frac{1}{2}.$$

The Birkhoff-Adams Theorem yields:

$$v_k^{\pm} = \alpha^k k^{\beta} e^{\delta_{\pm}\sqrt{k}} \left(1 + O\left(\frac{1}{\sqrt{k}}\right)\right), \ k \to \infty,$$

where $\beta = -\frac{1}{4}$, $\delta_+ = 2\sqrt{\frac{a_0(\lambda-\frac{1}{2})}{2c_1 c_2}} = -\delta_-$. If the value δ_+ is pure imaginary, then clearly the vectors $v^{\pm}$ do not belong to $l^2(\mathbb{N})$.

In order to prove the assertion of the lemma in relation to $w^{\pm}$, note that in our calculations we use only the first two orders of the asymptotical expansions for $P_1(k)$ and $P_2(k)$. These coincide with the ones for $R_1(k)$ and $R_2(k)$. Thus, the solutions of recurrence relations (2.4) and (2.6) coincide in their main orders, which completes the proof. $\qquad\Box$

Now we are able to solve the recurrence relation (2.1) combining the solutions of recurrence relations (2.4) and (2.6).

Lemma 2.2. *Recurrence relation (2.1) has two linearly independent solutions u_n^+ and u_n^- with the following asymptotics as $k \to \infty$:*

1.
$$\begin{cases} u^{\pm}_{2k-1} = \alpha^{k}_{\pm} k^{\beta_{\pm}} \left(1 + O\left(\frac{1}{k}\right)\right), \\ u^{\pm}_{2k} = -(c_1 + \alpha_{\pm}c_2)\alpha^{k}_{\pm} k^{\beta_{\pm}} \left(1 + O\left(\frac{1}{k}\right)\right), \end{cases}$$

if $\left|\frac{c_1^2 + c_2^2 - 1}{c_1 c_2}\right| \neq 2$, *where the values* α_+, α_-, β_+, β_- *are taken from the statement of Lemma 2.1.*

2.
$$\begin{cases} u^{\pm}_{2k-1} = \alpha^{k} e^{\delta_{\pm}\sqrt{k}} k^{-\frac{1}{4}} \left(1 + O\left(\frac{1}{\sqrt{k}}\right)\right), \\ u^{\pm}_{2k} = -(c_1 + \alpha c_2)\alpha^{k} e^{\delta_{\pm}\sqrt{k}} k^{-\frac{1}{4}} \left(1 + O\left(\frac{1}{\sqrt{k}}\right)\right), \end{cases}$$

if $\left|\frac{c_1^2 + c_2^2 - 1}{c_1 c_2}\right| = 2$ *and* $\lambda \neq \frac{1}{2}$, *where the values* α, δ_+, δ_- *are taken from the statement of Lemma 2.1.*

Proof. It is clear, that any solution of recurrence relation (2.1) u gives two vectors, v and w, constructed of its odd and even components, which solve recurrence relations (2.4) and (2.6), respectively. Consequently, any solution of the recurrence relation (2.1) belongs to the linear space with the basis $\{V^+,\ V^-,\ W^+,\ W^-\}$, where

$$V^{\pm}_{2k-1} = v^{\pm}_k,\ V^{\pm}_{2k} = 0 \text{ and } W^{\pm}_{2k-1} = 0,\ W^{\pm}_{2k} = w^{\pm}_k.$$

This 4-dimensional linear space contains 2-dimensional subspace of solutions of recurrence relation (2.1). In order to obtain a solution u of (2.1), one has to obtain two conditions on the coefficients a_+, a_-, b_+, b_- such that $u = a_+ V^+ + a_- V^- + b_+ W^+ + b_- W^-$,

$$u_{2k-1} = a_+ v^+_k + a_- v^-_k,\ u_{2k} = b_+ w^+_k + b_- w^-_k. \tag{2.9}$$

Using Lemma 2.1, we substitute the asymptotics of this u into (2.1) where n is taken equal to $2k$,

$$\lambda_{2k-1} u_{2k-1} + (q_{2k} - \lambda)u_{2k} + \lambda_{2k} u_{2k+1} = 0.$$

As in Lemma 2.1, we have two distinct cases.

Consider the case $\left|\frac{c_1^2 + c_2^2 - 1}{c_1 c_2}\right| \neq 2$. Then

$$\left(c_1 \left[a_+ \left(\frac{\alpha_+}{\alpha_-}\right)^k k^{(\beta_+ - \beta_-)} + a_-\right] + \left[b_+ \left(\frac{\alpha_+}{\alpha_-}\right)^k k^{(\beta_+ - \beta_-)} + b_-\right]\right.$$
$$\left. + c_2 \left[a_+ \left(\frac{\alpha_+}{\alpha_-}\right)^k \alpha_+ k^{(\beta_+ - \beta_-)} + a_- \alpha_-\right]\right)(1 + o(1)) = 0 \text{ as } k \to \infty.$$

Therefore, for any number k greater than some big enough positive K one has:

$$[c_1 a_+ + b_+ + c_2 a_+ \alpha_+]\left(\frac{\alpha_+}{\alpha_-}\right)^k k^{(\beta_+ - \beta_-)} + [c_1 a_- + b_- + c_2 a_- \alpha_-] = 0,$$

hence $b_{\pm} = -(c_1 + \alpha_{\pm}c_2)a_{\pm}$. Thus (2.9) admits the following form:

$$u_{2k-1} = a_+ v^+_k + a_- v^-_k,$$
$$u_{2k} = -(c_1 + \alpha_+ c_2)a_+ w^+_k - (c_1 + \alpha_- c_2)a_- w^-_k.$$

It is clear now, that the vectors u^+ and u^- defined as follows:

$$u^+_{2k-1} = v^+_k, \quad u^+_{2k} = -(c_1 + \alpha_+ c_2)w^+_k,$$

$$u^-_{2k-1} = v^-_k, \quad u^-_{2k} = -(c_1 + \alpha_- c_2)w^-_k,$$

are two linearly independent solutions of the recurrence relation (2.1).

The second case here, $\left| \frac{c_1^2 + c_2^2 - 1}{c_1 c_2} \right| = 2$, can be treated in an absolutely analogous fashion. $\qquad\qquad\square$

Due to Gilbert-Pearson-Khan subordinacy theory [6], [12], we are now ready to prove our main result concerning the spectral structure of the operator J.

Theorem 2.3. *Depending on the modulation parameters c_1 and c_2, there are four distinct cases, describing the spectral structure of the operator J:*

(a) *If $\left| \frac{c_1^2 + c_2^2 - 1}{c_1 c_2} \right| < 2$, the spectrum is purely absolutely continuous with local multiplicity one almost everywhere on $\mathbb{R}$,*

(b) *If $\left| |c_1| - |c_2| \right| = 1$ and $c_1 c_2 \neq 0$, the spectrum is purely absolutely continuous with local multiplicity one almost everywhere on $(-\infty; \frac{1}{2})$ and pure point on $(\frac{1}{2}; +\infty)$,*

(c) *If $|c_1| + |c_2| = 1$ and $c_1 c_2 \neq 0$, the spectrum is purely absolutely continuous with local multiplicity one almost everywhere on $(\frac{1}{2}; +\infty)$ and pure point on $(-\infty; \frac{1}{2})$,*

(d) *If $\left| \frac{c_1^2 + c_2^2 - 1}{c_1 c_2} \right| > 2$, the spectrum is pure point.*

The four cases described above are illustrated by Figure 1.

Proof. Without loss of generality, assume that $c_1, c_2 > 0$. Changing the sign of c_1 or c_2 leads to an unitarily equivalent operator.

Consider subordinacy properties of generalized eigenvectors [12].

If $\frac{|c_1^2 + c_2^2 - 1|}{c_1 c_2} > 2$, we have $|\alpha_-| < 1 < |\alpha_+|$. By Lemma 2.2, u_- is a subordinate solution and lies in $l^2(\mathbb{N})$. Thus, every real λ can either be an eigenvalue or belong to the resolvent set of the operator J.

If $\frac{|c_1^2 + c_2^2 - 1|}{c_1 c_2} < 2$, we have Re $\alpha_+ = $ Re α_-, Re $\beta_+ = $ Re β_-, $|u_n^+| \sim |u_n^-|$ as $n \to \infty$, and there is no subordinate solution for all real λ. The spectrum of J in this situation is purely absolutely continuous.

If $\frac{c_1^2 + c_2^2 - 1}{c_1 c_2} = 2$, which is equivalent to $|c_1 - c_2| = 1$, then either $\lambda > \frac{1}{2}$ or $\lambda < \frac{1}{2}$. If $\lambda > \frac{1}{2}$, then $|\alpha| = 1$, $\delta_+ = -\delta_- > 0$ and u_- is subordinate and lies in $l^2(\mathbb{N})$, hence λ can either be an eigenvalue or belong to the resolvent set. If $\lambda < \frac{1}{2}$, then $|\alpha| = 1$, both δ_+ and δ_- are pure imaginary, $|u_n^+| \sim |u_n^-|$ as $n \to \infty$, no subordinate solution exists and ultimately λ belongs to purely absolutely continuous spectrum.

If $\frac{c_1^2 + c_2^2 - 1}{c_1 c_2} = -2$, which is equivalent to $c_1 + c_2 = 1$, the subcases $\lambda > \frac{1}{2}$ and $\lambda < \frac{1}{2}$ change places, which completes the proof. $\qquad\square$

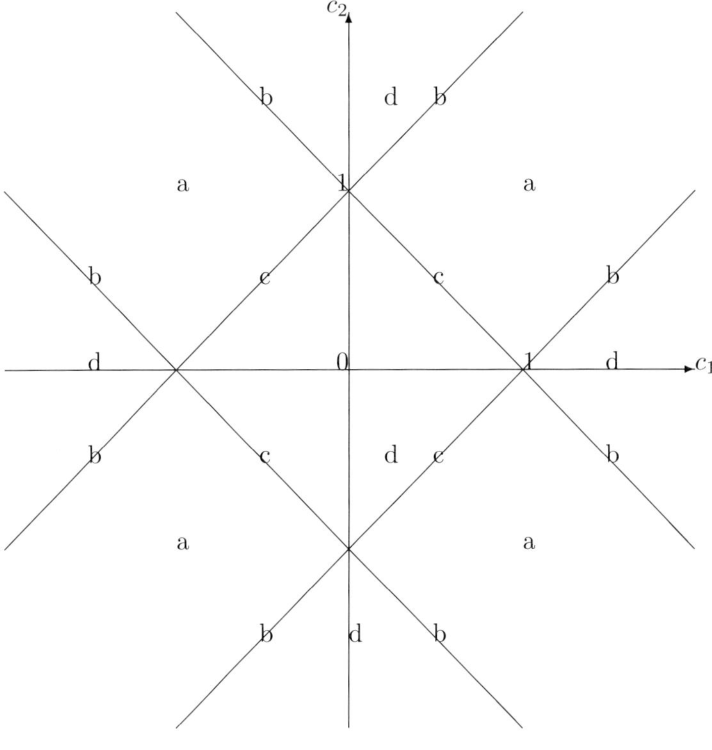

FIGURE 1

These results elaborate the domain structure, described in Section 1: we have obtained the information on the spectral structure of the operator J when the modulation parameters are on the boundaries of regions.

3. Criterion of semiboundedness and discreteness of the spectrum

We start with the following theorem which constitutes a criterion of semiboundedness of the operator J.

Theorem 3.1. *Let $c_1 c_2 \neq 0$.*

1. *If $|c_1| + |c_2| > 1$, then the operator J is not semibounded.*
2. *If $|c_1| + |c_2| \leq 1$, then the operator J is semibounded from below.*

Proof. Due to Theorem 2.3, there are four distinct cases of the spectral structure of the operator J, depending on the values of parameters c_1 and c_2 (see Figure 1).

The case (a), i.e., $\left| \frac{c_1^2 + c_2^2 - 1}{c_1 c_2} \right| < 2$, is trivial, since $\sigma_{\mathrm{ac}}(J) = \mathbb{R}$.

We are going to prove the assertion in the case (d), i.e., $\left| \frac{c_1^2 + c_2^2 - 1}{c_1 c_2} \right| > 2$, using the result of Janas and Naboko [8]. According to them, semiboundedness of the operator J depends on the location of the point zero relative to the spectrum of the periodic operator J_{per} ([8], see also Section 1).

It is easy to see, that the absolutely continuous spectrum of the operator J_{per} in our case consists of two intervals,

$$\sigma_{\mathrm{ac}}(J_{\mathrm{per}}) = [\lambda_{-+}; \lambda_{--}] \bigcup [\lambda_{+-}; \lambda_{++}],$$

where $\lambda_{\pm+} = 1 \pm (|c_1| + |c_2|)$, $\lambda_{\pm-} = 1 \pm ||c_1| - |c_2||$ and

$$\lambda_{-+} < \lambda_{--} < 1 < \lambda_{+-} < \lambda_{++}.$$

As it was established in [8], if the point zero lies in the gap between the intervals of the absolutely continuous spectrum of the operator J_{per}, then the operator J is not semibounded. If, on the other hand, the point zero lies to the left of the spectrum of the operator J_{per}, then the operator J is semibounded from below. A direct application of this result completes the proof in the case (d).

We now pass over to the cases (b) and (c), i.e., $\left| \frac{c_1^2 + c_2^2 - 1}{c_1 c_2} \right| = 2$, $c_1 c_2 \neq 0$. This situation is considerably more complicated, since the point zero lies right on the edge of the absolutely continuous spectrum of the operator J_{per}. We consider the cases (b) and (c) separately.

(b): We have to prove, that the operator J is not semibounded. By Theorem 2.3, $\sigma_{\mathrm{ac}}(J) = (-\infty; \frac{1}{2}]$, thus the operator J is not semibounded from below. Now consider the quadratic form of the operator, taken on the canonical basis element e_n. We have

$$(Je_n, e_n) = q_n \to +\infty, \ n \to \infty,$$

thus the operator J is not semibounded.

(c): We will show that the operator J is semibounded from below. To this end, we estimate its quadratic form: for any $u \in D(J)$ ($D(J)$ being the domain of the operator J) one has

$$(Ju, u) = \sum_{n=1}^{\infty} n|u_n|^2 + \sum_{k=1}^{\infty} c_1(2k-1)(u_{2k-1}\overline{u_{2k}}$$

$$+ \overline{u_{2k-1}}u_{2k}) + \sum_{k=1}^{\infty} c_2(2k)(u_{2k}\overline{u_{2k+1}} + \overline{u_{2k}}u_{2k+1}).$$

Using the Cauchy inequality [1] and taking into account, that $|c_1| + |c_2| = 1$, we ultimately arrive at the estimate

$$(Ju, u) \geq \sum_{n=1}^{\infty} n|u_n|^2 - \sum_{k=1}^{\infty}(|c_1|(2k-1)|u_{2k-1}|^2 + |c_1|(2k-1)|u_{2k}|^2)$$

$$- \sum_{k=1}^{\infty}(|c_2|(2k)|u_{2k}|^2 + |c_2|(2k)|u_{2k+1}|^2) =$$

$$= \sum_{k=1}^{\infty} (|c_1||u_{2k}|^2 + |c_2||u_{2k-1}|^2) \geq \min\{|c_1|, |c_2|\} \|u\|^2 > 0, \quad (3.1)$$

which completes the proof. $\qquad\square$

The remainder of the present section is devoted to the proof of discreteness of the operator's pure point spectrum in the case (c) of Theorem 2.3, i.e., when $|c_1| + |c_2| = 1$, $c_1 c_2 \neq 0$.

By Theorem 2.3, in this situation the absolutely continuous spectrum covers the interval $[\frac{1}{2}; +\infty)$ and the remaining part of the spectrum, if it is present, is of pure point type. The estimate (3.1) obtained in the proof of the previous theorem implies that there is no spectrum in the interval $(-\infty; \min\{|c_1|, |c_2|\})$. We will prove that nonetheless if $|c_1| \neq |c_2|$, the pure point spectral component of the operator J is non-empty.

It is clear, that if $|c_1| = |c_2| = \frac{1}{2}$, the spectrum of the operator J in the interval $(-\infty; \frac{1}{2})$ is empty and the spectrum in the interval $(\frac{1}{2}; +\infty)$ is purely absolutely continuous. This situation together with its generalization towards Jacobi matrices with zero row sums was considered by Dombrowski and Pedersen in [3] and absolute continuity of the spectrum was established.

Theorem 3.2. *In the case* (c), *i.e., when* $|c_1| + |c_2| = 1$, $c_1 c_2 \neq 0$, *under an additional assumption* $|c_1| \neq |c_2|$ *the spectrum of the operator* J *in the interval* $(-\infty; \frac{1}{2})$ *is non-empty.*

Proof. Without loss of generality, assume that $0 < c_1$, $c_2 < 1$. Changing the sign of c_1, c_2 or both leads to an unitarily equivalent operator.

Consider the quadratic form of the operator $J - \frac{1}{2}I$ for $u \in D(J)$.

$$\left(\left(J - \frac{1}{2}I \right) u, u \right) = \sum_{n=1}^{\infty} \left[q_n |u_n|^2 + \lambda_n (u_n \overline{u_{n+1}} + \overline{u_n} u_{n+1}) - \frac{1}{2}|u_n|^2 \right]$$

$$= \sum_{n=1}^{\infty} \left[n|u_n|^2 + c_n n(|u_{n+1} + u_n|^2 - |u_{n+1}|^2 - |u_n|^2) - \frac{1}{2}|u_n|^2 \right].$$

Shifting the index n by 1 in the term $c_n n |u_{n+1}|^2$ and then using the 2-periodicity of the sequence $\{c_n\}$, we have

$$\left(\left(J - \frac{1}{2}I \right) u, u \right)$$

$$= \sum_{n=1}^{\infty} [c_n n |u_{n+1} + u_n|^2] + \sum_{n=1}^{\infty} \left[n|u_n|^2 - c_n n |u_n|^2 - c_{n+1}(n-1)|u_n|^2 - \frac{1}{2}|u_n|^2 \right]$$

$$= \sum_{n=1}^{\infty} [c_n n |u_{n+1} + u_n|^2] + \sum_{n=1}^{\infty} \left(\frac{c_{n+1} - c_n}{2} \right) |u_n|^2 =$$

$$= \sum_{k=1}^{\infty} [c_1(2k-1)|u_{2k-1} + u_{2k}|^2 + c_2(2k)|u_{2k} + u_{2k+1}|^2]$$

$$- \left(\frac{c_1 - c_2}{2}\right) \sum_{k=1}^{\infty} [|u_{2k-1}|^2 - |u_{2k}|^2].$$

We need to find a vector $u \in D(J)$ which makes this expression negative. The following lemma gives a positive answer to this problem via an explicit construction and thus completes the proof.

Lemma 3.3. *For $0 < c_1, c_2 < 1$, $c_1 + c_2 = 1$ there exists a vector $u \in l_{\mathrm{fin}}(\mathbb{N})$ such that*

$$\sum_{k=1}^{\infty} [c_1(2k-1)|u_{2k-1} + u_{2k}|^2 + c_2(2k)|u_{2k} + u_{2k+1}|^2]$$

$$< \left(\frac{c_1 - c_2}{2}\right) \sum_{k=1}^{\infty} [|u_{2k-1}|^2 - |u_{2k}|^2] \tag{3.2}$$

Proof. We consider the cases $c_1 > c_2$ and $c_1 < c_2$ separately. Below we will see, that the latter can be reduced to the former.

1. $c_1 - c_2 > 0$. In this case, we will choose a vector v from $l_{\mathrm{fin}}(\mathbb{N})$ with nonnegative components such that if the vector u is defined by $u_{2k-1} = v_k$, $u_{2k} = -v_{k+1}$, the condition (3.2) holds true. In terms of such v, the named condition admits the following form:

$$c_1 \sum_{k=1}^{\infty} [(2k-1)(v_k - v_{k+1})^2] < \left(\frac{c_1 - c_2}{2}\right) v_1^2. \tag{3.3}$$

2. $c_1 - c_2 < 0$. In this case, we will choose a vector $w \in l_{\mathrm{fin}}(\mathbb{N})$ with non-negative components and the value t such that if the vector u is defined by $u_{2k} = -w_k$, $u_{2k-1} = w_k$, $u_1 = tw_1$, the condition (3.2) holds true. In terms of w and t, condition (3.2) admits the form

$$c_2 \sum_{k=1}^{\infty} [(2k)(w_k - w_{k+1})^2] < \left(-\frac{t^2}{2} + 2c_1 t + \frac{1 - 4c_1}{2}\right) w_1^2. \tag{3.4}$$

Take t such that the expression in the brackets on the right-hand side of the latter inequality is positive. This choice is possible, if we take take the maximum of the parabola $y(t) = -\frac{t^2}{2} + 2c_1 t + \frac{1-4c_1}{2}$, located at the point $t_0 = 2c_1$. Then the incquality (3.4) admits the form

$$c_2 \sum_{k=1}^{\infty} [(2k)(w_k - w_{k+1})^2] < \frac{(c_1 - c_2)^2}{2} w_1^2. \tag{3.5}$$

We will now explicitly construct a vector $v^{(N)} \in l_{\mathrm{fin}}(\mathbb{N})$ such that it satisfies both (3.3) and (3.5) for sufficiently large numbers of N. Consider the sequence

$v_n^{(N)} = \sum_{k=n}^{N} \frac{1}{k}$ for $n \leq N$ and put $v_n^{(N)} = 0$ for $n > N$. It is clear, that as $N \to +\infty$,

$$(v_1^{(N)})^2 = \left(\sum_{k=1}^{N} \frac{1}{k} \right)^2 \sim (\ln N)^2,$$

and

$$\sum_{k=1}^{\infty} [(2k-1)(v_k^{(N)} - v_{k+1}^{(N)})^2] \sim \sum_{k=1}^{\infty} [(2k)(v_k^{(N)} - v_{k+1}^{(N)})^2] \sim 2\ln N = o\left((v_1^{(N)})^2 \right),$$

which completes the proof of Lemma 3.3. $\quad\square\square$

We are now able to prove the discreteness of the pure point spectral component of the operator J in the case (c) of Theorem 2.3, which is non-empty due to Theorem 3.2.

Theorem 3.4. *In the case* (c), *i.e., when* $|c_1| + |c_2| = 1$ *and* $c_1 c_2 \neq 0$, *under an additional assumption* $|c_1| \neq |c_2|$ *the spectrum of the operator* J *in the interval* $(-\infty; \min\{|c_1|, |c_2|\})$ *is empty, the spectrum in the interval* $[\min\{|c_1|, |c_2|\}; \frac{1}{2})$ *is discrete, and the following estimate holds for the number of eigenvalues* λ_n *in the interval* $(-\infty; \frac{1}{2} - \varepsilon)$, $\varepsilon > 0$:

$$\#\{\lambda_n : \lambda_n < \frac{1}{2} - \varepsilon\} \leq \frac{1}{\varepsilon}.$$

Proof. According to the Glazman Lemma [1], dimension of the spectral subspace, corresponding to the interval $(-\infty; \frac{1}{2} - \varepsilon)$, is less or equal to the co-dimension of any subspace $H_\varepsilon \subset l^2(\mathbb{N})$ such that

$$(Ju, u) \geq \left(\frac{1}{2} - \varepsilon \right) \|u\|^2 \tag{3.6}$$

for any $u \in D(J) \cap H_\varepsilon$. Consider subspaces l_N^2 of vectors with zero first N components, i.e., $l_N^2 := \{u \in l^2(\mathbb{N}) : u_1 = u_2 = \cdots = u_N = 0\}$. For any ε, $0 < \varepsilon < \frac{1}{2}$ we will find a number $N(\varepsilon)$ such that for any vector from $H_\varepsilon = l_{N(\varepsilon)}^2$ the inequality (3.6) is satisfied. We consider ε such that $0 < \varepsilon < \frac{1}{2}$ only, since the spectrum is empty in the interval $(-\infty; 0]$ (see the estimate (3.1)). The co-dimension of the subspace $l_{N(\varepsilon)}^2$ is $N(\varepsilon)$, so this value estimates from above the number of eigenvalues in the interval $(-\infty; \frac{1}{2} - \varepsilon)$.

Consider the quadratic form of the operator J for $u \in D(J)$:

$$(Ju, u) = \sum_{n=1}^{\infty} n[|u_n|^2 + c_n 2 \operatorname{Re}(u_n \overline{u_{n+1}})]$$

$$\geq \sum_{n=1}^{\infty} n \left[|u_n|^2 - |c_n| \left(I_n |u_n|^2 + \frac{1}{I_n} |u_{n+1}|^2 \right) \right]$$

$$= |u_1|^2 (1 - |c_1| I_1) + \sum_{n=2}^{\infty} n |u_n|^2 \left[1 - |c_n| I_n - |c_{n-1}| \frac{1}{I_{n-1}} \frac{n-1}{n} \right],$$

where we have used the Cauchy inequality [1] $2\,\mathrm{Re}\,(u_n\overline{u_{n+1}}) \le I_n|u_n|^2 + \frac{1}{I_n}|u_{n+1}|^2$ with the sequence $I_n > 0$, $n \in \mathbb{N}$ which we will fix as $I_n := 1 - \frac{\phi}{n}$ in order that the expression

$$1 - |c_n|I_n - |c_{n-1}|\frac{1}{I_{n-1}}\left(1 - \frac{1}{n}\right) \tag{3.7}$$

takes its simplest form. This choice cancels out the first order with respect to n. The value of ϕ in the interval $0 < \phi < 1$ will be fixed later on. We have:

$$\frac{1}{I_{n-1}} = 1 + \frac{\phi}{n} + \phi_n, \tag{3.8}$$

where $\phi_n = O\left(\frac{1}{n^2}\right)$, $n \to \infty$. Moreover, as can be easily seen,

$$\phi_n = \frac{\phi(\phi+1)}{n(n-1-\phi)}.$$

After substituting the value of ϕ_n into (3.8) and then into (3.7) we obtain:

$$1 - |c_n|I_n - |c_{n-1}|\frac{1}{I_{n-1}}\left(1 - \frac{1}{n}\right) = \frac{1}{n}(\phi|c_n| + (1-\phi)|c_{n-1}|) + \theta_n \tag{3.9}$$

with $\theta_n := |c_{n-1}|\left(\frac{\phi}{n^2} - \phi_n\left(1 - \frac{1}{n}\right)\right) = O\left(\frac{1}{n^2}\right)$ as $n \to \infty$.

Choose ϕ in order to make the right-hand side of expression (3.9) symmetric with respect to the modulation parameters c_1 and c_2: $\phi = \frac{1}{2}$. Then

$$1 - |c_n|I_n - |c_{n-1}|\frac{1}{I_{n-1}}\left(1 - \frac{1}{n}\right) = \frac{1}{2n} + \theta_n.$$

Consequently,

$$(Ju, u) \ge |u_1|^2\left(1 - \frac{|c_1|}{2}\right) + \sum_{n=2}^{\infty}\left(\frac{1}{2} + \theta_n n\right)|u_n|^2.$$

Since $n\theta_n \to 0$ as $n \to \infty$, we can choose $N(\varepsilon)$ such that for any $n > N(\varepsilon)$ the condition $n\theta_n > -\varepsilon$ holds. Thus condition (3.6) will be satisfied for all vectors from $D(J) \cap l^2_{N(\varepsilon)}$, since their first components are zeros.

The discreteness of the pure point spectrum is proved. We pass on to the proof of the estimate for $N(\varepsilon)$. We start with θ_n:

$$|\theta_n| \le \left|\frac{\phi}{n^2} - \frac{\phi(\phi+1)}{n(n-1-\phi)}\frac{n-1}{n}\right| = \frac{1}{2n^2}\left|1 - \frac{3}{2}\frac{n-1}{n-\frac{3}{2}}\right|.$$

We have

$$n > 2 \Rightarrow \left\{2 > \frac{n-1}{n-\frac{3}{2}} > 1\right\} \Rightarrow \left\{n|\theta_n| < \frac{1}{4n}\right\}.$$

Taking $N(\varepsilon) = \frac{1}{\varepsilon}$, $0 < \varepsilon < \frac{1}{2}$, we see that for any $n > N(\varepsilon) > 2$ the condition $n\theta_n > -\varepsilon$ holds. Thus, the condition (3.6) is satisfied for all vectors from $D(J) \cap l^2_{N(\varepsilon)}$, which completes the proof. $\qquad\square$

4. The degenerate case

Now we consider the case, when one of the modulation parameters turns to zero (we call this case degenerate). Formally speaking, we cannot call such matrix a Jacobi one, but this limit case is of certain interest for us, supplementing the whole picture.

Theorem 4.1. *If $c_1 c_2 = 0$, $c \neq 0$ (denoting $c := \max\{|c_1|, |c_2|\}$), then the spectrum of the operator J is the closure of the set of eigenvalues λ_n:*

$$\sigma(J) = \overline{\{\lambda_n,\ n \in \mathbb{N}\}}.$$

The set of eigenvalues is

$$\{\lambda_n,\ n \in \mathbb{N}\} \begin{cases} \{\lambda_n^+,\ \lambda_n^-,\ n \in \mathbb{N}\}, & \text{if } c_1 \neq 0,\ c_2 = 0 \\ \{1,\ \tilde{\lambda}_n^+,\ \tilde{\lambda}_n^-,\ n \in \mathbb{N}\}, & \text{if } c_1 = 0,\ c_2 \neq 0, \end{cases}$$

where eigenvalues $\lambda_n^\pm$, $\tilde{\lambda}_n^\pm$ have the following asymptotics:

$$\lambda_n^+,\ \tilde{\lambda}_n^+ = 2(1+c)n + O(1),\ n \to \infty,$$

$$\lambda_n^- = 2(1-c)n + \left(c - \frac{1}{2}\right) - \frac{1}{16cn} + O\left(\frac{1}{n^2}\right),\ n \to \infty$$

$$\tilde{\lambda}_n^- = 2(1-c)n + \left(2c - \frac{3}{2}\right) - \frac{1}{16cn} + O\left(\frac{1}{n^3}\right),\ n \to \infty.$$

Proof. When one of the parameters c_1 or c_2 is zero, the infinite matrix consists of 2x2 (or 1x1) blocks. Thus, the operator J is an orthogonal sum of finite 2x2 (or 1x1) matrices J_n, $J = \bigoplus_{n=1}^{\infty} J_n$. Then, the spectrum of the operator J is the closure of the sum of spectrums of these matrices, $\sigma(J) = \overline{\bigcup_{n=1}^{\infty} \sigma(J_n)}$. Let us calculate $\sigma(J_n)$.

If $c_1 \neq 0$, $c_2 = 0$, then

$$J_n \begin{pmatrix} 2n - 1 & c_1(2n-1) \\ c_1(2n-1) & 2n \end{pmatrix}$$

and $\sigma(J_n) = \{\lambda_n^+,\ \lambda_n^-\}$, where $\lambda_n^\pm = \frac{4n-1\pm\sqrt{4c^2(2n-1)^2+1}}{2}$ and it is easy to see that

$$\lambda_n^\pm = 2(1\pm c)n - \left(\frac{1}{2} \pm c\right) \pm \frac{1}{16cn} + O\left(\frac{1}{n^2}\right),\ n \to \infty.$$

If $c_1 = 0$, $c_2 \neq 0$, then $J_1 = 1$, $\sigma(J_1) = \{1\}$,

$$J_n \begin{pmatrix} 2n - 2 & c_2(2n-2) \\ c_2(2n-2) & 2n - 1 \end{pmatrix},\ n \geq 2$$

and $\sigma(J_n) = \{\tilde{\lambda}_n^+,\ \tilde{\lambda}_n^-\}$, $n \geq 1$, where $\tilde{\lambda}_n^\pm = \frac{4n-3\pm\sqrt{4c^2(2n-2)^2+1}}{2}$ and it is easy to see that

$$\tilde{\lambda}_n^\pm = 2(1\pm c)n - \left(\frac{3}{2} \pm 2c\right) \pm \frac{1}{16cn} + O\left(\frac{1}{n^2}\right),\ n \to \infty,$$

which completes the proof. $\qquad\square$

Remark 4.2. From the last theorem it follows that as $n \to \infty$, λ_n^+, $\tilde{\lambda}_n^+ \to +\infty$. As for λ_n^- and $\tilde{\lambda}_n^-$, their asymptotic behavior depends on the parameter c:

If $c > 1$, then λ_n^-, $\tilde{\lambda}_n^- \to -\infty$.

If $c = 1$, then λ_n^-, $\tilde{\lambda}_n^- \to \frac{1}{2}$.

Finally, if $0 < c < 1$, then λ_n^-, $\tilde{\lambda}_n^- \to +\infty$.

Hence, if $0 < c \leq 1$, the operator J is semibounded from below, and if $c > 1$, the operator J is not semibounded. This clearly corresponds to results, obtained in Section 3.

When we move along the side of the boundary square (see Figure 1, case (c)) towards one of the points $\{D_j\}_{j=1}^4 = \{(1;0);\ (0;1);\ (-1;0);\ (0;-1)\}$, the absolutely continuous spectrum covers the interval $[\frac{1}{2};+\infty)$. At the same time, at each limit point D_j, $j = 1,\ 2,\ 3,\ 4$, the spectrum of J becomes pure point, which demonstrates yet another phenomenon of the spectral phase transition. Moreover, note that the spectrum at each limit point consists of two series of eigenvalues, one going to $+\infty$, another accumulating to the point $\lambda = \frac{1}{2}$, both points prior to the spectral phase transition having been the boundaries of the absolutely continuous spectrum.

Remark 4.3. The proof of discreteness of the spectrum in the case (c) of Theorem 2.3 essentially involves the semiboundedness property of the operator J. In the case (b) one does not have the advantage of semiboundedness and due to that reason the proof of discreteness supposedly becomes much more complicated.

Remark 4.4. The choice $q_n = n$ was determined by the possibility to apply the Birkhoff-Adams technique. It should be mentioned that much more general situation $q_n = n^\alpha$, $0 < \alpha < 1$ may be considered on the basis of the generalized discrete Levinson Theorem. Proper approach has been developed in [11], see also [5]. One can apply similar method in our situation. Another approach which is also valid is so-called Jordan box case and is presented in [7].

Acknowledgements

The author expresses his deep gratitude to Prof. S.N. Naboko for his constant attention to this work and for many fruitful discussions of the subject and also to Dr. A.V. Kiselev for his help in preparation of this paper.

References

[1] N.I. Akhiezer, I.M. Glazman, *Theory of linear operators in Hilbert space. (2nd edition)* Dover, New York, 1993.

[2] Yu.M. Berezanskii, *Expansions in eigenfunctions of selfadjoint operators. (Russian)* Naukova Dumka, Kiev, 1965.

[3] J. Dombrowski, S. Pedersen, *Absolute continuity for unbounded Jacobi matrices with constant row sums.* J. Math. Anal. Appl., vol. **267** (2002), no. 2, pp. 695–713.

[4] S.N. Elaydi, *An Introduction to Difference Equations.* Springer-Verlag, New York, 1999.

[5] D. Damanik, S.N. Naboko, *A first order phase transition in a class of unbounded Jacobi matrices: critical coupling.* (to appear in Journal Approximation Theory)

[6] D. Gilbert, D. Pearson, *On subordinacy and analysis of the spectrum of one dimensional Schrödinger operators.* J. Math. Anal. Appl., vol. **128** (1987), pp. 30–56.

[7] J. Janas, *The asymptotic analysis of generalized eigenvectors of some Jacobi operators. Jordan box case.* (to appear in J. Difference Eq. Appl.)

[8] J. Janas, S.N. Naboko, *Criteria for semiboundedness in a class of unbounded Jacobi operators.* translation in St. Petersburg Math. J., vol. **14** (2003), no. 3, pp. 479–485.

[9] J. Janas, S.N. Naboko, *Multithreshold spectral phase transition examples for a class of unbounded Jacobi matrices.* Oper. Theory Adv. Appl., vol. **124** (2001), pp. 267–285.

[10] J. Janas, S.N. Naboko, *Spectral analysis of selfadjoint Jacobi matrices with periodically modulated entries.* J. Funct. Anal., vol. **191** (2002), no. 2 , pp. 318–342.

[11] J. Janas, S.N. Naboko, E. Sheronova, *Asymptotic behaviour of generalizes eigenvectors of Jacobi matrices in Jordan box case.* (submitted to Rocky Mount. Math. J.)

[12] S. Khan, D. Pearson, *Subordinacy and spectral theory for infinite matrices.* Helv. Phys. Acta, vol. **65** (1992), pp. 505–527.

Sergey Simonov
Department of Mathematical Physics
Institute of Physics
St. Petersburg University
Ulianovskaia 1
198904, St. Petergoff
St. Petersburg, Russia
e-mail: `sergey_simonov@mail.ru`

Operator Theory:
Advances and Applications, Vol. 174, 205–246

On Connection Between Factorizations of Weighted Schur Functions and Invariant Subspaces

Alexey Tikhonov

Abstract. We study operator-valued functions of weighted Schur classes over multiply-connected domains. There is a correspondence between functions of weighted Schur classes and so-called "conservative curved" systems introduced in the paper. In the unit disk case the fundamental relationship between invariant subspaces of the main operator of a conservative system and factorizations of the corresponding Schur class function (characteristic function) is well known. We extend this connection to weighted Schur classes. With this aim we develop new notions and constructions and make suitable changes in standard theory.

Mathematics Subject Classification (2000). Primary 47A48; Secondary 47A45, 47A56, 47A15.

Keywords. factorization, conservative system, invariant subspace, functional model, Schur class.

0. Introduction

It is well known [1, 2] that there is a one-to-one correspondence between (simple) unitary colligations

$$\mathfrak{A} = \begin{pmatrix} T & N \\ M & L \end{pmatrix} \in \mathcal{L}(H \oplus \mathfrak{N}, H \oplus \mathfrak{M}), \quad \mathfrak{A}^*\mathfrak{A} = I, \ \mathfrak{A}\mathfrak{A}^* = I$$

and operator-valued functions $\Theta(z)$ of the Schur class

$$S := \{\Theta \in H^\infty(\mathbb{D}, \mathcal{L}(\mathfrak{N}, \mathfrak{M})) : ||\Theta||_\infty \leq 1\}.$$

Here $H, \mathfrak{N}, \mathfrak{M}$ are separable Hilbert spaces and $\mathcal{L}(\mathfrak{N}, \mathfrak{M})$ is the space of all bounded linear operators acting from $\mathfrak{N}$ to $\mathfrak{M}$. The mapping $\mathfrak{A} \mapsto \Theta$ defined by the formula $\Theta(z) = L^* + zN^*(I - zT^*)^{-1}M^*$, $|z| < 1$ is one of the directions of the above mentioned correspondence. The operator-valued function $\Theta(z)$ is called the characteristic function of the unitary colligation $\mathfrak{A}$. Its property $||\Theta||_\infty \leq 1$ is a consequence of the unitary property of the colligation $\mathfrak{A}$.

The reverse direction of the correspondence is realized via functional model [1, 2], whose essential ingredients are Hardy's spaces H^2 and H_-^2 (see [3]). These two aspects of the theory (unitary colligations and Schur class functions) are equipollent: both have simple, clear and independent descriptions and we can easily change a point of view from unitary colligations to Schur class functions and back. This context gives a nice opportunity to connect operator theory and function theory in a very deep and fruitful manner [4].

One of the cornerstones of this theory is the link (see [1, 2]) between factorizations of characteristic function $\Theta(z)$ and invariant subspaces of operator T, which goes back to [5] and [6]. The most simple way to explain this connection is to look at it from the point of view of systems theory and to employ the well-known correspondence between unitary colligations $\mathfrak{A}$ and conservative linear discrete time-invariant systems $\Sigma = (T, M, N, L; H, \mathfrak{N}, \mathfrak{M})$ (see, e.g., [7])

$$\begin{cases} x(n+1) &= Tx(n) + Nu(n), \quad x(n) \in H, \ u(n) \in \mathfrak{N}, \\ y(n) &= Mx(n) + Lu(n), \quad y(n) \in \mathfrak{M}, \ n \geq 0. \end{cases}$$

The conservative property (the property of energy balance) of Σ corresponds to the unitary property of colligation $\mathfrak{A}$. If we send a sequence $\{u(n)\}$ into the system Σ with initial state $x(0) = 0$, we get the output sequence $\{y(n)\}$. For their "Laplace" transformations, we have $\hat{y}(z) = \hat{\Upsilon}(z)\hat{u}(z)$, where $\hat{u}(z) = \sum_{n=0}^\infty z^n u(n)$, $\hat{y}(z) = \sum_{n=0}^\infty z^n y(n)$, and $\hat{\Upsilon}(z) = L + zM(I - zT)^{-1}N$. Note that the transfer function $\hat{\Upsilon}(z)$ of system Σ is equal to $\Theta^\sim(z) := \Theta(\bar{z})^*$, where $\Theta(z)$ is the characteristic function of the unitary colligation $\mathfrak{A}$.

Sending the output of a system $\Sigma_2 = (T_2, M_2, N_2, L_2; H_2, \mathfrak{N}, \mathfrak{L})$ into the input of a system $\Sigma_1 = (T_1, M_1, N_1, L_1; H_1, \mathfrak{L}, \mathfrak{M})$, we obtain the cascade system $\Sigma_{21} := \Sigma_2 \cdot \Sigma_1 = (T_{21}, M_{21}, N_{21}, L_{21}; H_{21}, \mathfrak{N}, \mathfrak{M})$. It is clear that the transfer function $\hat{\Upsilon}_{21}(z) = \Theta_{21}^\sim(z)$ of the system Σ_{21} is the product of the transfer functions of systems Σ_1 and Σ_2. It is easily shown that

$$\Sigma_{21} = \left(\begin{pmatrix} T_1 & N_1 M_2 \\ 0 & T_2 \end{pmatrix}, \ (M_1, L_1 M_2), \ \begin{pmatrix} N_1 L_2 \\ N_2 \end{pmatrix}, \ L_2 L_1 \right),$$

where $H_{21} = H_1 \oplus H_2$. The subspace H_1 is invariant under the operator T_{21} and therefore, if we fix $\Theta_{21}(z)$, one may hope to study invariant subspaces of the operator T_{21} using this approach. Unfortunately, there are some pitfalls for this: the operator T_{21} can vary when we run over all factorizations of $\Theta_{21}(z)$. More precisely, its variable part is the unitary component T_{21u} from the decomposition $T_{21} = T_{21s} \oplus T_{21u}$ into completely non-unitary and unitary parts [1]. In the context of conservative systems the corresponding decomposition is $\Sigma = \Sigma_s \oplus \Sigma_u$, where $H_s = H_c \vee H_o$, $H_u = H \ominus H_s$, $H_c = \vee_{n \geq 0} T^n N(\mathfrak{N})$, $H_o = \vee_{n \geq 0} (T^*)^n M^*(\mathfrak{M})$. Here Σ_s and Σ_u are the simple and "purely unitary" parts of the system Σ, respectively. A system Σ is called simple if $H = H_s$. A system $(T, 0, 0, 0; H, \{0\}, \{0\})$ is called "purely unitary" if T is unitary.

B. Sz.-Nagy and C. Foiaş established the following criterion (see [1, 2]): *the product of consecrative systems $\Sigma_{21} = \Sigma_2 \cdot \Sigma_1$ is simple if and only if the corresponding factorization* $\Theta_{21}(z) = \Theta_2(z)\Theta_1(z)$ *is regular.* Recall that the product $\Theta_{21}(z) = \Theta_2(z)\Theta_1(z)$ of Schur class functions is regular [2] if

$$\operatorname{Ran}\left(I - \Theta_2^*(z)\Theta_2(z)\right)^{1/2} \cap \operatorname{Ran}\left(I - \Theta_1(z)\Theta_1^*(z)\right)^{1/2} = \{0\}, \quad \text{a.e. } z \in \mathbb{T}.$$

This condition is equivalent to the standard definition of regularity from [1].

Moreover, B. Sz.-Nagy and C. Foiaş described (Theorems VII.1.1 and VII.4.3 in [1]) the order preserving one-to-one correspondence between regular factorizations of a characteristic function and invariant subspaces of the corresponding model operator. The order relation for invariant subspaces is the ordinary inclusion. The order relation for factorizations is $\Theta_2\Theta_1 \prec \Theta_2'\Theta_1'$, where we write $\Theta_2\Theta_1 \prec \Theta_2'\Theta_1'$ if there exists a Schur class function θ such that $\Theta_2 = \Theta_2'\theta$ and $\Theta_1' = \theta\Theta_1$. Extension of this correspondence between factorizations and invariant subspaces to the case of weighted Schur classes is the main aim of the present paper.

We shall consider operator-valued functions (or rather, triplets of operator-valued functions) of *weighted Schur classes*

$$S_\Xi := \left\{ (\Theta^+, \Xi_+, \Xi_-) : \Theta^+ \in H^\infty(G_+, \mathcal{L}(\mathfrak{N}_+, \mathfrak{N}_-)), \right. \tag{Cfn}$$
$$\left. \forall\, \zeta \in C \;\; \forall\, n \in \mathfrak{N}_+ \;\; ||\Theta^+(\zeta)n||_{-,\zeta} \leq ||n||_{+,\zeta} \right\},$$

where $\mathfrak{N}_\pm$ are separable Hilbert spaces; G_+ is a finite-connected domain of the complex plane $\mathbb{C}$ bounded by a rectifiable Carleson curve C, $G_- = \mathbb{C}\backslash \operatorname{clos} G_+$ and $\infty \in G_-$; $\Xi_\pm$ are operator-valued weights such that $\Xi_\pm, \Xi_\pm^{-1} \in L^\infty(C, \mathcal{L}(\mathfrak{N}_\pm))$, $\Xi_\pm(\zeta) \geq 0$, $\zeta \in C$, and $||n||_{\pm,\zeta} := (\Xi_\pm(\zeta)n, n)^{1/2}$, $n \in \mathfrak{N}_\pm$. We shall also use the parallel notation $\Theta \in \operatorname{Cfn}$ whenever $\Theta \in S_\Xi$.

First, we recall the construction of free *functional model* of Sz.-Nagy-Foiaş type (see [8, 9, 16]). Let $\Pi = (\pi_+, \pi_-)$ be a pair of operators $\pi_\pm \in \mathcal{L}(L^2(C, \mathfrak{N}_\pm), \mathcal{H})$ such that

$$
\begin{array}{llll}
(\text{i})_1 & (\pi_\pm^*\pi_\pm)z = z(\pi_\pm^*\pi_\pm); & (\text{i})_2 & \pi_\pm^*\pi_\pm >> 0; \\[4pt]
(\text{ii})_1 & (\pi_-^\dagger\pi_+)z = z(\pi_-^\dagger\pi_+); & (\text{ii})_2 & P_-(\pi_-^\dagger\pi_+)P_+ = 0; \\[4pt]
(\text{iii}) & \operatorname{Ran}\pi_+ \vee \operatorname{Ran}\pi_- = \mathcal{H}, & &
\end{array} \tag{Mod}
$$

where $\mathfrak{N}_\pm, \mathcal{H}$ are separable Hilbert spaces; $A >> 0$ means that $\exists\, c > 0$ such that $\forall u\; (Au, u) \geq c(u, u)$; the (nonorthogonal) projections $P_\pm$ are uniquely determined by conditions $\operatorname{Ran} P_\pm = E^2(G_\pm, \mathfrak{N}_\pm)$ and $\operatorname{Ker} P_\pm = E^2(G_\mp, \mathfrak{N}_\pm)$ (since the curve C is a Carleson curve, the projections $P_\pm$ are bounded); the spaces $E^2(G_\pm, \mathfrak{N}_\pm)$ are Smirnov's spaces [3] of vector-valued functions with values in $\mathfrak{N}_\pm$; the operators $\pi_\pm^\dagger$ are adjoint to $\pi_\pm$ if we regard $\pi_\pm : L^2(C, \Xi_\pm) \to \mathcal{H}$ as operators acting from weighted L^2 spaces with operator-valued weights $\Xi_\pm = \pi_\pm^*\pi_\pm$. In this interpretation $\pi_\pm$ are isometries. For such pairs $\Pi = (\pi_+, \pi_-)$, we shall say that Π is a free functional model and write $\Pi \in \operatorname{Mod}$.

Note that, in our approach, we strive to hold analyticity in both the domains G_+ and G_- with the aim to reserve possibility to exploit techniques typical for boundary values problems (singular integral operators, the Riemann-Hilbert problem, the stationary scattering theory, including smooth methods of T.Kato). Thus we will use both Smirnov's spaces $E^2(G_\pm)$, which are analogues of the Hardy spaces H^2 and H^2_-. The requirement of analyticity in both the domains $G_\pm$ conflicts with orthogonality: in general, the decomposition $L^2(C) = E^2(G_+) \dotplus E^2(G_-)$ is not orthogonal. Note that, starting from [10], the use of the combination "analyticity in G_+ and orthogonality" (without analyticity in G_-) is a mainstream of development in the multiply-connected case. In this paper we sacrifice the orthogonality and hold analyticity in both the domains $G_\pm$. Therefore at this point we fork with the traditional way of generalization of Sz.-Nagy-Foiaş theory [11, 12, 13]. Nevertheless, our requirements are substantial as well and descend from applications (see [8, 14, 15]: in [8] we studied the duality of spectral components for trace class perturbations of normal operators with spectrum on a curve; the functional model from [14] goes back to the paper [15], which is devoted to spectral analysis of linear neutral functional differential equations).

The operator $\pi_-^\dagger \pi_+$ can be regarded as an analytic operator-valued function. Namely, $\Theta^+(z) := (\pi_-^\dagger \pi_+)(z)$, $z \in G_+$. In this connection, we shall say that the triplet of operator-valued functions

$$\Theta = (\pi_-^\dagger \pi_+, \pi_+^* \pi_+, \pi_-^* \pi_-) \in S_\Xi \,. \tag{MtoC}$$

is the *characteristic function* of model $\Pi = (\pi_+, \pi_-)$. Besides, (MtoC) defines the transformation $\Theta = \mathcal{F}_{cm}(\Pi)$. Conversely, for a given $\Theta \in S_\Xi$, it is possible to construct (up to unitary equivalence) a functional model $\Pi \in \mathrm{Mod}$ such that $\Theta = (\pi_-^\dagger \pi_+, \pi_+^* \pi_+, \pi_-^* \pi_-)$, i.e., there exists the inverse transformation $\mathcal{F}_{mc} := \mathcal{F}_{cm}^{-1}$ (see Proposition 1.1).

At this moment we should look for a suitable generalization of conservative systems (=unitary colligations). We define curved conservative systems in terms of the functional model. Let $\Pi \in \mathrm{Mod}$. We put

$$\widehat{\Sigma} = \mathcal{F}_{sm}(\Pi) := (\widehat{T}, \widehat{M}, \widehat{N}, \widehat{\Theta}_u, \widehat{\Xi}; \mathcal{K}_\Theta, \mathfrak{N}_+, \mathfrak{N}_-)$$

with

$$
\begin{aligned}
\widehat{T} &\in \mathcal{L}(\mathcal{K}_\Theta)\,, & \widehat{T}f &:= \mathcal{U}f - \pi_+ \widehat{M}f\,, \quad f \in \mathcal{K}_\Theta\,; \\
\widehat{M} &\in \mathcal{L}(\mathcal{K}_\Theta, \mathfrak{N}_+)\,, & \widehat{M}f &:= \frac{1}{2\pi i} \int_C (\pi_+^\dagger f)(z)\, dz\,; \\
\widehat{N} &\in \mathcal{L}(\mathfrak{N}_-, \mathcal{K}_\Theta)\,, & \widehat{N}n &:= P_\Theta \pi_- n\,, \quad n \in \mathfrak{N}_-\,; \\
\widehat{\Xi} &:= (\pi_+^* \pi_+, \pi_-^* \pi_-)\,;
\end{aligned}
\tag{MtoS}
$$

where $\mathcal{K}_\Theta := \mathrm{Ran}\, P_\Theta$, $P_\Theta := (I - \pi_+ P_+ \pi_+^\dagger)(I - \pi_- P_- \pi_-^\dagger)$; the normal operator $\mathcal{U}$, which spectrum is absolutely continuous and lies on C, is uniquely determined by conditions $\mathcal{U}\pi_\pm = \pi_\pm z$; the "unitary part" $\widehat{\Theta}_u$ corresponds to the unitary constant Θ_u^0 from pure-unitary decomposition [1, 2] of Schur class function $\Theta^0(w) = \Theta_p^0(w) \oplus \Theta_u^0$, $w \in \mathbb{D}$. The contractive-valued function $\Theta^0(w)$ is the lifting (see, e.g., [10]) of the (possible multiple-valued) character-automorphic operator-valued function $(\chi_- \Theta^+ \chi_+^{-1})(z)$ to the universal cover space $\mathbb{D}$ of the domain G_+, where $\chi_\pm$ are outer (possible multiple-valued) character-automorphic operator-valued functions such that $\chi_\pm^* \chi_\pm = \Xi_\pm = \pi_\pm^* \pi_\pm$. In the sequel, we shall refer $\widehat{\Sigma}$ as the model system and the operator $\widehat{T}$ as the model (or main) operator. Note also that (MtoS) defines the transformation $\widehat{\Sigma} = \mathcal{F}_{sm}(\Pi)$.

A coupling of operators and Hilbert spaces $\Sigma = (T, M, N, \Theta_u, \Xi; H, \mathfrak{N}, \mathfrak{M})$ is called a *conservative curved system* if there exists a functional model Π with $\mathfrak{N}_+ = \mathfrak{N}$ and $\mathfrak{N}_- = \mathfrak{M}$, a Hilbert space $\mathcal{K}_u$, a normal operator $\widehat{T}_u \in \mathcal{L}(\mathcal{K}_u), \sigma(\widehat{T}_u) \subset C$ and an invertible operator $X \in \mathcal{L}(H, \mathcal{K}_\Theta \oplus \mathcal{K}_u)$ such that

$$\Sigma = (T, M, N, \Theta_u, \Xi; H, \mathfrak{N}, \mathfrak{M}) \overset{X}{\sim} (\widehat{\Sigma} \oplus \widehat{\Sigma}_u), \tag{Sys}$$

where $\widehat{\Sigma} = \mathcal{F}_{sm}(\Pi)$ and $\widehat{\Sigma}_u = (\widehat{T}_u, 0, 0, 0; \mathcal{K}_u, \{0\}, \{0\})$. We write $\Sigma_1 \overset{X}{\sim} \Sigma_2$ if

$$XT_1 = T_2 X, \quad M_1 = M_2 X, \quad N_1 X = N_2, \quad \Theta_{1u} = \Theta_{2u}, \quad \Xi_1 = \Xi_2.$$

The spaces $\mathcal{K}_\Theta$ and $\mathcal{K}_u$ play roles of the simple and "normal" subspaces of the system $\widehat{\Sigma} \oplus \widehat{\Sigma}_u$, respectively. A curved conservative system Σ is called simple if

$$\rho(T) \cap G_+ \neq \emptyset \qquad \text{and} \qquad \bigcap_{z \in \rho(T)} \operatorname{Ker} M(T - z)^{-1} = \{0\}.$$

For unitary colligations under the condition $\rho(T) \cap \mathbb{D} \neq \emptyset$, this definition is equivalent to the standard one from [2]. Note that there appear some troubles if we extend the standard definition (simple subspace = controllable subspace $\vee$ observable subspace) straightforwardly.

In the case when $G_+ = \mathbb{D}$ and $\Xi_\pm \equiv I$, a conservative curved system $\Sigma = (T, M, N, \Theta_u, \Xi; H, \mathfrak{N}, \mathfrak{M})$ is exactly conservative, the block-matrix $\mathfrak{A} = \begin{pmatrix} T & N \\ M & \Theta^+(0)^* \end{pmatrix}$ is a unitary colligation and the operator-valued function $\Theta^+ = \pi_-^\dagger \pi_+$ can be expressed as $\Theta^+(z) = L^* + zN^*(I - zT^*)^{-1}M^*, |z| < 1$. For general simple-connected domains we lose the unitary property but we can regard a system $\Sigma = (T, M, N)$ as the result of certain transformation (deformation) of some unitary colligation $\mathfrak{A}$ (see [8, 9]). Another reason to call our systems as "curved conservative" is the fact that the characteristic function of such a system is a weighted Schur function.

Thus we have defined the notion of conservative curved systems. Note that linear similarity (instead of unitary equivalence for unitary colligations) is a natural kind of equivalence for conservative curved systems and duality is a substitute for orthogonality. The following diagram shows relationships between models, characteristic functions, and conservative curved systems

$$\mathrm{Cfn} \underset{\mathcal{F}_{cm}}{\overset{\mathcal{F}_{mc}}{\rightleftarrows}} \mathrm{Mod} \overset{\mathcal{F}_{sm}}{\longrightarrow} \mathrm{Sys} \ . \tag{dgr}$$

As we can now see, characteristic functions and conservative curved systems are not on equal terms: first of them play leading role because the definition of conservative curved system depends on the functional model, which, in turn, is uniquely determined by its characteristic function. But, surprisingly, the notion of conservative curved system is comparatively autonomous. Though we define such systems in terms of the functional model, many properties and operations with conservative curved systems can be formulated intrinsically and do not refer explicitly to the functional model. One of the aims of this paper is to "measure" a degree of this autonomy with the point of view of the correspondence "factorizations of characteristic function $\leftrightarrow$ invariant subspaces".

If we are going to follow the way described above for conservative systems, we need to introduce transfer functions. For a curved conservative system Σ, we define the *transfer function*

$$\Upsilon = (\Upsilon(z), \Theta_u, \Xi), \quad \text{where} \quad \Upsilon(z) := M(T-z)^{-1}N. \qquad \text{(Tfn)} + \text{(StoT)}$$

At the same time we have defined the transformation $\Upsilon = \mathcal{F}_{ts}(\Sigma)$. Then, using the functional model, the transformation $\mathcal{F}_{tc} := \mathcal{F}_{ts} \circ \mathcal{F}_{sm} \circ \mathcal{F}_{mc}$ can be computed as

$$\Upsilon(z) = (\mathcal{F}_{tc}(\Theta))(z) = \begin{cases} \Theta_+^-(z) - \Theta^+(z)^{-1}, & z \in G_+ \cap \rho(T); \\ -\Theta_-^-(z), & z \in G_-, \end{cases} \qquad \text{(CtoT)}$$

where the operator-valued functions $\Theta_\pm^-(z)$ are defined by the formulas

$$\Theta_\pm^-(z)\,n := (P_\pm \Theta^- n)(z), \quad z \in G_\pm, \quad n \in \mathfrak{N}_-\,;$$

$$\Theta^-(\zeta) := (\pi_+^\dagger \pi_-)(\zeta) = \Xi_+(\zeta)^{-1}\Theta^+(\zeta)^*\Xi_-(\zeta), \quad \zeta \in C.$$

In the case when $G_+ = \mathbb{D}$ and $\Xi_\pm \equiv I$ we get $\Theta^-(\zeta) = \Theta^+(1/\bar\zeta)^*, |\zeta| = 1$ and therefore, $\Theta_+^-(z) = \Theta^+(0)^*, |z| < 1; \; \Theta_-^-(z) = \Theta^+(1/\bar z)^* - \Theta^+(0)^*, |z| > 1$.

In connection with (CtoT), note that the spectrum of a model operator coincides with the spectrum of its characteristic function, i.e., $z \in G_+ \cap \rho(T) \Leftrightarrow \exists \Theta^+(z)^{-1}$.

Thus we arrive at the complete diagram

$$
\begin{array}{ccc}
& \overset{\mathcal{F}_{cm}}{\underset{\mathcal{F}_{mc}}{\rightleftarrows}} & \\
\text{Mod} & & \text{Cfn} \\
{\scriptstyle \mathcal{F}_{sm}} \downarrow & & \downarrow {\scriptstyle \mathcal{F}_{tc}} \\
\text{Sys} & \underset{\mathcal{F}_{ts}}{\longrightarrow} & \text{Tfn}
\end{array}
\qquad \text{(Dgr)}
$$

Unfortunately, we have obtained almost nothing for our purpose: to study the correspondence "factorizations $\leftrightarrow$ invariant subspaces". The main difficulty is to invert the arrows $\mathcal{F}_{tc}$ and $\mathcal{F}_{ts}$. In the case of the unit circle the transfer function can be calculated as $\Upsilon(z) = \Theta^+(0)^* - \Theta^+(1/\bar z)^*, |z| > 1$ and, conversely, one can easily recover the characteristic function $\Theta^+(z)$ from the transfer function $\Upsilon(z)$ (see [17] for this case and for the case of simple connected domains). But, in general, especially for multiply-connected domains, this is a considerable problem: to construct $\Theta \in \text{Cfn}$ such that $\Upsilon = \mathcal{F}_{tc}(\Theta)$. Note that the condition $\Upsilon(z) = M(T-z)^{-1}N \in N(G_+ \cup G_-)$ (that is, $\Upsilon(z)$ is an operator-valued function of Nevanlinna class: $\Upsilon(z) = 1/\delta(z)\,\Omega(z)$, where $\delta \in H^\infty(G_+ \cup G_-)$ and $\Omega \in H^\infty(G_+ \cup G_-, \mathcal{L}(\mathfrak{N}_-, \mathfrak{N}_+)))$ is sufficient for uniqueness of Θ. Moreover, under this assumption, we can recover the corresponding characteristic function and it is possible to find an intrinsic description of conservative curved systems. Note that we reap the benefit of the functional model when we are able to determine that some set of operators (T, M, N) is a conservative curved system [8, 14, 9]. The author plans to address all these problems elsewhere.

Thus we distinguish notions of characteristic and transfer functions and there are no simple enough (and suitable in the study of factorizations) relationships between them. These circumstances dictate that we have to use only the partial diagram (dgr) and to ignore other objects and transformations related to transfer functions from the complete diagram (Dgr). Note also that we study the correspondence "factorizations of *characteristic* function $\leftrightarrow$ invariant subspaces of operator T" in contrast to the correspondence studied in [7]: "factorizations of *transfer* function $\leftrightarrow$ invariant subspaces". At this point we fork with [7].

The paper is organized as follows. In Section 1 we deal with the fragment

$$\mathrm{Cfn} \xrightleftharpoons[\mathcal{F}_{cm}]{\mathcal{F}_{mc}} \mathrm{Mod} :$$ in the context of functional models we develop the constructions

corresponding to factorizations of characteristic functions. If we restrict ourselves to *regular* factorizations, we can keep on to exploit the functional model Mod. But to handle arbitrary factorizations and to obtain a pertinent definition of the product of conservative curved systems we need some generalization of Mod. Moreover, the order relation $\Theta_2\Theta_1 \prec \Theta_2'\Theta_1'$ implies the factorizations like $\Theta_2'\theta\Theta_1$ and therefore we need a functional model suited to handle factorizations with three or more multipliers. With this aim we introduce the notion of n-model Mod_n and extend the transformations $\mathcal{F}_{mc}$ and $\mathcal{F}_{cm}$ to this context. In the rest of the section we study geometric properties of n-models in depth and do this mainly because they form a solid foundation for our definition of the product of curved conservative systems in the next section.

At this moment it is unclear how to define the product of conservative curved systems. As a first approximation we can consider the following construction. Let

$$\Sigma_1 \sim \widehat{\Sigma}_1 = \mathcal{F}_{sm}(\mathcal{F}_{mc}(\Theta_1)) \quad \text{and} \quad \Sigma_2 \sim \widehat{\Sigma}_2 = \mathcal{F}_{sm}(\mathcal{F}_{mc}(\Theta_2)).$$

Then a candidate for their product is $\widehat{\Sigma}_{21} = \mathcal{F}_{sm}(\mathcal{F}_{mc}(\Theta_2\Theta_1))$, where $\mathcal{F}_{mc}(\Theta_2\Theta_1)$ is 3-model corresponding to the factorization $\Theta_{21} = \Theta_2 \cdot \Theta_1$. Our aim is to define the product $\Sigma_2 \cdot \Sigma_1$ by explicit formulas without referring to the functional model. In Section 2 we suggest such a definition and study basic properties of it. The main one among those properties is the property that the product of conservative curved systems $\Sigma_2 \cdot \Sigma_1$ is a conservative curved system too (Theorem A). The geometrical properties of n-model established in Section 1 play a crucial role in our reasoning.

In Section 3 we establish a correspondence between two notions of regularity: the first of them is the regularity of the product of conservative curved systems; the second one is the notion of regular factorization of operator-valued functions [1, 2], which we extend to the weighted Schur classes. We obtain the desired correspondence indirectly: first, we introduce yet another notion – the notion of regularity for models and then establish separately the correspondences Cfn $\leftrightarrow$ Mod and Sys $\leftrightarrow$ Mod.

In Section 4 we study the transformation $\mathcal{F}_{ic}$ defined therein, which takes a factorization $\Theta = \Theta_2\Theta_1$ of characteristic function to the invariant subspace of the model operator of the system $\widehat{\Sigma}_{21} = \mathcal{F}_{sm}(\mathcal{F}_{mc}(\Theta_2\Theta_1))$. We show that this mapping is surjective. Combining this property of $\mathcal{F}_{ic}$ with the criterion of regularity from Section 3, we establish the main result of the paper: *there is an order preserving one-to-one correspondence between regular factorizations of a characteristic function and invariant subspaces of the resolvent of the corresponding model operator.* In conclusion we translate results obtained for model operators into the language of conservative curved systems.

For the convenience of readers, some proofs of attendant assertions have been placed in the Appendix.

Note that the multiply connected specific appears essentially only in the proofs of Proposition 4.1 and 4.2. So, at first a reader can study the paper assuming that the domain G_+ is simply connected. On the other hand, the multiply connected specific influenced on our choice of other proofs throughout the paper. Note that, for simple connected domains, some of them can be reduced to the case of the unit disk (see, e.g., [9, 17]).

The author is grateful to the referee for numerous helpful suggestions improving paper's readability.

1. Geometric properties of n-model

We start with the definition of an *n-characteristic function*, which formalizes products of weighted Schur class functions like the following $\theta_{n-1} \cdots \theta_2\theta_1$: in fact, we merely rearrange them $\Theta_{ij} := \theta_{i-1} \cdots \theta_j$.

Definition. *Let* Ξ_k, $k = \overline{1,n}$ *be operator-valued weights such that* $\Xi_k, \Xi_k^{-1} \in L^\infty(C, \mathcal{L}(\mathfrak{N}_k))$, $\Xi_k(\zeta) \geq 0$, $\zeta \in C$. *A set of analytic in* G_+ *operator-valued functions* $\Theta = \{\Theta_{ij} : i \geq j\}$ *is called an n-characteristic function if* $\Theta_{ij} \in S_{\Xi_{ij}}$ *and* $\forall\, i \geq j \geq k$ $\Theta_{ik} = \Theta_{ij}\Theta_{jk}$, *where* $\Xi_{ij} = (\Xi_i, \Xi_j)$.

We assume also that $\Theta_{kk} := I$ and denote by Cfn_n the class of all n-characteristic functions. In the sequel, we shall usually identify a 3-characteristic function $\Theta = \{\Theta_{31}, \Theta_{32}, \Theta_{21}\}$ with the factorization of Schur class function $\theta = \theta_2 \cdot \theta_1$, where $\theta = (\Theta_{31}, \Xi_1, \Xi_3)$, $\theta_1 = (\Theta_{21}, \Xi_1, \Xi_2)$, $\theta_2 = (\Theta_{32}, \Xi_2, \Xi_3)$. Besides, it is clear how to define the product of n-characteristic functions $\Theta = \Theta'' \cdot \Theta'$: assuming that $\Xi'_{n'} = \Xi''_1$, we need only to renumber multipliers, for instance, $\Theta_{ij} = \Theta''_{i-n'+1,1}\Theta'_{n'j}$, $i \geq n' \geq j$.

In the context of functional models the corresponding notion is the notion of *n-functional model.*

Definition. *An* n-*tuple* $\Pi = (\pi_1, \ldots, \pi_n)$ *of operators* $\pi_k \in \mathcal{L}(L^2(C, \mathfrak{N}_k), \mathcal{H})$ *is called an* n-*model if*

$$
\begin{array}{llll}
\text{(i)} & \forall\, k & (\pi_k^* \pi_k) z = z(\pi_k^* \pi_k); & \pi_k^* \pi_k \gg 0; \\[4pt]
\text{(ii)} & \forall\, j \geq k & (\pi_j^\dagger \pi_k) z = z(\pi_j^\dagger \pi_k); & P_-(\pi_j^\dagger \pi_k) P_+ = 0; \\[4pt]
\text{(iii)} & \forall\, i \geq j \geq k & \pi_i^\dagger \pi_k = \pi_i^\dagger \pi_j \pi_j^\dagger \pi_k; & \\[4pt]
\text{(iv)} & & \mathcal{H}_{\pi_n \vee \cdots \vee \pi_1} = \mathcal{H}\,. &
\end{array}
\qquad (\mathrm{Mod}_n)
$$

Here $\mathcal{H}_{\pi_n \vee \cdots \vee \pi_1} := \vee_{k=1}^n \operatorname{Ran} \pi_k$. The definition (Mod_n) is an extension of the definition (Mod): namely, $\mathrm{Mod} = \mathrm{Mod}_2$. It is readily seen that $\Theta = \{\pi_i^\dagger \pi_j\}_{i \geq j}$ is an n-characteristic function with weights $\Xi_k = \pi_k^* \pi_k$ and therefore we have defined the transformation $\mathcal{F}_{cm} : \mathrm{Mod}_n \to \mathrm{Cfn}_n$. The existence of the "inverse" transformation $\mathcal{F}_{mc}$ follows from

Proposition 1.1. *Suppose* $\Theta \in \mathrm{Cfn}_n$. *Then* $\exists\, \Pi \in \mathrm{Mod}_n$ *such that* $\Theta = \mathcal{F}_{cm}(\Pi)$. *If also* $\Theta = \mathcal{F}_{cm}(\Pi')$, *then there exists a unitary operator* $X : \mathcal{H}_{\pi_n \vee \cdots \vee \pi_1} \to \mathcal{H}'_{\pi_n \vee \cdots \vee \pi_1}$ *such that* $\pi_k' = X \pi_k$.

Proof. We put $\mathcal{H} = \oplus_{k=1}^n \mathcal{H}_k^\Delta$, where $\mathcal{H}_k^\Delta = \operatorname{clos} \Delta_{k\,k+1\,k} L^2(C, \mathfrak{N}_k)$, $k = \overline{1, n-1}$, $\mathcal{H}_n^\Delta = L^2(C, \mathfrak{N}_n)$, $\Delta_{k\,k+1\,k} := (I - \Theta_{k+1\,k}^\dagger \Theta_{k+1\,k})^{1/2}$, and $\Theta_{k+1\,k}^\dagger$ is adjoint to the operator $\Theta_{k+1\,k} : L^2(C, \Xi_k) \to L^2(C, \Xi_{k+1})$. Let ν_k, $k = \overline{1, n}$ be the operators of embedding of $\mathcal{H}_k^\Delta$ into $\mathcal{H}$ and

$$
\pi_n := \nu_n\,, \qquad \pi_k := \pi_{k+1} \Theta_{k+1\,k} + \nu_k \Delta_{k\,k+1\,k}\,, \qquad k = \overline{1, n-1}\,.
$$

It can easily be calculated that

$$
\pi_k = \nu_n \Theta_{nk} + \nu_{n-1} \Delta_{n-1\,n\,n-1} \Theta_{n-1\,k} + \cdots + \nu_j \Delta_{j\,j+1\,j} \Theta_{jk} + \cdots + \nu_k \Delta_{k\,k+1\,k}\,.
$$

From this identity we get $\pi_i^\dagger \pi_j = \Theta_{ij}$, $i \geq j$.

The existence and unitary property of X follows from the identity

$$
\| \pi_1 u_1 + \cdots + \pi_n u_n \|^2 = \sum_{i,j=1}^n (\pi_j^\dagger \pi_i u_i,\, u_j)_{L^2(C, \Xi_j)}
$$

$$
= \sum_{i,j=1}^n (\pi'^\dagger_j \pi_i' u_i,\, u_j)_{L^2(C, \Xi_j)} = \| \pi_1' u_1 + \cdots + \pi_n' u_n \|^2\,. \qquad \square
$$

The construction of Proposition 1.1 is simplified if all functions Θ_{ij} are two-sided Ξ-inner. In this case, $\mathcal{H} = L^2(C)$ and $\pi_k = \Theta_{nk}$.

We can consider an equivalence relation $\sim$ in Mod_n. We write $\Pi \sim \Pi'$ if there exists a unitary operator $X : \mathcal{H}_{\pi_n \vee \cdots \vee \pi_1} \to \mathcal{H}'_{\pi_n \vee \cdots \vee \pi_1}$ such that $\pi_k' = X \pi_k$. It is clear that the transformation $\mathcal{F}_{cm}$ induces a transformation $\mathcal{F}_{cm}^\sim : \mathrm{Mod}_n^\sim \to \mathrm{Cfn}_n$ such that $\mathcal{F}_{cm}^\sim(\Pi^\sim) = \mathcal{F}_{cm}(\Pi)$, $\Pi \in \Pi^\sim$. By Proposition 1.1, there exists the inverse transformation $\mathcal{F}_{mc}^\sim : \mathrm{Cfn}_n \to \mathrm{Mod}_n^\sim$. But, in the sequel, we shall usually ignore this equivalence relation and use merely the transformations $\mathcal{F}_{cm}$ and $\mathcal{F}_{mc}$.

The product of any n-models Π', Π'' with the only restriction $\pi_n'^* \pi_n' = \pi_1''^* \pi_1''$ is defined (up to unitary equivalence) as $\Pi = \Pi'' \cdot \Pi' := \mathcal{F}_{mc}(\mathcal{F}_{cm}(\Pi'') \cdot \mathcal{F}_{cm}(\Pi'))$.

214   A. Tikhonov

Using the construction of Proposition 1.1, we can uniquely determine the normal operator $\mathcal{U} = XzX^{-1} \in \mathcal{L}(\mathcal{H}_{\pi_n \vee \cdots \vee \pi_1})$ with absolutely continuous spectrum $\sigma(\mathcal{U}) \subset C$ such that $\mathcal{U}\pi_k = \pi_k z$, where $X : \mathcal{H}_{\hat{\pi}_n \vee \cdots \vee \hat{\pi}_1} \to \mathcal{H}_{\pi_n \vee \cdots \vee \pi_1}$ is a unitary operator such that $\pi_k = X\hat{\pi}_k$; the operators $\hat{\pi}_k$ are constructed for the n-characteristic function $\Theta = \mathcal{F}_{cm}(\Pi)$ by the same way as in Proposition 1.1.

Taking into account the existence of the operator $\mathcal{U}$, note that $\mathcal{F}_{sc}(\Theta) = \mathcal{F}_{sc}(\Theta_{n1}) \oplus \widehat{\Sigma}_u$, where $\mathcal{F}_{sc} := \mathcal{F}_{sm} \circ \mathcal{F}_{mc}$ and the system $\widehat{\Sigma}_u = (\widehat{T}_u, 0, 0, 0)$ is "purely normal" with $\widehat{T}_u = \mathcal{U} \,|\, (\mathcal{H}_{\pi_n \vee \cdots \vee \pi_1} \ominus \mathcal{H}_{\pi_n \vee \pi_1})$.

Let $\Pi \in \mathrm{Mod}_n$. Now we define our building bricks: orthoprojections $P_{\pi_i \vee \cdots \vee \pi_j}$ onto $\mathcal{H}_{\pi_i \vee \cdots \vee \pi_j}$ and projections $q_{i\pm} := \pi_i P_{\pm} \pi_i^{\dagger}$.

Lemma 1.2. *For* $i \geq j \geq k \geq l \geq m$

1) $q_{i-} q_{j+} = 0$;

2) $q_{i+} + q_{i-} = \pi_i \pi_i^{\dagger} = P_{\pi_i}$;

3) $P_{\pi_i \vee \cdots \vee \pi_j}(I - \pi_k \pi_k^{\dagger}) P_{\pi_l \vee \cdots \vee \pi_m} = 0$;

4) $P_{\pi_l \vee \cdots \vee \pi_m}(I - \pi_k \pi_k^{\dagger}) P_{\pi_i \vee \cdots \vee \pi_j} = 0$.

Proof. Statement 1) is a direct consequences of (ii) from (Mod_n). Statement 2) is obvious. Statement 3) is equivalent to the relation

$$\forall f, g \in \mathcal{H} \qquad ((I - \pi_k \pi_k^{\dagger}) P_{\pi_l \vee \cdots \vee \pi_m} f, P_{\pi_i \vee \cdots \vee \pi_j} g) = 0.$$

The latter can be rewritten in the form

$$((I - \pi_k \pi_k^{\dagger}) \pi_{l'} u, \pi_{i'} v) = 0, \quad j \leq i' \leq i, \quad m \leq l' \leq l$$

and is true because of (iii) from (Mod_n). Statement 4) can be obtained from Statement 3) by conjugation. $\qquad\square$

We also define the projections

$$P_{(ij)} := P_{\pi_i \vee \cdots \vee \pi_j}(I - q_{j+})(I - q_{i-}), \quad i \geq j.$$

It is easily shown that $P_{(ii)} = 0$ and

$$P_{(ij)} = (I - q_{j+}) P_{\pi_i \vee \cdots \vee \pi_j}(I - q_{i-}) = (I - q_{j+})(I - q_{i-}) P_{\pi_i \vee \cdots \vee \pi_j}.$$

Indeed, $P_{\pi_i \vee \cdots \vee \pi_j} - P_{\pi_i}$ is orthoprojection onto $\mathcal{H}_{\pi_i \vee \cdots \vee \pi_j} \ominus \mathcal{H}_{\pi_i}$, $P_{\pi_i} q_{i\pm} = q_{i\pm} P_{\pi_i}$ and $(P_{\pi_i \vee \cdots \vee \pi_j} - P_{\pi_i}) q_{i\pm} = q_{i\pm}(P_{\pi_i \vee \cdots \vee \pi_j} - P_{\pi_i}) = 0$. The same is hold for π_j. Then,

$$\begin{aligned}
P_{(ij)}^2 &= P_{\pi_i \vee \cdots \vee \pi_j}(I - q_{j+})(I - q_{i-})(I - q_{j+})(I - q_{i-}) P_{\pi_i \vee \cdots \vee \pi_j} \\
&= P_{\pi_i \vee \cdots \vee \pi_j}[(I - q_{j+})(I - q_{j+})(I - q_{i-}) - (I - q_{j+}) q_{i-}(I - q_{i-})] P_{\pi_i \vee \cdots \vee \pi_j} \\
&= P_{\pi_i \vee \cdots \vee \pi_j}[(I - q_{j+})(I - q_{i-})] P_{\pi_i \vee \cdots \vee \pi_j} = P_{(ij)}.
\end{aligned}$$

Note also that $P_{(ij)} = \Theta_{nj} P_{-} \Theta_{ij}^{-1} P_{+} \Theta_{ni}^{-1}$ whenever all functions Θ_{ij} are two-sided Ξ-inner (recall that then we can choose $\pi_k = \Theta_{nk}$).

Lemma 1.3. *For $i \geq j \geq k \geq l$, one has*

$$1)\ P_{(ij)}q_{k+} = 0; \qquad 2)\ q_{i-}P_{(jk)} = 0; \qquad 3)\ P_{(ij)}P_{(kl)} = 0;$$

$$4)\ P_{(ik)}P_{(jk)} = P_{(jk)}; \quad 5)\ P_{(ij)}P_{(ik)} = P_{(ij)}; \quad 6)\ P_{(jk)}P_{(ij)} = 0.$$

We also define the subspaces

$$\mathcal{K}_{(ij)} := \operatorname{Ran} P_{(ij)}, \qquad\qquad \mathcal{H}_{ij} := \mathcal{H}_{\pi_i \vee \cdots \vee \pi_j}$$

$$\mathcal{H}_{ij+} := \mathcal{H}_{ij} \cap \operatorname{Ker} q_{i-}, \qquad\qquad \mathcal{D}_{j+} := \operatorname{Ran} q_{j+}.$$

It is easy to prove that $\mathcal{H}_{ij+} \cap \operatorname{Ker} P_{(ij)} = \mathcal{D}_{j+}$. Indeed, let $f \in \mathcal{H}_{ij+} \cap \operatorname{Ker} P_{(ij)}$. Then

$$f = (I - P_{(ij)})f = f - (I - q_{j+})(I - q_{i-})P_{\pi_i \vee \cdots \vee \pi_j}f$$
$$= f - (I - q_{j+})(I - q_{i-})f = f - (I - q_{j+})f = q_{j+}f \in \mathcal{D}_{j+}.$$

Conversely, let $f \in \mathcal{D}_{j+}$. Then $f = q_{j+}f \in \mathcal{H}_{ij}$ and therefore $q_{i-}f = q_{i-}q_{j+}f - 0$, that is, $f \in \mathcal{H}_{ij+}$. Hence we have

$$P_{(ij)}f = (I - q_{j+})(I - q_{i-})P_{\pi_i \vee \cdots \vee \pi_j}f = (I - q_{j+})(I - q_{i-})f = (I - q_{j+})f = 0$$

and $f \in \mathcal{H}_{ij+} \cap \operatorname{Ker} P_{(ij)}$.

Translating the assertions of the above lemmas into the language of geometry, we obtain

$$\mathcal{K}_{(ij)} \subset \mathcal{H}_{ij+}, \qquad \mathcal{K}_{(jk)} \subset \mathcal{K}_{(ik)}, \qquad \mathcal{H}_{jk+} \subset \mathcal{H}_{il+}, \qquad i \geq j \geq k \geq l.$$

Indeed, let $f \in \mathcal{K}_{(ij)}$. Then $f = P_{(ij)}f \in \mathcal{H}_{ij}$ and $g_{i-}f = g_{i-}(I-q_{j+})(I-q_{i-})f = g_{i-}(I-q_{i-})f = 0 \Rightarrow f \in \operatorname{Ker} q_{i-}$. The inclusion $\mathcal{K}_{(jk)} \subset \mathcal{K}_{(ik)}$ is a straightforward consequence of Lemma 1.3(4). Let $f \in \mathcal{H}_{jk+}$. Then

$$g_{i-}f = g_{i-}(I - g_{j-})P_{\pi_j \vee \cdots \vee \pi_k}f = g_{i-}[(I - \pi_j \pi_j^{\dagger}) + g_{j+}]P_{\pi_j \vee \cdots \vee \pi_k}f = 0.$$

and therefore $f \in \mathcal{H}_{il+}$.

Let $1 = m_1 \leq \cdots \leq m_i \leq \cdots \leq m_N = n$. We define the operators

$$P_{[m_i m_j]} := P_{(m_{j+1} m_j)}(I - P_{(m_{j+2} m_{j+1})}) \cdots (I - P_{(m_i m_{i-1})})$$
$$+ P_{(m_{j+2} m_{j+1})}(I - P_{(m_{j+3} m_{j+2})}) \cdots (I - P_{(m_i m_{i-1})})$$
$$+ \cdots + P_{(m_{i-1} m_{i-2})}(I - P_{(m_i m_{i-1})}) + P_{(m_i m_{i-1})}, \qquad i \geq j.$$

Note that our notation is ambiguous: the projection $P_{[m_i m_j]}$ depends on the whole chain $m_j \leq \cdots \leq m_i$ but not only on two numbers m_j and m_i. The following properties of operators $P_{[m_i m_j]}$ are straightforward consequences of Lemma 1.3.

Proposition 1.4. *For $i \geq j \geq k \geq l$,*

1) $P_{[m_i m_j]}q_{m_k+} = 0;$

2) $q_{m_i-}P_{[m_j m_k]} = 0;$

3) $P_{[m_i m_j]}P_{[m_k m_l]} = 0.$

Further, since $I - P_{[m_i m_j]} = (I - P_{(m_{j+1} m_j)})(I - P_{(m_{j+2} m_{j+1})}) \ldots (I - P_{(m_i m_{i-1})})$, we get the following recursion relation

$$P_{[m_i m_k]} = P_{[m_j m_k]}(I - P_{[m_i m_j]}) + P_{[m_i m_j]}, \qquad i \geq j \geq k.$$

Since $P_{[m_{j+1} m_j]} = P_{(m_{j+1} m_j)}$, we obtain by induction that the operator $P_{[m_i m_j]}$ is a projection and

$$\mathcal{K}_{[m_i m_j]} = \mathcal{K}_{(m_i m_{i-1})} \dotplus \cdots \dotplus \mathcal{K}_{(m_{j+1} m_j)},$$

where $\mathcal{K}_{[m_i m_j]} := \operatorname{Ran} P_{[m_i m_j]}$. We use the notation $H = H' \dotplus H''$ if there exists a projection P' such that $H' = \operatorname{Ran} P'$, $H'' = \operatorname{Ker} P'$. Besides, we have

$$\mathcal{K}_{[m_i m_j]} \subset \mathcal{H}_{m_i m_j +}, \qquad \mathcal{D}_{m_j +} \subset \operatorname{Ker} P_{[m_i m_j]}.$$

The first inclusion follows straightforwardly from Proposition 1.4(2). The second one is a consequence of Proposition 1.4(1).

The following proposition affirms a more delicate property of projections $P_{[m_i m_j]}$.

Proposition 1.5. *One has* $\mathcal{H}_{m_i m_j +} \cap \operatorname{Ker} P_{[m_i m_j]} = \mathcal{D}_{m_j +}$, $i \geq j$.

Remark. Since $P_{[31]} = P_{(32)} + P_{(21)}(I - P_{(32)}) = P_{(32)} + P_{(21)}$, by Proposition 1.5 and the corollary of Lemma (iii), we obtain the following identities

$$(P_{(32)} + P_{(21)}) \, P_{(31)} \, (P_{(32)} + P_{(21)}) = P_{(32)} + P_{(21)}$$

and

$$P_{(31)} \, (P_{(32)} + P_{(21)}) \, P_{(31)} = P_{(31)}.$$

This means that

$$(P_{(31)} | \mathcal{K}_{[31]})^{-1} = (P_{(21)} + P_{(32)}) | \mathcal{K}_{(31)}, \qquad ((P_{(21)} + P_{(32)}) | \mathcal{K}_{(31)})^{-1} = P_{(31)} | \mathcal{K}_{[31]}.$$

Example. Let $w = \varphi(z) = z + \varepsilon z^2$, $|\varepsilon| < 1/2$, $G_+ = \varphi(\mathbb{D})$, $C = \varphi(\mathbb{T})$. We put

$$\theta(w) = \frac{2w}{1 + \sqrt{1 + 4\varepsilon w}}, \qquad w \in G_+, \qquad \Theta_{ij}(w) = \theta(w)^{i-j}, \qquad 1 \leq j \leq i \leq n,$$

and $\Xi_i(w) = 1$, $w \in C$. It can easily be checked that $|\theta(w)| = 1$, $w \in C$. Then $P_{(ij)} = P_{(ij)}^{(n)} = \theta^{n-j} P_- \theta^{j-i} P_+ \theta^{i-n}$. For the functions

$$f_k^{ij}(w) = \theta(w)^{n-j} w^{-k}, \qquad k = \overline{1, i-j},$$

we have $f_k^{ij} \in \mathcal{K}_{(ij)}^{(n)} = \operatorname{Ran} P_{(ij)}^{(n)}$. By [9], $\mathcal{K}_{(ij)}^{(n)}(\varepsilon) = P_{(ij)}^{(n)}(\varepsilon)\mathcal{K}_{(ij)}^{(n)}(0)$ and $\mathcal{K}_{(ij)}^{(n)}(0) = P_{(ij)}^{(n)}(0)\mathcal{K}_{(ij)}^{(n)}(\varepsilon)$. Since $\dim \mathcal{K}_{(ij)}^{(n)}(0) = i - j$, the functions $f_k^{ij}(w)$ form a basis of the subspace $\mathcal{K}_{(ij)}^{(n)}$. Note also that $P_{ij}^{(n)} = \theta^{n-i} P_{i-j+1,1}^{(i-j+1)} \theta^{i-n}$ and therefore $\mathcal{K}_{(ij)}^{(n)} = \theta^{n-i} \mathcal{K}_{(i-j+1,1)}^{(i-j+1)}$.

Consider particular cases. In the case of $n = 3$ we have

$$f_1^{31} = \theta^2/w, \; f_2^{31} = \theta^2/w^2 \qquad \text{and} \qquad f_1^{21} = \theta^2/w, \; f_1^{32} = \theta/w.$$

Hence, $\mathcal{K}_{(21)} \subset \mathcal{K}_{(31)}$, $\mathcal{K}_{(32)} \nsubseteq \mathcal{K}_{(31)}$ and $\mathcal{K}_{(32)} \dotplus \mathcal{K}_{(21)} \neq \mathcal{K}_{(31)}$.

In the case of $n = 5$ it can be calculated that

$$P_{(21)} f_1^{53} = -\varepsilon^2 f_1^{21} \qquad \text{and} \qquad P_{(21)} f_2^{53} = 2\varepsilon^3 f_1^{21}.$$

Therefore, $P_{(21)} P_{(53)} \neq 0$.

Our calculations was based on the formula

$$P_{(kl)} f_p^{ij} = \theta^{n-l} w^{l-j-p} u_{l-j-p,j-l}(w) - 2^{l-k} \theta^{n-l} P_- \frac{w^{l-j-p} u_{k-j-p,j-k}(w)}{(1 + \sqrt{1 + 4\varepsilon w})^{l-k}},$$

where $1 \le i,j,k,l \le n$, $p = \overline{1, i-j}$, $i \ge j$, $k \ge l$, and

$$u_{q,r}(w) := 2^{-r} w^{-q} P_- w^q (1 + \sqrt{1 + 4\varepsilon w})^r.$$

It can easily be checked that $u_{q,r}(w) \equiv 0$, $q \ge 0$. For $q < 0$, we make use of the Residue Theorem. Calculating $P_+ w^q (1 + \sqrt{1 + 4\varepsilon w})^r$ and interpreting the projection P_+ as the boundary values of the Cauchy integral operator, we get

$$u_{-1,r}(w) = 1;$$

$$u_{-2,r}(w) = 1 + r\varepsilon w;$$

$$u_{-3,r}(w) = \tfrac{1}{2}(2 + 2r\varepsilon w + r(r-3)\varepsilon^2 w^2);$$

$$u_{-4,r}(w) = \tfrac{1}{6}(2 + 6r\varepsilon w + 3r(r-3)\varepsilon^2 w^2 + 3r(r-4)(r-5)\varepsilon^3 w^3).$$

2. Product of conservative curved systems

Definition. *Let* $\Sigma_k = (T_k, M_k, N_k, \Theta_{ku}, \Xi_k; H_k, \mathfrak{N}_{k+}, \mathfrak{N}_{k-})$, $k = 1,2$ *be conservative curved systems,* $G_{1+} = G_{2+}$, $\mathfrak{N}_{1-} = \mathfrak{N}_{2+}$, *and* $\Xi_{1-} = \Xi_{2+}$. *We define the product of them as*

$$\Sigma_{21} = \Sigma_2 \cdot \Sigma_1 := (T_{21}, M_{21}, N_{21}, \Theta_{21u}, \Xi_{21}; H_{21}, \mathfrak{N}_{1+}, \mathfrak{N}_{2-})$$

with

$$\Theta_{21u} = \Theta_{2u}\Theta_{1u}, \qquad \Xi_{21} = (\Xi_{1+}, \Xi_{2-}), \qquad H_{21} = H_1 \oplus H_2,$$

$$T_{21} = \begin{pmatrix} T_1 & N_1 M_2 \\ 0 & T_2 \end{pmatrix}, \qquad M_{21} = (M_1, M_2^{21}), \qquad N_{21} = \begin{pmatrix} M_{*1}^{21*} \\ N_2 \end{pmatrix},$$

$$M_2^{21} f_2 = -\frac{1}{2\pi i} \int_C \Theta_1^-(\zeta) \, [M_2(T_2 - \cdot)^{-1} f_2]_-(\zeta) \, d\zeta, \qquad f_2 \in H_2, \qquad \text{(Prod)}$$

$$M_{*1}^{21} f_1 = -\frac{1}{2\pi i} \int_{\overline{C}} \Theta_{*2}^-(\zeta) \, [N_1^*(T_1^* - \cdot)^{-1} f_1]_-(\zeta) \, d\zeta, \qquad f_1 \in H_1,$$

where $[M_2(T_2 - \cdot)^{-1} f_2]_-$ *and* $[N_1^*(T_1^* - \cdot)^{-1} f_1]_-$ *are the boundary limits of* $M_2(T_2 - z)^{-1} f_2$ *and* $N_1^*(T_1^* - z)^{-1} f_1$ *from the domains* G_- *and* $\overline{G}_- := \{\bar{z}: z \in G_-\}$, *respectively;* $\Theta_{*2}^- = \Theta_2^{-\sim}$ *(see* (CtoT) *for the definition of* Θ^- *).*

Note that we can consider the product $\Sigma_2 \cdot \Sigma_1$ without the assumption that Σ_1, Σ_2 are conservative curved systems. We need only to assume additionally that $\forall f_2 \in H_2 : M_2(T_2 - z)^{-1} f_2 \in E^2(G_-)$ and $\forall f_1 \in H_1 : N_1^*(T_1^* - z)^{-1} f_1 \in E^2(\bar{G}_-)$. For conservative curved systems these assumptions are always satisfied (it follows from the definition of conservative curved system).

We start to justify the definition with the observation that in case of unitary colligations we get the standard algebraic definition [2]: $M_2^{21} = \Theta_1^+(0)^* M_2$ and $N_1^{21} := M_{*1}^{21*} = N_1 \Theta_2^+(0)^*$ (see the Introduction). Indeed, since in this case $\Theta_+^- \equiv \Theta^+(0)^* = L$ and $M_2(T_2 - z)^{-1} f_2 \in E^2(G_-)$, we obtain

$$M_2^{21} f_2 = -\frac{1}{2\pi i} \int_C \Theta_1^-(\zeta)\, [M_2(T_2 - \cdot)^{-1} f_2]_-(\zeta)\, d\zeta$$

$$= -\frac{1}{2\pi i} \int_C \Theta_{1+}^-(\zeta)\, [M_2(T_2 - \cdot)^{-1} f_2]_-(\zeta)\, d\zeta = L_1 M_2\,.$$

By a similar computation, we get $N_1^{21} = N_1 L_2$. Besides, we have

Proposition 2.1.
1) $\Sigma_1 \sim \Sigma_1', \Sigma_2 \sim \Sigma_2' \Rightarrow \Sigma_2 \cdot \Sigma_1 \sim \Sigma_2' \cdot \Sigma_1'$;
2) $(\Sigma_2 \cdot \Sigma_1)^* = \Sigma_1^* \cdot \Sigma_2^*$.

Here $\Sigma^* := (T^*, N^*, M^*, \Theta_u^\sim, \Xi_*)$, $\Xi_{*\pm} = \Xi_\mp^{\sim -1}$.

Further, we shall say that a triplet of operators (T, M, N) is a realization of a transfer function $\Upsilon = \mathcal{F}_{tc}(\Theta)$ if $\Upsilon(z) = M(T - z)^{-1} N$.

Proposition 2.2. *Suppose that triplets (T_1, M_1, N_1) and (T_2, M_2, N_2) are realizations of transfer functions $\Upsilon_1 = \mathcal{F}_{tc}(\Theta_1)$ and $\Upsilon_2 = \mathcal{F}_{tc}(\Theta_2)$, respectively. Suppose also that $\forall f_1 \in H_1 : N_1^*(T_1^* - z)^{-1} f_1 \in E^2(\bar{G}_-)$ and $\forall f_2 \in H_2 : M_2(T_2 - z)^{-1} f_2 \in E^2(G_-)$. Then the triplet (T_{21}, M_{21}, N_{21}) defined by (Prod) is a realization of the transfer function $\Upsilon_{21} = \mathcal{F}_{tc}(\Theta_2 \Theta_1)$.*

Thus we have obtained important properties of product of systems. But the main question whether the product $\Sigma_2 \cdot \Sigma_1$ of conservative curved systems Σ_1, Σ_2 is a conservative curved system too leaves unexplained. The following proposition answers this question. It also answers a question about author's motivation of the definition (Prod): in fact, the connection between the product of systems and the product of models established in the proposition sheds genuine light on our definition (Prod).

Proposition 2.3. *Suppose $\Pi_1, \Pi_2 \in \mathrm{Mod}$, $\Pi = \Pi_2 \cdot \Pi_1$, $\Sigma_1 = \mathcal{F}_{sm}(\Pi_1)$, $\Sigma_2 = \mathcal{F}_{sm}(\Pi_2)$, $\Sigma_{21} = \Sigma_2 \cdot \Sigma_1$, and $\widehat{\Sigma} = \mathcal{F}_{sm}(\Pi)$. Then $\Sigma_{21} \sim \widehat{\Sigma}$.*

We hope that it will cause no confusion if we use the same symbol $\mathcal{F}_{ms}$ for the transformations $\mathcal{F}_{ms} : \mathrm{Mod} \to \mathrm{Sys}$ and $\mathcal{F}_{ms} : \mathrm{Mod}_n \to \mathrm{Sys}$: the latter one is defined by (MtoS) as well (with $\pi_+ = \pi_1$ and $\pi_- = \pi_n$).

Proof. Let

$$\Sigma_1 = (T_1, M_1, N_1), \quad \Sigma_2 = (T_2, M_2, N_2), \quad \Sigma_{21} = (T_{21}, M_{21}, N_{21})$$

Let also $\Pi = (\pi_1, \pi_2, \pi_3)$, $\widehat{\Sigma} = (\widehat{T}, \widehat{M}, \widehat{N}) = \mathcal{F}_{sm}(\Pi)$, and

$$\widehat{\Sigma}_1 = (\widehat{T}_1, \widehat{M}_1, \widehat{N}_1) = \mathcal{F}_{sm}(\pi_1, \pi_2), \quad \widehat{\Sigma}_2 = (\widehat{T}_2, \widehat{M}_2, \widehat{N}_2) = \mathcal{F}_{sm}(\pi_2, \pi_3).$$

It is obvious that the systems $\widehat{\Sigma}_k$ and Σ_k, $k = 1, 2$ are unitarily equivalent.

Since there are no simple and convenient expressions for operators $\widehat{T}^*$, $\widehat{M}^*$, $\widehat{N}^*$ in terms of the model Π, we need to employ the dual model $\Pi_* = (\pi_{*+}, \pi_{*-})$, where $\pi_{*\mp} \in \mathcal{L}(L^2(\bar{C}, \mathfrak{N}_{\mp}), \mathcal{H})$ are defined by the conditions

$$(f, \pi_{*\mp} v)_{\mathcal{H}} = \langle \pi_{\pm}^{\dagger} f, v \rangle_C, \ f \in \mathcal{H}, \ v \in L^2(\bar{C}, \mathfrak{N}_{\mp}), \ \text{where}$$

$$\langle u, v \rangle_C := \tfrac{1}{2\pi i} \int_C (u(z), v(\bar{z}))_{\mathfrak{N}} \, dz, \ u \in L^2(C, \mathfrak{N}), \ v \in L^2(\bar{C}, \mathfrak{N}).$$

Then we can define the dual objects $\widehat{T}_*, \widehat{M}_*, \widehat{N}_*$ corresponding to the subspace $\mathcal{K}_{*\Theta} = \operatorname{Ran} P_{*\Theta} \subset \mathcal{H}$. Note that $P_{*\Theta} = P_{\Theta}^*$ and $(\widehat{T}_*, \widehat{M}_*, \widehat{N}_*) \sim (\widehat{T}^*, \widehat{N}^*, \widehat{M}^*)$.

Since $P_{\Theta}^* \neq P_{\Theta}$, we have $\mathcal{K}_{\Theta} \neq \mathcal{K}_{*\Theta}$. Besides, as is known from Section 1, $\mathcal{K}_{\Theta} = \mathcal{K}_{(31)} \neq \mathcal{K}_{(32)} \dotplus \mathcal{K}_{(21)}$ and therefore the main challenge of the proposition is to handle all these subspaces coordinately. In [8, 9], the author noticed that it was convenient to use the pair of operators $W, W_* \in \mathcal{L}(H, \mathcal{H})$ for a model and the dual one simultaneously. We extend this construction to 3-models. By [8, 9], there exist operators $W_k, W_{*k} \in \mathcal{L}(H_k, \mathcal{H})$, $k = 1, 2$ such that $W_{*k}^* W_k = I$, $W_k W_{*k}^* = P_k$, and

$$\widehat{T}_k W_k = W_k T_k, \qquad \widehat{M}_k W_k = M_k, \qquad \widehat{N}_k = W_k N_k,$$

$$\widehat{T}_{*k} W_{*k} = W_{*k} T_k^*, \qquad \widehat{M}_{*k} W_{*k} = N_k^*, \qquad \widehat{N}_{*k} = W_{*k} M_k^*,$$

where $P_1 = P_{(21)}$, $P_2 = P_{(32)}$ are projections related to the 3-model $\Pi = (\pi_1, \pi_2, \pi_3)$. Define $W_{21} := (W_1, W_2)$ and $W_{*21} := (W_{*1}, W_{*2})$. By Lemma 1.3, $P_{(32)} P_{(21)} = P_{(21)} P_{(32)} = 0$. This implies

$$W_{*21}^* W_{21} = \operatorname{diag}(I, I) \qquad \text{and} \qquad W_{21} W_{*21}^* = P_{(21)} + P_{(32)}.$$

We put

$$\widehat{T}'_{21} = W_{21} T_{21} W_{*21}^*, \qquad \widehat{M}'_{21} = M_{21} W_{*21}^*, \qquad \widehat{N}'_{21} = W_{21} N_{21},$$

$$\widehat{T}'_{*21} = W_{*21} T_{21}^* W_{21}^*, \qquad \widehat{M}'_{*21} = N_{21}^* W_{21}^*, \qquad \widehat{N}'_{*21} = W_{*21} M_{21}^*,$$

and (see the remark after Proposition 1.5)

$$\widehat{T}_{21} = (P_{(21)} + P_{(32)}) \widehat{T} P_{(31)}, \qquad \widehat{M}_{21} = \widehat{M} P_{(31)}, \qquad \widehat{N}_{21} = (P_{(21)} + P_{(32)}) \widehat{N},$$

$$\widehat{T}_{*21} = (P_{(21)}^* + P_{(32)}^*) \widehat{T}_* P_{(31)}^*, \qquad \widehat{M}_{*21} = \widehat{M}_* P_{(31)}^*, \qquad \widehat{N}_{21} = (P_{(21)}^* + P_{(32)}^*) \widehat{N}_*.$$

Our aim is to show that $(\widehat{T}'_{21}, \widehat{M}'_{21}, \widehat{N}'_{21}) = (\widehat{T}_{21}, \widehat{M}_{21}, \widehat{N}_{21})$. If this identity holds, we get

$$\widehat{T} W = W T_{21}, \qquad \widehat{M} W = M_{21}, \qquad \widehat{N} = W N_{21},$$

220   A. Tikhonov

where $W = P_{(31)}W_{21}$, $W_* = P_{(31)}^*W_{*21}$. Thus, $\Sigma_{21} \overset{W}{\sim} \widehat{\Sigma}$ and the proposition is proved. Note also that $W_*^*W = I$ and $WW_*^* = P_{(31)}$.

We check the desired identities by computations within the functional model. The identities

$$\widehat{T}_{21}' = \begin{pmatrix} \widehat{T}_1 & \widehat{N}_1\widehat{M}_2 \\ 0 & \widehat{T}_2 \end{pmatrix}, \quad \widehat{M}_{21}' = \left(\widehat{M}_1, \frac{-1}{2\pi i}\int_C \Theta_1^-(z)[\widehat{M}_2(\widehat{T}_2 - \varsigma)^{-1} f]_-(z)\,dz \right)$$

can be obtained by a straightforward calculation. Indeed, we have

$$\widehat{T}_{21}' = W_{21}T_{21}W_{*21}^* = \begin{pmatrix} W_1, & W_2 \end{pmatrix} \begin{pmatrix} T_1 & N_1M_2 \\ 0 & T_2 \end{pmatrix} \begin{pmatrix} W_{*1}^* \\ W_{*2}^* \end{pmatrix}$$

$$= \begin{pmatrix} W_1T_1W_{*1}^* & W_1N_1M_2W_{*2}^* \\ 0 & W_2T_2W_{*2}^* \end{pmatrix} = \begin{pmatrix} \widehat{T}_1 & \widehat{N}_1\widehat{M}_2 \\ 0 & \widehat{T}_2 \end{pmatrix}$$

and

$$\widehat{M}_{21}'f = M_{21}W_{*21}^*f = \begin{pmatrix} M_1, & M_2^{21} \end{pmatrix} \begin{pmatrix} W_{*1}^* \\ W_{*2}^* \end{pmatrix} f = M_1W_{*1}^*f_{21} + M_2^{21}W_{*2}^*f_{32}$$

$$= \widehat{M}_1 f_{21} - \frac{1}{2\pi i}\int_C \Theta_1^-(z)\,[M_2(T_2 - \cdot)^{-1}W_{*2}^*f_{32}]_-(z)\,dz$$

$$= \widehat{M}_1 f_{21} - \frac{1}{2\pi i}\int_C \Theta_1^-(z)\,[\widehat{M}_2(\widehat{T}_2 - \cdot)^{-1}f_{32}]_-(z)\,dz,$$

where $f = f_{21} + f_{32} \in \mathcal{K}_{(21)}\dot{+}\mathcal{K}_{(32)}$.

On the other hand, using Lemma 1.3, Proposition 1.4, and the inclusions $\mathcal{U}\mathcal{D}_{1+} \subset \mathcal{D}_{1+}$, $\mathcal{U}\mathcal{H}_{31+} \subset \mathcal{H}_{31+}$, we get

$$\widehat{T}_{21}f = (P_{(21)} + P_{(32)})\widehat{T}P_{(31)}f = (P_{(21)} + P_{(32)})P_{(31)}\mathcal{U}P_{(31)}f$$

$$= (P_{(21)} + P_{(32)})P_{(31)}\mathcal{U}f = (P_{(21)} + P_{(32)})\mathcal{U}f$$

$$= P_{(21)}\mathcal{U}f_{21} + P_{(32)}\mathcal{U}f_{21} + P_{(21)}\mathcal{U}f_{32} + P_{(32)}\mathcal{U}f_{32}$$

$$= \widehat{T}_1 f_{21} + 0 + P_{(21)}\mathcal{U}f_{32} + \widehat{T}_2 f_{32}$$

$$= \widehat{T}_1 f_{21} + P_{(21)}(I - P_{(32)})\mathcal{U}f_{32} + \widehat{T}_2 f_{32}$$

$$= \widehat{T}_1 f_{21} + P_{(21)}(\mathcal{U}f_{32} - \mathcal{U}f_{32} + \pi_2\widehat{M}_2 f_{32}) + \widehat{T}_2 f_{32}$$

$$= \widehat{T}_1 f_{21} + \widehat{N}_1\widehat{M}_2 f_{32} + \widehat{T}_2 f_{32},$$

where $f = f_{21} + f_{32} \in \mathcal{K}_{(21)}\dot{+}\mathcal{K}_{(32)}$. Thus we have $\widehat{T}_{21}' = \widehat{T}_{21}$. Further, if we recall Lemma 1.2, we obtain

$$\widehat{M}_{21}f = \widehat{M}P_{(31)}f = \widehat{M}f_{21} + \widehat{M}(I - \pi_1 P_+ \pi_1^\dagger)f_{32}$$

$$= \frac{1}{2\pi i}\int_C (\pi_1^\dagger f_{21})(z)\,dz + \frac{1}{2\pi i}\int_C [\pi_1^\dagger(I - \pi_1 P_+ \pi_1^\dagger)f_{32}](z)\,dz$$

$$= \widehat{M}_1 f_{21} + \frac{1}{2\pi i} \int_C (P_- \pi_1^\dagger f_{32})(z)\, dz = \widehat{M}_1 f_{21} + \frac{1}{2\pi i} \int_C (\pi_1^\dagger f_{32})(z)\, dz$$

$$= \widehat{M}_1 f_{21} + \frac{1}{2\pi i} \int_C (\pi_1^\dagger \pi_2 \pi_2^\dagger f_{32})(z)\, dz + \frac{1}{2\pi i} \int_C (\pi_1^\dagger (I - \pi_2 \pi_2^\dagger) f_{32})(z)\, dz$$

$$= \widehat{M}_1 f_{21} + \frac{1}{2\pi i} \int_C \Theta_1^-(z)(\pi_2^\dagger f_{32})(z)\, dz + 0$$

$$= \widehat{M}_1 f_{21} - \frac{1}{2\pi i} \int_C \Theta_1^-(z)\, [\widehat{M}_2(\widehat{T}_2 - \zeta)^{-1} f_{32}]_-(z)\, dz\,.$$

Therefore, $\widehat{M}'_{21} = \widehat{M}_{21}$. Similarly, $\widehat{M}'_{*21} = \widehat{M}_{*21}$. We can obtain the residuary identity $\widehat{N}'_{21} = \widehat{N}_{21}$ if we make use of the duality relations

$$(\widehat{M}'_{21} f', n) = (f', \widehat{N}'_{*21} n)\,, \qquad (\widehat{N}'_{21} m, g') = (m, \widehat{M}'_{*21} g')$$

and

$$(\widehat{M}_{21} f, n) = (f, \widehat{N}_{*21} n)\,, \qquad (\widehat{N}_{21} m, g) = (m, \widehat{M}_{*21} g)\,,$$

where $f' \in \mathcal{K}_{[31]} = \mathcal{K}_{(32)} \dotplus \mathcal{K}_{(21)}$, $g' \in \mathcal{K}_{*[31]} = \mathcal{K}_{*(32)} \dotplus \mathcal{K}_{*(21)}$, $f \in \mathcal{K}_{(31)}$, $g \in \mathcal{K}_{*(31)}$, $n \in \mathfrak{N}_1$, and $m \in \mathfrak{N}_3$. Therefore we have

$$(\widehat{N}'_{21} m, g) = (m, \widehat{M}'_{*21} g) = (m, \widehat{M}_{*21} g) = (\widehat{N}_{21} m, g)\,. \qquad \square$$

Remark. Note that we do not claim that $\mathcal{F}_{sm}(\Pi_2)\mathcal{F}_{sm}(\Pi_1) = \mathcal{F}_{sm}(\Pi_2 \Pi_1)$. The statement and the proof of Proposition 2.3 is a good illustration to our previous remark in the Introduction that the linear similarity (but not unitary equivalence) is the natural kind of equivalence for conservative curved systems.

The following theorem is a direct consequence of Proposition 2.3.

Theorem A. *Let $\widehat{\Sigma}_1 = \mathcal{F}_{sc}(\Theta_1)$, $\widehat{\Sigma}_2 = \mathcal{F}_{sc}(\Theta_2)$ and $\widehat{\Sigma}_{21} = \mathcal{F}_{sc}(\Theta_{21})$, where $\Theta_1, \Theta_2; \Theta_{21} = \Theta_2 \Theta_1 \in \mathrm{Cfn}$. Suppose that $\Sigma_1 \sim (\widehat{\Sigma}_1 \oplus \Sigma_{1u})$ and $\Sigma_2 \sim (\widehat{\Sigma}_2 \oplus \Sigma_{2u})$, where the systems Σ_{1u} and Σ_{2u} are "purely normal" systems. Then there exists a "purely normal" system $\Sigma_u = (T_u, 0, 0, 0)$ such that $\Sigma_2 \cdot \Sigma_1 \sim (\widehat{\Sigma}_{21} \oplus \Sigma_u)$.*

Proof. By Proposition 2.3, $\widehat{\Sigma}_2 \cdot \widehat{\Sigma}_1 \sim \mathcal{F}_{sc}(\Theta_2 \cdot \Theta_1) = \widehat{\Sigma}_{21} \oplus \widehat{\Sigma}_u$. Then we have

$$\Sigma_2 \cdot \Sigma_1 \sim (\widehat{\Sigma}_2 \oplus \Sigma_{2u}) \cdot (\widehat{\Sigma}_1 \oplus \Sigma_{1u}) \sim (\widehat{\Sigma}_2 \cdot \widehat{\Sigma}_1) \oplus \Sigma_{1u} \oplus \Sigma_{2u}$$

By Proposition 2.1, $\widehat{\Sigma}_2 \cdot \widehat{\Sigma}_1 \sim (\widehat{\Sigma}_{21} \oplus \widehat{\Sigma}_u)$. Therefore, $\Sigma_2 \cdot \Sigma_1 \sim (\widehat{\Sigma}_{21} \oplus \Sigma_u)$, where $\Sigma_u = \widehat{\Sigma}_u \oplus \Sigma_{1u} \oplus \Sigma_{2u}$. $\qquad \square$

Thus we see that the definition of product of conservative curved systems (Prod) is tightly linked to functional model though we do not refer to it explicitly. On the other hand, its formal independence from functional model characterizes the comparative autonomy of conservative curved system well enough. Moreover, we have explicit formulas for $\Sigma_2 \cdot \Sigma_1$ and the product depends only on the factors Σ_2, Σ_1 and their characteristic functions. Theoretically, the dependence on characteristic

222 A. Tikhonov

functions is undesirable, but, in author's opinion, we cannot count on having more than we have.

Now we turn to the associativity of multiplication of systems.

Proposition 2.4. *One has* $\Sigma_3 \cdot (\Sigma_2 \cdot \Sigma_1) \sim (\Sigma_3 \cdot \Sigma_2) \cdot \Sigma_1$, *where* $\Sigma_k \in \mathrm{Sys}$, $k = \overline{1,3}$.

Proof. Let $\Sigma_k = \mathcal{F}_{sc}(\Theta_k)$, $\Pi_k = \mathcal{F}_{mc}(\Theta_k)$, $\Pi = \Pi_3 \cdot \Pi_2 \cdot \Pi_1$ and $\Sigma = \mathcal{F}_{sm}(\Pi)$. For the functional model Π, we consider the following subspaces $\mathcal{K}_1 = \mathcal{K}_{(21)}$, $\mathcal{K}_2 = \mathcal{K}_{(32)}$, $\mathcal{K}_3 = \mathcal{K}_{(43)}$, $\mathcal{K}_{21} = \mathcal{K}_{(31)}$, $\mathcal{K}_{32} = \mathcal{K}_{(42)}$, $\mathcal{K}_{321} = \mathcal{K}_{(41)}$, and dual to them. We will denote by $W_1 \colon H_1 \to \mathcal{K}_1$, $W_2 \colon H_2 \to \mathcal{K}_2$, $W_3 \colon H_3 \to \mathcal{K}_3$ the operators that realize similarities of the systems Σ_k, $k = 1,2,3$ with the corresponding systems $\widehat{\Sigma}_k$ in the model Π (see the proof of Proposition 2.3). Denote by W_{*k}, $k = 1,2,3$ the dual operators. As in the proof of Proposition 2.3, we get that the operator $W_{21} = P_{(31)}(W_1, W_2)$ realizes similarity $\Sigma_2 \cdot \Sigma_1 \sim \widehat{\Sigma}_{21}$. Similarly, the operator $W_{32} = P_{(42)}(W_2, W_3)$ realizes similarity $\Sigma_3 \cdot \Sigma_2 \sim \widehat{\Sigma}_{32}$. By the same argument, we get that the operator $W_{3(21)} = P_{(41)}(W_{21}, W_3)$ realizes similarity $\Sigma_3 \cdot (\Sigma_2 \cdot \Sigma_1) \sim \widehat{\Sigma}_{321}$ and the operator $W_{(32)1} = P_{(41)}(W_1, W_{32})$ realizes similarity $(\Sigma_3 \cdot \Sigma_2) \cdot \Sigma_1 \sim \widehat{\Sigma}_{321}$. Thus, the operators

$$W_{3(21)} = P_{(41)}(P_{(31)}(W_1, W_2), W_3), \quad W_{(32)1} = P_{(41)}(W_1, P_{(42)}(W_2, W_3))$$

realize the similarities $\Sigma_3 \cdot (\Sigma_2 \cdot \Sigma_1) \sim \widehat{\Sigma}_{321}$ and $(\Sigma_3 \cdot \Sigma_2) \cdot \Sigma_1 \sim \widehat{\Sigma}_{321}$, respectively. Therefore, $\Sigma_3 \cdot (\Sigma_2 \cdot \Sigma_1) \sim (\Sigma_3 \cdot \Sigma_2) \cdot \Sigma_1$. $\qquad\square$

Recall that the operator $P_{[41]} = P_{(21)}(I - P_{(32)})(I - P_{(43)}) + P_{(23)}(I - P_{(32)}) + P_{(43)} = P_{(21)}(I - P_{(43)}) + P_{(32)} + P_{(43)}$ is a projection in $\mathcal{H}$ onto the subspace $\mathcal{K}_{(21)} \dotplus \mathcal{K}_{(32)} \dotplus \mathcal{K}_{(43)}$ and its components $P_{(21)}(I - P_{(43)})$, $P_{(32)}$, $P_{(43)}$ are commuting projections onto the subspaces $\mathcal{K}_{(21)}$, $\mathcal{K}_{(32)}$, $\mathcal{K}_{(43)}$, respectively. Then we have

$$\begin{aligned}
W_{3(21)} &= P_{(41)}(P_{(31)}(W_1, W_2), W_3) \\
&= P_{(41)}(P_{(31)}(P_{(21)}(I - P_{(43)}) + P_{(32)}) + P_{(43)})(W_1, W_2, W_3)
\end{aligned}$$

and

$$\begin{aligned}
W_{(32)1} &= P_{(41)}(W_1, P_{(42)}(W_2, W_3)) \\
&= P_{(41)}(P_{(21)}(I - P_{(43)}) + P_{(42)}(P_{(32)}) + P_{(43)}))(W_1, W_2, W_3).
\end{aligned}$$

Thus, $W_{3(21)} = Y(W_1, W_2, W_3)$ and $W_{(32)1} = Z(W_1, W_2, W_3)$, where

$$Y, Z : \mathcal{K}_{(21)} \dotplus \mathcal{K}_{(32)} \dotplus \mathcal{K}_{(43)} \to \mathcal{K}_{(41)},$$
$$(W_1, W_2, W_3) : H_1 \oplus H_2 \oplus H_3 \to \mathcal{K}_{(21)} \dotplus \mathcal{K}_{(32)} \dotplus \mathcal{K}_{(43)},$$

and

$$Y = P_{(41)}(P_{(31)}(P_{(21)}(I - P_{(43)}) + P_{(32)}) + P_{(43)}),$$
$$Z = P_{(41)}(P_{(21)}(I - P_{(43)}) + P_{(42)}(P_{(32)} + P_{(43)})).$$

As is shown in the Appendix, $Z^{-1} = [P_{(21)} + (P_{(32)} + P_{(43)})P_{(42)}]|\mathcal{K}_{(41)}$ and $Z^{-1}Y = (I + P_{(21)}P_{(43)})|\mathcal{K}_{(21)}\dot{+}\mathcal{K}_{(32)}\dot{+}\mathcal{K}_{(43)}$. Further, it can easily be checked that

$$(W_1, W_2, W_3)^{-1} = \begin{pmatrix} W_{*1}^* P_{(21)}(I - P_{(43)}) \\ W_{*2}^* P_{(32)} \\ W_{*3}^* P_{(43)} \end{pmatrix}.$$

Thus we obtain

$$W_{(32)1}^{-1} W_{3(21)} = (W_1, W_2, W_3)^{-1} Z^{-1} Y (W_1, W_2, W_3)$$

$$= \begin{pmatrix} I & 0 & W_{*1}^* P_{(21)} P_{(43)} W_3 \\ 0 & I & 0 \\ 0 & 0 & I \end{pmatrix}$$

and therefore $W_{(32)1}^{-1} W_{3(21)} \neq I$ because, in general, we have no the property $P_{(21)} P_{(43)} = 0$. Thus the identity $\Sigma_3 \cdot (\Sigma_2 \cdot \Sigma_1) = (\Sigma_3 \cdot \Sigma_2) \cdot \Sigma_1$ does not hold.

Example. We continue the example from Section 1. Consider the systems

$$\Sigma_1 = \Sigma_2 = \mathcal{F}_{sc}(\theta) = ((0), (1), (1)), \ \Sigma_3 = \mathcal{F}_{sc}(\theta^2) = \left(\begin{pmatrix} 0 & 1 \\ 0 & 0 \end{pmatrix}, (1, 0), \begin{pmatrix} 2\varepsilon \\ 1 \end{pmatrix} \right).$$

Then we can easily calculate that

$$\Sigma_3 \cdot (\Sigma_2 \cdot \Sigma_1) = \left(\begin{pmatrix} 0 & 1 & \varepsilon & 0 \\ 0 & 0 & 1 & 0 \\ 0 & 0 & 0 & 1 \\ 0 & 0 & 0 & 0 \end{pmatrix}, (1, \varepsilon, -\varepsilon^2, 2\varepsilon^3), \begin{pmatrix} \varepsilon^3 \\ -\varepsilon^2 \\ 2\varepsilon \\ 1 \end{pmatrix} \right)$$

and

$$(\Sigma_3 \cdot \Sigma_2) \cdot \Sigma_1 = \left(\begin{pmatrix} 0 & 1 & \varepsilon & -\varepsilon^2 \\ 0 & 0 & 1 & 0 \\ 0 & 0 & 0 & 1 \\ 0 & 0 & 0 & 0 \end{pmatrix}, (1, \varepsilon, 0, 0), \begin{pmatrix} \varepsilon^3 \\ -\varepsilon^2 \\ 2\varepsilon \\ 1 \end{pmatrix} \right).$$

Thus, $\Sigma_3 \cdot (\Sigma_2 \cdot \Sigma_1) \neq (\Sigma_3 \cdot \Sigma_2) \cdot \Sigma_1$. The matrix

$$X = \begin{pmatrix} 1 & 0 & -\varepsilon^2 & 2\varepsilon^3 \\ 0 & 1 & 0 & 0 \\ 0 & 0 & 1 & 0 \\ 0 & 0 & 0 & 1 \end{pmatrix}$$

realizes the similarity $\Sigma_3 \cdot (\Sigma_2 \cdot \Sigma_1) \sim (\Sigma_3 \cdot \Sigma_2) \cdot \Sigma_1$.

3. Regular factorizations

We start with extension of the notion of regularity to the class of n-characteristic functions.

Definition. *We shall say that an n-characteristic function Θ is regular (and write $\Theta \in \mathrm{Cfn}_n^{\mathrm{reg}}$) if $\forall\, i \geq j \geq k$*

$$\mathrm{Ran}\,(I - \Theta_{ij}^{\dagger}(z)\Theta_{ij}(z))^{1/2} \cap \mathrm{Ran}\,(I - \Theta_{jk}(z)\Theta_{jk}^{\dagger}(z))^{1/2} = \{0\}\,, \quad a.e. \ \ z \in C\,.$$

Note that it suffices to check these conditions for $k = 1$, $i = n$, $j = \overline{1,n}$ (it follows from [1], Lemma VII.4.1).

Taking into account the fact that we identify n-characteristic function with factorization of Schur class function, we obtain the definition of regularity for factorization of Schur class function. In particular, we have the corresponding definition in the case when $n = 3$. If additionally we assume that $\Xi_k \equiv 1$, $k = \overline{1,3}$, we arrive at the standard definition [1, 2] (see the Introduction).

In the context of functional models the corresponding notion is the following.

Definition. *Let* $\mathrm{Mod}_n^{\mathrm{reg}} := \{\Pi \in \mathrm{Mod}_n : \mathrm{Ran}\,\pi_1 \vee \mathrm{Ran}\,\pi_n = \mathcal{H}\}$. *We shall say that an n-model* $\Pi \in \mathrm{Mod}_n$ *is regular if* $\Pi \in \mathrm{Mod}_n^{\mathrm{reg}}$.

We are going to show that these two notions of regularity (for n-characteristic functions and for n-models) agree. With that end in mind we employ the construction of Proposition 1.1. It is easy to show that for any two contractive operators $A_{21}\colon \mathcal{N}_1 \to \mathcal{N}_2$ and $A_{32}\colon \mathcal{N}_2 \to \mathcal{N}_3$ there exist three isometries $V_1\colon \mathcal{N}_1 \to \mathcal{H}$, $V_2\colon \mathcal{N}_2 \to \mathcal{H}$, and $V_3\colon \mathcal{N}_3 \to \mathcal{H}$ such that

$$A_{21} = V_2^* V_1\,, \qquad A_{32} = V_3^* V_2\,, \qquad A_{32} A_{21} = V_3^* V_1\,.$$

Note that we do not need to assume as in Proposition 1.1 that the operators A_{21} and A_{32} are operator-valued functions of weighted Schur class: it suffices to assume that they are merely contractive operators. Evolving this approach, we obtain the following Lemmas 3.1 and 3.2. Let V_1, V_2, V_3 be isometries; $A_{21} = V_2^* V_1$, $A_{32} = V_3^* V_2$, $A_{31} = V_3^* V_1$; $\mathcal{E}_1 = \mathrm{Ran}\, V_1$, $\mathcal{E}_2 = \mathrm{Ran}\, V_2$, and $\mathcal{E}_3 = \mathrm{Ran}\, V_3$.

Lemma 3.1. *The following conditions are equivalent:*
1) $A_{31} = A_{32} A_{21}$;
2) $V_3^* V_1 - V_3^* V_2 V_2^* V_1 = 0$;
3) $((\mathcal{E}_1 \vee \mathcal{E}_2) \ominus \mathcal{E}_2) \perp ((\mathcal{E}_3 \vee \mathcal{E}_2) \ominus \mathcal{E}_2)$.

Lemma 3.2. *Assume that* $V_3^* V_1 - V_3^* V_2 V_2^* V_1 = 0$. *Then the following conditions are equivalent:*
1) $\mathrm{Ran}\,(I - A_{32}^* A_{32})^{1/2} \cap \mathrm{Ran}\,(I - A_{21} A_{21}^*)^{1/2} = \{0\}$;
2) $\mathrm{clos}\,\mathrm{Ran}\,(I - V_3 V_3^*)V_1 = \mathrm{clos}\,\mathrm{Ran}\,(I - V_2 V_2^*)V_1 \oplus \mathrm{clos}\,\mathrm{Ran}\,(I - V_3 V_3^*)V_2$;
3) $\mathcal{E}_2 \subset \mathcal{E}_1 \vee \mathcal{E}_3$.

These two lemmas allow us to translate factorization problems into geometrical language and now we can say about purely geometrical nature of the notion of regularity. Note that this fact is the underlying basis of the generalization of Sz.-Nagy-Foiaş's regularity criterion in [18], where the authors dropped the condition of analyticity.

The following assertion is a straightforward consequence of the lemmas.

Proposition 3.3. *One has* $\Pi = \mathcal{F}_{mc}(\Theta) \in \mathrm{Mod}_n^{\mathrm{reg}} \iff \Theta \in \mathrm{Cfn}_n^{\mathrm{reg}}$.

We have defined the notions of regularity for Cfn_n and Mod_n. Now we pass over to curved conservative systems looking for a counterpart of the regularity in this context. In the Introduction we have defined the notion of *simple* curved conservative systems. For this notion, we have the following list of properties.

Proposition 3.4.
1) *Let* $\widehat{\Sigma} = \mathcal{F}_{sm}(\Pi)$, $\Pi \in \mathrm{Mod}$ *and* $\rho(\widehat{T}) \cap G_+ \neq \emptyset$. *Then the system* $\widehat{\Sigma}$ *is simple;*
2) *If* $\Sigma \sim \Sigma'$, *then* Σ *is simple* $\Leftrightarrow$ Σ' *is simple;*
3) *If a system* Σ *is simple, then the system* Σ^* *is simple too;*
4) *If* $\Sigma \overset{X}{\sim} \Sigma'$, $\Sigma \overset{X'}{\sim} \Sigma'$ *and the system* Σ *is simple, then* $X = X'$;
5) *If the system* $\Sigma = \Sigma_2 \cdot \Sigma_1$ *is simple and* $\rho(T) \cap \rho(T_1) \cap G_+ \neq \emptyset$, *then the systems* Σ_1 *and* Σ_2 *are simple.*

Definition. *The product of systems* $\Sigma_{21} = \Sigma_2 \cdot \Sigma_1$ *is called regular if the system* Σ_{21} *is simple.*

Now we are ready to establish a correspondence between notions of regularity for systems and models.

Proposition 3.5. *Let* $\widehat{\Sigma}_1 = \mathcal{F}_{sm}(\Pi_1)$, $\widehat{\Sigma}_2 = \mathcal{F}_{sm}(\Pi_2)$. *Suppose* $\Sigma_1 \sim \widehat{\Sigma}_1$, $\Sigma_2 \sim \widehat{\Sigma}_2$, $\Sigma_{21} = \Sigma_2 \cdot \Sigma_1$, *and* $\rho(T_{21}) \cap G_+ \neq \emptyset$. *Then the product* $\Sigma_2 \cdot \Sigma_1$ *is regular* $\Leftrightarrow$ *the product* $\Pi_2 \cdot \Pi_1$ *is regular.*

Proof. Without loss of generality (see Propositions 2.3 and 3.4) it can be assumed that
$$\Sigma_{21} = \widehat{\Sigma} = \mathcal{F}_{sm}(\Pi_2 \cdot \Pi_1) \quad \text{and} \quad T_{21} = \widehat{T}.$$
As above (see the proof of Proposition 3.4), we get
$$\mathcal{K}_u = \bigcap_{z \in \rho(T)} \mathrm{Ker}\, \widehat{M}(\widehat{T} - z)^{-1} = \{f \in \mathcal{K}_{(31)} : \pi_+^\dagger f = 0, \, \pi_-^\dagger f = 0\}$$

and therefore $\mathcal{K}_u \subset (\mathrm{Ran}\, \pi_+ \vee \mathrm{Ran}\, \pi_-)^\perp$. On the hand, if $f \in (\mathrm{Ran}\, \pi_+ \vee \mathrm{Ran}\, \pi_-)^\perp$, then $P_{(31)}f = f$ and $f \in \mathcal{K}_{(31)}$. Thus, $\mathcal{K}_u = (\mathrm{Ran}\, \pi_+ \vee \mathrm{Ran}\, \pi_-)^\perp$. It remains to note that the product $\Sigma_2 \cdot \Sigma_1$ is regular iff $\mathcal{K}_u = \{0\}$ and the product $\Pi_2 \cdot \Pi_1$ is regular iff $\mathrm{Ran}\, \pi_+ \vee \mathrm{Ran}\, \pi_- = \mathcal{H}$ (recall that $\pi_+ = \pi_3$ and $\pi_- = \pi_1$). $\qquad\square$

Combining Propositions 3.3 and 3.5, we arrive at

Proposition 3.6 (Criterion of regularity). *Let* $\widehat{\Sigma}_1 = \mathcal{F}_{sc}(\Theta_1)$, $\widehat{\Sigma}_2 = \mathcal{F}_{sc}(\Theta_2)$. *Suppose* $\Sigma_1 \sim \widehat{\Sigma}_1$, $\Sigma_2 \sim \widehat{\Sigma}_2$, $\Sigma_{21} = \Sigma_2 \cdot \Sigma_1$, *and* $\rho(T_{21}) \cap G_+ \neq \emptyset$. *Then the product* $\Sigma_2 \cdot \Sigma_1$ *is regular* $\Leftrightarrow$ *the factorization* $\Theta_2 \cdot \Theta_1$ *is regular.*

Thus we obtain the correspondence between regular factorizations of characteristic functions and regular products of systems.

Remark. It can easily be shown that the inner-outer factorization [1] of Schur class functions is regular (see [1]). Hence, using the criterion of regularity, one can prove that the product of colligations with C_{11} and C_{00} contractions is regular. It is possible to extend this result to the case of weighted Schur functions employing the generalization of regularity criterion (Proposition 3.6). Note that, for J-contractive analytic operator functions, J-inner-outer factorization is regular too [19]. However, since in this situation we have no such a geometrical functional model (and such a geometrical description of regularity) as is the Sz.-Nagy-Foiaş model for contractions, we have to establish directly the regularity of the product of "absolutely continuous" and "singular" colligations (analogous of C_{11} and C_{00} contractions). The uniqueness of J-inner-outer factorizations is the most important consequence of this regularity (see [19]).

4. Factorizations and invariant subspaces

The most remarkable feature of the product of systems is its connection with invariant subspaces. We see that the subspace H_1 in the definition (Prod) is invariant under the operator T_{21} (and under its resolvent $(T_{21} - z)^{-1}$, $z \in G_-$). In the context of functional model this implies that the subspace $\mathcal{K}_{(21)}$ is invariant under the operator $\widehat{T}$ (see Proposition 2.3). Following B.Sz.-Nagy and C.Foiaş, we shall work within the functional model and use the model as a tool for studying the correspondence "factorizations $\leftrightarrow$ invariant subspaces". Let $\Theta \in \mathrm{Mod}_3$, $\Pi = \mathcal{F}_{mc}(\Theta) = (\pi_1, \pi_2, \pi_3) \in \mathrm{Mod}_3$. We define the transformation $L = \mathcal{F}_{ic}(\Theta)$ as a mapping that takes each 3-characteristic function Θ (which we identify with factorization of Schur class function) to the invariant subspace $L := \mathcal{K}_{(21)} = \mathrm{Ran}\, P_{(21)}$. To study the transformation $\mathcal{F}_{ic}$ (and its ingenuous extension to n-characteristic functions), we need to make some preliminary work.

Let $\Pi \in \mathrm{Mod}_n$. Consider the chain of subspaces $\mathcal{H}_{11+} \subset \cdots \subset \mathcal{H}_{n1+}$ (see the definition of $\mathcal{H}_{ij+}$ after Lemma 1.3). These subspaces are invariant under the resolvent $(\mathcal{U} - z)^{-1}$, $z \in G_-$. The inverse is also true accurate up to the "normal" part of the chain.

Proposition 4.1. *Suppose* $\mathcal{U} \in \mathcal{L}(\mathcal{H})$ *is a normal operator,* $\sigma(\mathcal{U}) \subset C$, *and* $\mathcal{H}_{1+} \subset \cdots \subset \mathcal{H}_{n+}$ *is a chain of invariant under* $(\mathcal{U} - z)^{-1}$, $z \in G_-$ *subspaces. Then there exists an n-model* $\Pi \in \mathrm{Mod}_n$ *such that* $\mathcal{H}_{k1+} \subset \mathcal{H}_{k+}$, $k = \overline{1, n}$ *and the subspaces* $\mathcal{H}_{uk} := \mathcal{H}_{k+} \ominus \mathcal{H}_{k1+}$ *reduce the operator* $\mathcal{U}$. *If an n-model* $\Pi' \in \mathrm{Mod}_n$

satisfies the same conditions, then $\mathcal{H}'_{k+} = \mathcal{H}_{k+}$ *and* $\exists \psi_k$ *such that* $\psi_k, \psi_k^{-1} \in H^\infty(G_+, \mathcal{L}(\mathfrak{N}_k))$ *and* $\pi'_k = \pi_k \psi_k$. *Besides, we have* $\mathcal{H}_{u1} \subset \cdots \subset \mathcal{H}_{un}$.

Proof. Consider the Wold type decompositions $\mathcal{H}_{k+} = \mathcal{H}^{\mathrm{pur}}_{k+} \oplus \mathcal{H}^{\mathrm{nor}}_{k+}$ with respect to the normal operator $\mathcal{U}$, $\sigma(\mathcal{U}) \subset C$ (see [10]). The operators $\mathcal{U}|\mathcal{H}^{\mathrm{pur}}_{k+}$ are the pure subnormal parts of $\mathcal{U}|\mathcal{H}_{k+}$ and $\mathcal{U}|\mathcal{H}^{\mathrm{nor}}_{k+}$ are normal operators. These decompositions are unique. We set

$$\mathcal{E}_{k+} = \mathcal{H}^{\mathrm{pur}}_{k+}, \qquad \mathcal{E}_k = \vee_{z \notin C}(\mathcal{U} - z)^{-1}\mathcal{E}_{k+}, \qquad \mathcal{E}_{k-} = \mathcal{E}_k \ominus \mathcal{E}_{k+}.$$

Obviously, $\mathcal{E}_{k-} \subset \mathcal{H}^\perp_{k+}$ and the operator $\mathcal{U}^*|\mathcal{E}_{k-}$ is the pure subnormal part of $\mathcal{U}^*|\mathcal{H}^\perp_{k+}$. For $i \geq j \geq k$, we have $\mathcal{E}_{i-} \perp \mathcal{E}_{j+}$, $\mathcal{E}_{k+} \subset \mathcal{H}_{j+}$, and $\mathcal{E}_{i-} \subset \mathcal{H}^\perp_{j+}$. Hence,

$$\mathcal{E}_k \subset \vee_{z \notin C}(\mathcal{U} - z)^{-1}\mathcal{H}_{j+} \qquad \text{and} \qquad \mathcal{E}_i \subset \vee_{z \notin C}(\mathcal{U}^* - \bar{z})^{-1}\mathcal{H}^\perp_{j+}.$$

This implies that

$$\mathcal{E}_j \oplus ((\mathcal{E}_k \vee \mathcal{E}_j) \ominus \mathcal{E}_j) = \mathcal{E}_k \vee \mathcal{E}_j \subset \mathcal{E}_j \vee \mathcal{H}_{j+} = \mathcal{E}_j \oplus \mathcal{H}^{\mathrm{nor}}_{j+}.$$

Therefore we get

$$(\mathcal{E}_k \vee \mathcal{E}_j) \ominus \mathcal{E}_j \subset \mathcal{H}^{\mathrm{nor}}_j \qquad \text{and} \qquad \mathcal{E}_{j+} \oplus ((\mathcal{E}_k \vee \mathcal{E}_j) \ominus \mathcal{E}_j) \subset \mathcal{H}_j.$$

In the same way, $\mathcal{E}_{j-} \oplus ((\mathcal{E}_i \vee \mathcal{E}_j) \ominus \mathcal{E}_j) \subset \mathcal{H}^\perp_j$. And finally,

$$((\mathcal{E}_i \vee \mathcal{E}_j) \ominus \mathcal{E}_j) \perp ((\mathcal{E}_k \vee \mathcal{E}_j) \ominus \mathcal{E}_j.$$

We need to make use of the following lemma.

Lemma. *Suppose* $\mathcal{U} \in \mathcal{L}(\mathcal{H})$ *is a normal operator,* $\sigma(\mathcal{U}) \subset C$, $\mathcal{E}_+ \subset \mathcal{H}$ *and* $\mathcal{U}|\mathcal{E}_+$ *is a pure subnormal operator. Then there exists an operator* $\pi \in \mathcal{L}(L^2(C, \mathfrak{N}), \mathcal{H})$ *such that* $\mathrm{Ran}\,\pi = \vee_{\lambda \notin C}(\mathcal{U} - \lambda)^{-1}\mathcal{E}_+$, $\mathrm{Ker}\,\pi = \{0\}$, $\pi E^2(G_+, \mathfrak{N}) = \mathcal{E}_+$ *and* $\mathcal{U}\pi = \pi z$.

Proof. Without loss of generality we can assume that $\mathcal{H} = \vee_{\lambda \notin C}(\mathcal{U} - \lambda)^{-1}\mathcal{E}_+$. By [10], there exists a unitary operator $Y_0 \in \mathcal{L}(E^2_\alpha(G_+, \mathfrak{N}), \mathcal{E}_+)$ such that $\mathcal{U}Y_0 = Y_0 z$, where $E^2_\alpha(G_+, \mathfrak{N})$ is the Smirnov space of character-automorphic functions (see the proof of Proposition 1.5). By Mlak's lifting theorem [20], the operator Y_0 can be extended to the space $L^2(C, \mathfrak{N})$ lifting the intertwining condition. This extension will be denoted by $\pi_0 \in \mathcal{L}(L^2(C, \mathfrak{N}), \mathcal{H})$. So, we have $\mathcal{U}\pi_0 = \pi_0 z$. Similarly, there exists an extension $X_0 \in \mathcal{L}(\mathcal{H}, L^2(C, \mathfrak{N}))$ of the operator Y_0^{-1} such that $X_0 \mathcal{U} = z X_0$. Thus, $X_0 \pi_0 | E^2_\alpha(G_+, \mathfrak{N}) = I | E^2_\alpha(G_+, \mathfrak{N})$. Since $L^2(C, \mathfrak{N}) = \vee_{\lambda \notin C}(z - \lambda)^{-1} E^2_\alpha(G_+, \mathfrak{N})$, we get $X_0 \pi_0 = I$. Likewise, since $\pi_0 X_0 | \mathcal{E}_+ = I | \mathcal{E}_+$ and $\mathcal{H} = \vee_{\lambda \notin C}(\mathcal{U} - \lambda)^{-1}\mathcal{E}_+$, we get $\pi_0 X_0 = I$ and therefore $\pi_0^{-1} = X_0 \in \mathcal{L}(\mathcal{H}, L^2(C, \mathfrak{N}))$.

According to [10], the "bundle" shift $z|F^2_\alpha(G_+, \mathfrak{N})$ is similar to the trivial shift $z|E^2(G_+, \mathfrak{N})$. The similarity is realized by operator-valued function $\chi \in L^\infty(C, \mathcal{L}(\mathfrak{N}))$ such that $\chi^{-1} \in L^\infty(C, \mathcal{L}(\mathfrak{N}))$ and $\chi E^2(G_+, \mathfrak{N}) = E^2_\alpha(G_+, \mathfrak{N})$. Then we put $\pi := \pi_0 \chi$. $\square$

Since $\mathcal{U}|\mathcal{E}_{j+}$ are the pure subnormal parts of $\mathcal{U}|\mathcal{H}_j$, there exist operators $\pi_j \in \mathcal{L}(L^2(C, \mathfrak{N}_j), \mathcal{H})$ such that $\mathrm{Ran}\,\pi_j = \mathcal{E}_j$, $\pi_j E^2(G_+) = \mathcal{E}_{j+}$, and $\mathcal{U}\pi_j = \pi_j z$.

In terms of operators π_j we rewrite the relations obtained earlier. The relation $\mathcal{E}_{i-}\perp\mathcal{E}_{j+}$ implies $P_-(\pi_i^\dagger\pi_j)P_+ = 0$ and the orthogonality $((\mathcal{E}_i \vee \mathcal{E}_j)\ominus\mathcal{E}_j)\perp((\mathcal{E}_k \vee \mathcal{E}_j)\ominus\mathcal{E}_j)$ means that $\mathrm{Ran}(I-\pi_j\pi_j^\dagger)\pi_i \perp \mathrm{Ran}(I-\pi_j\pi_j^\dagger)\pi_k$. Hence, $\pi_i^\dagger(I-\pi_j\pi_j^\dagger)\pi_k = 0$ and $\pi_i^\dagger\pi_k = \pi_i^\dagger\pi_j\pi_j^\dagger\pi_k$. Thus the n-tuple $\Pi = (\pi_1,\ldots,\pi_n)$ is an n-model. We put $\mathcal{H}_{j1+} = \mathcal{H}_{\pi_j\vee\cdots\vee\pi_1} \cap \mathrm{Ker}\,\pi_j P_-\pi_j^\dagger$. Then,

$$\mathcal{H}_{j1+} = \mathrm{Ran}\,P_{\pi_j\vee\cdots\vee\pi_1} \cap \mathrm{Ran}\,(I - \pi_j P_-\pi_j^\dagger) = \mathrm{Ran}\,P_{\pi_j\vee\cdots\vee\pi_1}(I - \pi_j P_-\pi_j^\dagger)$$

$$= \mathrm{Ran}\,P_{\pi_j\vee\cdots\vee\pi_1}((I - \pi_j\pi_j^\dagger) + \pi_j P_+\pi_j^\dagger) = \mathcal{E}_{j+} \oplus \mathrm{Ran}(I - \pi_j\pi_j^\dagger)P_{\pi_j\vee\cdots\vee\pi_1}$$

$$= \mathcal{E}_{j+} \oplus \vee_{k=1}^{j-1} \mathrm{clos}\,\mathrm{Ran}(I - \pi_j\pi_j^\dagger)\pi_k = \mathcal{E}_{j+} \oplus (\vee_{k=1}^{j-1}((\mathcal{E}_k \vee \mathcal{E}_j) \ominus \mathcal{E}_j))\,.$$

Hence we get $\mathcal{H}_{j1+} \subset \mathcal{H}_{j+} = \mathcal{E}_{j+} \oplus \mathcal{H}_{j+}^{\mathrm{nor}}$ and

$$\mathcal{H}_{uj} = \mathcal{H}_{j+} \ominus \mathcal{H}_{j1+} = \mathcal{H}_{j+}^{\mathrm{nor}} \ominus (\vee_{k=1}^{j-1}((\mathcal{E}_k \vee \mathcal{E}_j) \ominus \mathcal{E}_j))\,.$$

It is obvious that the subspaces $\mathcal{H}_{uj}$ reduce the operator $\mathcal{U}$.

Assume that $\mathcal{H}'_{j1+} = \mathcal{H}_{\pi'_j\vee\cdots\vee\pi'_1} \cap \mathrm{Ker}\,\pi'_j P_-\pi'^\dagger_j$, $\mathcal{H}'_{j1+} \subset \mathcal{H}_{j+}$ and the subspaces $\mathcal{H}_{j+} \ominus \mathcal{H}'_{j1+}$ reduce the operator $\mathcal{U}$, where $\Pi' = (\pi'_1,\ldots,\pi'_n) \in \mathrm{Mod}_n$. Then we have the generalized Wold decompositions [10] $\mathcal{H}_{j+} = \mathcal{E}'_{j+} \oplus (\vee_{k=1}^{j-1}((\mathcal{E}'_k \vee \mathcal{E}'_j) \ominus \mathcal{E}'_j)) \oplus (\mathcal{H}_{j+} \ominus \mathcal{H}'_{j1+})$. Since these decompositions are unique, we obtain $\mathcal{E}'_{j+} = \mathcal{E}_{j+}$, $\mathcal{E}'_j = \mathcal{E}_j$ and, by induction, $\mathcal{H}'_{j1+} = \mathcal{H}_{j1+}$. Then, $\pi'_j = \pi_j\psi_j$, where $\psi_j = \pi_j^\dagger\pi'_j$, $\psi_j^{-1} = \pi'^\dagger_j\pi_j \in H^\infty(G_+,\mathcal{L}(\mathfrak{N}_j))$.

Since $\mathcal{H}_{uj} \perp (\mathcal{E}_{j+}\oplus((\mathcal{E}_k\vee\mathcal{E}_j)\ominus\mathcal{E}_j))$ and $\mathcal{H}_{uj} \subset \mathcal{H}_{j+}$, we get $\mathcal{H}_{uj} \perp (\mathcal{E}_k\vee\mathcal{E}_j)$. For $i > j$, we have $\mathcal{E}_{i-} \subset \mathcal{H}_{j+}^\perp \subset \mathcal{H}_{uj}^\perp$. Hence, $\mathcal{H}_{uj} \perp \mathcal{E}_i$ and $\mathcal{H}_{uj} \perp \mathcal{H}_{\pi_n\vee\cdots\vee\pi_1}$. Since $\mathcal{H}_{j1+} \subset \mathcal{H}_{\pi_n\vee\cdots\vee\pi_1}$, $\mathcal{H}_{uj} \subset \mathcal{H}_{\pi_n\vee\cdots\vee\pi_1}^\perp$ and $\mathcal{H}_{j+} = \mathcal{H}_{uj} \oplus \mathcal{H}_{j1+}$, we have $\mathcal{H}_{uj} = \mathcal{H}_{\pi_n\vee\cdots\vee\pi_1}^\perp \cap \mathcal{H}_{j+}$ and therefore $\mathcal{H}_{u1} \subset \cdots \subset \mathcal{H}_{un}$. $\qquad\square$

Let $\theta \in \mathrm{Cfn}$. We fix θ and define $\mathrm{Mod}_n^\theta := \{\Pi \in \mathrm{Mod}_n : \pi_n^\dagger\pi_1 = \theta\}$. Then we can consider the chain of subspaces $\mathcal{F}_{im}^\theta(\Pi) := (\mathcal{K}_{(11)} \subset \mathcal{K}_{(21)} \subset \cdots \subset \mathcal{K}_{(n1)})$, where $\mathcal{K}_{(k1)} = \mathrm{Ran}\,P_{(k1)}$. The subspaces $\mathcal{K}_{(k1)}$ are invariant under the operator $\widehat{T}$ and this observation motivates the following definition.

Let $\theta = \pi_-^\dagger\pi_+$, where the operators $\pi_\pm \in \mathcal{L}(L^2(\Xi_\pm),\mathcal{H})$ are isometries. Let $\mathcal{U} \in \mathcal{L}(\mathcal{H})$ be a normal operator such that $\mathcal{U}\pi_\pm = \pi_\pm z$ and $\sigma(\mathcal{U}) \subset C$. Let also $\mathcal{K} = \mathrm{Ran}\,P$, $P = (I - \pi_+ P_+\pi_+^\dagger)(I - \pi_- P_-\pi_-^\dagger)$, and $T = P\mathcal{U}|\mathcal{K}$.

Definition. *A chain of subspaces* $L = (L_1 \subset L_2 \subset \cdots \subset L_n)$ *is called n-invariant if* $L_n \subset \mathcal{K}$, $(T-z)^{-1}L_k \subset L_k$, $z \in G_-$, $k = \overline{1,n}$, *and the subspaces* L_1, $\mathcal{K}\ominus L_n$ *reduce the operator* $\mathcal{U}$. *We will denote the class of all n-invariant chains by* Inv_n^θ.

In fact, we have already defined the transformation $\mathcal{F}_{im}^\theta \colon \mathrm{Mod}_n^\theta \to \mathrm{Inv}_n^\theta$, which takes each $\Pi \in \mathrm{Mod}_n^\theta$ to the n-invariant chain of subspaces $(\mathcal{K}_{(11)} \subset \mathcal{K}_{(21)} \subset \cdots \subset \mathcal{K}_{(n1)}) \in \mathrm{Inv}_n^\theta$. This transformation is surjective accurate up to the "normal" part of the chain.

Proposition 4.2. *Suppose a chain L is n-invariant. Then there exists an n-model $\Pi \in \operatorname{Mod}_n^\theta$ such that $\mathcal{K}_{(k1)} \subset L_k$, $k = \overline{1,n}$ and the subspaces $L_{uk} := L_k \ominus \mathcal{K}_{(k1)}$ reduce the operator $\mathcal{U}$. If an n-model $\Pi' \in \operatorname{Mod}_n$ satisfies the same conditions, then $\mathcal{K}'_{(k1)} = \mathcal{K}_{(k1)}$ and $\exists \psi_k$ such that $\psi_k, \psi_k^{-1} \in H^\infty(G_+, \mathcal{L}(\mathfrak{N}_k))$ and $\pi'_k = \pi_k \psi_k$. Besides, we have $L_{u1} \subset \cdots \subset L_{un}$.*

Proof. We put $\mathcal{H}_{k+} = L_k \dotplus \mathcal{D}_+$, where $\mathcal{D}_+ = \operatorname{Ran} q_+$, $q_+ = \pi_+ P_+ \pi_+^\dagger$. Then, for $z \in G_-$, we get $(\mathcal{U} - z)^{-1}\mathcal{D}_+ \subset \mathcal{D}_+ \subset \mathcal{H}_k$ and

$$(\mathcal{U} - z)^{-1}L_k \subset P(\mathcal{U} - z)^{-1}L_k \dotplus q_+(\mathcal{U} - z)^{-1}L_k \subset (T - z)^{-1}L_k \dotplus \mathcal{D}_+ \subset \mathcal{H}_k.$$

Therefore the chain $\mathcal{H}_{1+} \subset \cdots \subset \mathcal{H}_{n+}$ is invariant under $(\mathcal{U} - z)^{-1}$, $z \in G_-$. By Proposition 4.1, there exists an n-model $\Pi \in \operatorname{Mod}_n$ such that $\mathcal{H}_{k1+} \subset \mathcal{H}_{k+}$, $k = \overline{1,n}$ and the subspaces $\mathcal{H}_{uk} = \mathcal{H}_{k+} \ominus \mathcal{H}_{k1+}$ reduce the operator $\mathcal{U}$. Since L_1 reduces $\mathcal{U}$, we have that $\mathcal{H}_{1+} = \mathcal{D}_+ \dotplus L_1$ is the generalized Wold decomposition of $\mathcal{H}_{1+}$. Taking into account the uniqueness of Wold decomposition, we obtain $\pi_+ = \pi_1 \psi_1$. Comparing the Wold decompositions of the equal subspaces $\mathcal{K} \dotplus \mathcal{D}_+$ and $(\mathcal{K} \ominus L_n) \dotplus \mathcal{H}_{n+}$, we obtain $\pi_- = \pi_n \psi_n$. Thus we can assume without loss of generality (see the proof of Proposition 4.1) that $\pi_1 = \pi_+$ and $\pi_n = \pi_-$, i.e., $\Pi \in \operatorname{Mod}_n^\theta$.

Since $L_k \subset \mathcal{K}$, we have $L_k = (I - \pi_+ P_+ \pi_+^\dagger)\mathcal{H}_{k+}$. Taking into account that $\mathcal{K}_{(k1)} = (I - \pi_+ P_+ \pi_+^\dagger)\mathcal{H}_{k1+}$ and $\mathcal{H}_{k1+} \subset \mathcal{H}_{k+}$, we get $\mathcal{K}_{(k1)} \subset L_k$. Since $\mathcal{H}_{uk} = \mathcal{H}_{\pi_n \vee \cdots \vee \pi_1}^\perp \cap \mathcal{H}_{k+}$, we have $\mathcal{H}_{uk} = P\mathcal{H}_{uk} \subset P\mathcal{H}_{k+} = L_k$ and therefore $\mathcal{H}_{uk} \oplus \mathcal{K}_{(k1)} \subset L_k$. In fact, these two spaces are equal. Consider the operator $q'_+ = \pi'_+ P'_+ \pi'^*_+$, which is the orthogonal counterpart to $q_+ = \pi_+ P_+ \pi_+^\dagger$ (see the proof of Proposition 1.5). Put $L'_k := (I - q'_+)L_k$ and $\mathcal{K}'_{(k1)} := (I - q'_+)\mathcal{K}_{(k1)}$. By the corollary of Lemma (iii), $L_k = (I - q_+)L'_k$ and $\mathcal{K}_{(k1)} = (I - q_+)\mathcal{K}'_{(k1)}$. Further, we have

$$L'_k \ominus \mathcal{K}'_{(k1)} = (L'_k \oplus \mathcal{D}_+) \ominus (\mathcal{K}'_{(k1)} \oplus \mathcal{D}_+) = \mathcal{H}_{k+} \ominus \mathcal{H}_{k1+} = \mathcal{H}_{uk}.$$

Then

$$L_k = (I - q_+)L'_k = (I - q_+)(\mathcal{K}'_{(k1)} \oplus \mathcal{H}_{uk}) = \mathcal{K}_{(k1)} \dotplus \mathcal{H}_{uk} = \mathcal{K}_{(k1)} \oplus \mathcal{H}_{uk}$$

and therefore $L_k = \mathcal{K}_{(k1)} \oplus \mathcal{H}_{uk}$. Hence, $L_{uk} = L_k \ominus \mathcal{K}_{(k1)} = \mathcal{H}_{uk}$. Then, by Proposition 4.1, we have $L_{u1} \subset \cdots \subset L_{un}$.

Let $\Pi' \in \operatorname{Mod}_n^\theta$ be an n-model such that $\mathcal{K}'_{(k1)} \subset L_k$, $k = \overline{1,n}$ and the subspaces $L'_{uk} = L_k \ominus \mathcal{K}'_{(k1)}$ reduce the operator $\mathcal{U}$. Then $\mathcal{H}'_{k1+} = \mathcal{K}'_{(k1)} \dotplus \mathcal{D}_+ \subset L_k \dotplus \mathcal{D}_+ = \mathcal{H}_{k+}$ and the subspaces $\mathcal{H}_{k+} \ominus \mathcal{H}'_{k1+} = L'_{uk}$ reduce the operator $\mathcal{U}$. By Proposition 4.1, we get $\mathcal{H}'_{k1+} = \mathcal{H}_{k1+}$. Hence, $\mathcal{K}'_{(k1)} = (I - \pi_+ P_+ \pi_+^\dagger)\mathcal{H}'_{k1+} = (I - \pi_+ P_+ \pi_+^\dagger)\mathcal{H}_{k1+} = \mathcal{K}_{(k1)}$. $\qquad\square$

Remark. In the case of $n = 2$ this proposition is an analogue of the well-known decomposition of a contraction T into the orthogonal sum $T = T_{cnu} \oplus T_u$ of the completely non-unitary part T_{cnu} and the unitary part T_u (see [1]). In this connection, we will use the notation

$$\mathrm{Inv}_n^{\theta\,cnu} := \{(L_1 \subset L_2 \subset \cdots \subset L_n) \in \mathrm{Inv}_n^\theta : L_{un} = \{0\}\}$$

In this notation Proposition 4.2 means merely that $\mathrm{Ran}\,\mathcal{F}_{im}^\theta = \mathrm{Inv}_n^{\theta\,cnu}$. Note also that the condition $L_{un} = \{0\}$ is equivalent to the condition

$$\vee_{k=1}^n [\vee_{z\notin C}(\mathcal{U}-z)^{-1}([\vee_{z\notin C}(\mathcal{U}-z)^{-1}(L_k \dot{+} \mathcal{D}_+)] \ominus (L_k \dot{+} \mathcal{D}_+))] = \mathcal{H}.$$

Let us now return to the transformation $\mathcal{F}_{ic}$. Fix $\theta \in \mathrm{Cfn}$ and define $\mathrm{Mod}_n^\theta := \{\Theta \in \mathrm{Mod}\colon \Theta_{n1} = \theta\}$. Then we can consider the restriction $\mathcal{F}_{ic}|\,\mathrm{Mod}_3^\theta$, which takes each 3-characteristic function $\Theta \in \mathrm{Mod}_3^\theta$ to the invariant subspace $L := \mathcal{K}_{(21)} \subset \mathcal{H}$. The main difficulty to handle effectively factorizations of the function θ is the fact that the space $\mathcal{H}$ is variable and we cannot compare invariant subspaces when we run over all factorizations of θ. To avoid this effect we shall restrict ourselves to models $\Pi = (\pi_+, \pi_2, \pi_-)$ for which $\mathcal{H} = \mathcal{H}_{\pi_+ \vee \pi_-} = \mathrm{Ran}\,\pi_+ \vee \mathrm{Ran}\,\pi_-$, where $\pi_\pm \in \mathcal{L}(L^2(\Xi_\pm), \mathcal{H})$ are fixed isometries such that $\theta = \pi_-^\dagger \pi_+$. Then we obviously have $\mathrm{Ran}\,\pi_2 \subset \mathrm{Ran}\,\pi_+ \vee \mathrm{Ran}\,\pi_-$ and therefore $\Pi = (\pi_+, \pi_2, \pi_-) \in \mathrm{Mod}_3^{\mathrm{reg}}$. In this connection, we also define the subclasses

$$\mathrm{Cfn}_n^{\theta\,\mathrm{reg}} := \mathrm{Cfn}_n^\theta \cap \mathrm{Cfn}_n^{\mathrm{reg}}, \qquad \mathrm{Mod}_n^{\theta\,\mathrm{reg}} := \mathrm{Mod}_n^\theta \cap \mathrm{Mod}_n^{\mathrm{reg}}$$

and

$$\mathrm{Inv}_n^{\theta\,\mathrm{reg}} := \{L \in \mathrm{Inv}_n^\theta : \mathrm{Ran}\,\pi_+ \vee \mathrm{Ran}\,\pi_- = \mathcal{H}\}.$$

By Proposition 3.3, it can easily be shown that

$$\mathcal{F}_{im}^\theta(\Pi) \in \mathrm{Inv}_n^{\theta\,\mathrm{reg}} \iff \Pi \in \mathrm{Mod}_n^{\theta\,\mathrm{reg}}.$$

Besides, it is clear that $\mathrm{Inv}_n^{\theta\,\mathrm{reg}} \subset \mathrm{Inv}_n^{\theta\,cnu}$. Finally, we define the transformation

$$\mathcal{F}_{ic}^\theta : \mathrm{Cfn}_n^{\theta\,\mathrm{reg}} \to \mathrm{Inv}_n^{\theta\,\mathrm{reg}}$$

by the following procedure.

Definition. *Let $\theta \in \mathrm{Cfn}$. Fix isometries $\pi_\pm \in \mathcal{L}(L^2(\Xi_\pm), \mathcal{H})$ such that $\theta = \pi_-^\dagger \pi_+$. Let $\Theta \in \mathrm{Cfn}_n^{\theta\,\mathrm{reg}}$ and $\Pi = (\pi_1, \pi_2, \ldots, \pi_n) = \mathcal{F}_{mc}(\Theta)$. By Proposition 3.3, $\Pi \in \mathrm{Mod}_n^{\mathrm{reg}}$, i.e., $\mathcal{H} = \mathcal{H}_{\pi_1 \vee \pi_n} = \mathrm{Ran}\,\pi_1 \vee \mathrm{Ran}\,\pi_n$ and $\theta = \pi_n^\dagger \pi_1$. By Proposition 1.1, there exists a unique unitary operator $X\colon \mathcal{H}_{\pi_1 \vee \pi_n} \to \mathcal{H}_{\pi_+ \vee \pi_-}$ such that $\pi_+ = X\pi_1$ and $\pi_- = X\pi_n$. We put*

$$\mathcal{F}_{ic}^\theta(\Theta) := \mathcal{F}_{im}^\theta(X\Pi),$$

where $X\Pi = (X\pi_1, X\pi_2, \ldots, X\pi_n)$.

This definition of the fundamental transformation $\mathcal{F}_{ic}^\theta$ is rather indirect. As justification of it we note that even in the unit disk case the known approaches [1, 2, 16] are not simpler than our procedure. The following proposition is a straightforward consequence of Proposition 4.2.

Proposition 4.3. *One has*

1) $\operatorname{Ran}\mathcal{F}_{ic}^{\theta} = \operatorname{Inv}_{n}^{\theta\,\mathrm{reg}}$;

2) *If* $\mathcal{F}_{ic}^{\theta}(\Theta') = \mathcal{F}_{ic}^{\theta}(\Theta)$, $\Theta,\Theta' \in \operatorname{Cfn}_{n}^{\theta\,\mathrm{reg}}$, *then* $\Theta' \sim \Theta$, *where* $\sim$ *is equivalence relation:* $\Theta' \sim \Theta$ *if* $\exists\,\psi_{k}$, $k = \overline{2, n-1}$ *such that* $\psi_{k},\,\psi_{k}^{-1} \in H^{\infty}(G_{+},\mathcal{L}(\mathfrak{N}_{k}))$, $\Theta'_{ij} = \psi_{i}^{-1}\Theta_{ij}\psi_{j}$, *and* $\Xi'_{k} = \psi_{k}^{*}\Xi_{k}\psi_{k}$; $\psi_{1} = I$, $\psi_{n} = I$.

Thus, one can consider the quotient space $\operatorname{Cfn}_{n}^{\theta\,\mathrm{reg}\sim} := \operatorname{Cfn}_{n}^{\theta\,\mathrm{reg}}/_{\sim}$ and the corresponding one-to-one transformation $\mathcal{F}_{ic}^{\theta\sim} : \operatorname{Cfn}_{n}^{\theta\,\mathrm{reg}\sim} \to \operatorname{Inv}_{n}^{\theta\,\mathrm{reg}}$. Note that the functions ψ_{k} can be regarded as Ξ-unitary constants, i.e., $\psi_{k}^{\dagger} = \psi_{k}^{-1} \in H^{\infty}(G_{+},\mathcal{L}(\mathfrak{N}_{k}))$, where $\psi_{k}^{\dagger}$ are adjoint to $\psi_{k} \colon L^{2}(\Xi'_{k}) \to L^{2}(\Xi_{k})$.

Let us consider particular cases. In the case of $n = 3$ we obtain that the transformation $\mathcal{F}_{ic}^{\theta\sim} : \operatorname{Cfn}_{3}^{\theta\,\mathrm{reg}\sim} \to \operatorname{Inv}_{3}^{\theta\,\mathrm{reg}}$ is a one-to-one correspondence between regular factorizations of a characteristic function and invariant subspaces of the corresponding model operator.

Consider the case $n = 4$. Let $L = (L_{1}, L_{2}, L_{3}, L_{4}) \in \operatorname{Inv}_{4}^{\theta\,\mathrm{reg}}$. By Proposition 4.3, there exists $\Theta \in \operatorname{Cfn}_{4}^{\theta\,\mathrm{reg}}$ such that $L = \mathcal{F}_{ic}^{\theta}(\Theta)$. If we rename $L' = L_{2}$, $L'' = L_{3}$ (recall that $L_{1} = \{0\}$, $L_{4} = \mathcal{K}_{\theta}$) and $\theta = \Theta_{41}$, $\theta'_{1} = \Theta_{21}$, $\theta'_{2} = \Theta_{42}$, $\theta''_{1} = \Theta_{31}$, $\theta''_{2} = \Theta_{43}$, $\Xi_{+} = \Xi_{1}$, $\Xi' = \Xi_{2}$, $\Xi'' = \Xi_{3}$, $\Xi_{-} = \Xi_{4}$, then we have

$$\theta = \theta'_{2}\theta'_{1} = \theta''_{2}\theta''_{1} \quad \text{and} \quad \exists\,\vartheta \in S_{\Xi} \quad \text{such that} \quad \theta''_{1} = \vartheta\theta'_{1}, \quad \theta'_{2} = \theta''_{2}\vartheta. \qquad (\prec)$$

Certainly, $\vartheta = \Theta_{32}$ and $\Xi = (\Xi', \Xi'')$. We shall say that a factorization $\theta = \theta'_{2}\theta'_{1}$ precedes a factorization $\theta = \theta''_{2}\theta''_{1}$ and write $\theta'_{2}\theta'_{1} \prec \theta''_{2}\theta''_{1}$ if the condition $(\prec)$ is satisfied. Thus, $L' \subset L'' \implies \theta'_{2}\theta'_{1} \prec \theta''_{2}\theta''_{1}$.

Conversely, suppose that factorizations $\theta'_{2}\theta'_{1} = \theta''_{2}\theta''_{1}$ are regular and $\theta'_{2}\theta'_{1} \prec \theta''_{2}\theta''_{1}$. After backward renaming we have $\Theta \in \operatorname{Cfn}_{4}^{\theta\,\mathrm{reg}}$. Let $L = (L_{1}, L_{2}, L_{3}, L_{4}) = \mathcal{F}_{ic}^{\theta}(\Theta)$, $L' = \mathcal{F}_{ic}^{\theta}(\theta'_{2}\theta'_{1})$, and $L'' = \mathcal{F}_{ic}^{\theta}(\theta''_{2}\theta''_{1})$. Since the factorizations are regular, we have $L' = L_{2}$, $L'' = L_{3}$. Therefore, $\theta'_{2}\theta'_{1} \prec \theta''_{2}\theta''_{1} \implies L' \subset L''$. Thus we finally obtain

$$\mathcal{F}_{ic}^{\mathrm{reg}}(\Theta_{42}\Theta_{21}) \subset \mathcal{F}_{ic}^{\mathrm{reg}}(\Theta_{43}\Theta_{31}) \iff \Theta_{42}\Theta_{21} \prec \Theta_{43}\Theta_{31}.$$

It is easy to check that $\theta'_{2}\theta'_{1} \prec \theta''_{2}\theta''_{1}, \theta'_{2}\theta'_{1} \sim \vartheta'_{2}\vartheta'_{1}, \theta''_{2}\theta''_{1} \sim \vartheta''_{2}\vartheta''_{1} \implies \vartheta'_{2}\vartheta'_{1} \prec \vartheta''_{2}\vartheta''_{1}$, i.e., the order relation $\prec$ is well defined on the quotient space $\operatorname{Cfn}_{3}^{\theta\,\mathrm{reg}\sim}$. Taking all this into account, we arrive at the main result of the Section.

Theorem B. *There is an order preserving one-to-one correspondence $\mathcal{F}_{ic}^{\mathrm{reg}}$ between regular factorizations of a characteristic function (up to the equivalence relation) and invariant subspaces of the resolvent $(\widehat{T} - z)^{-1}$, $z \in G_{-}$ of the corresponding model operator.*

This theorem is an extension of the fundamental result from [1] (Theorems VII.1.1 and VII.4.3; see also [21] for some refinement).

Corollary. *Suppose that factorizations $\theta'_{2}\theta'_{1}$, $\theta''_{2}\theta''_{1}$ are regular, $\theta'_{2}\theta'_{1} \prec \theta''_{2}\theta''_{1}$ and $\theta''_{2}\theta''_{1} \prec \theta'_{2}\theta'_{1}$. Then $\theta'_{2}\theta'_{1} \sim \theta''_{2}\theta''_{1}$.*

Proof. Let $L' = \mathcal{F}_{ic}^{\mathrm{reg}}(\theta_2'\theta_1')$ and $L'' = \mathcal{F}_{ic}^{\mathrm{reg}}(\theta_2''\theta_1'')$. By Theorem B, we get $L' \subset L'' \subset L'$ and therefore $L' = L''$. Then, by Proposition 4.3, we have $\theta_2'\theta_1' \sim \theta_2''\theta_1''$. $\qquad\square$

Note that the corollary can be proved independently from Theorem B: the corresponding argumentation make use of Lemmas 3.1 and 3.2 and therefore we can drop the assumptions that θ_2', θ_2', θ_1'', θ_2'' are operator-valued functions (see Proposition 5.1 in the Appendix).

In conclusion we again consider curved conservative systems. The following assertion is just a translation of Proposition 4.2 into the language of systems.

Proposition 4.4. *Suppose* $\Sigma = (T, M, N) \in \mathrm{Sys}$ *and a subspace* L *is invariant under the resolvent* $(T - z)^{-1}$, $z \in G_-$. *Then there exist systems* $\Sigma_1, \Sigma_2 \in \mathrm{Sys}$ *and an operator* $X : H_1 \oplus H_2 \to H$ *such that* $\Sigma \overset{X}{\sim} \Sigma_2 \cdot \Sigma_1$ *and* $L = XH_1$.

Proof. Let $\Sigma \overset{Y}{\sim} \widehat{\Sigma} = \mathcal{F}_{sc}(\theta)$ and $\Pi = (\pi_1, \pi_3) = \mathcal{F}_{mc}(\theta)$. Then $\widehat{L} = YL$ is an invariant subspace for the model operator. By Theorem B, there exists a regular factorization $\theta = \theta_2 \cdot \theta_1$ such that $\widehat{L} = \mathcal{F}_{ic}^{\mathrm{reg}}(\theta_2 \cdot \theta_1) = \mathrm{Ran}\, P_{(21)}$. Besides, $\theta_1 = \pi_2^\dagger\pi_1$ and $\theta_2 = \pi_3^\dagger\pi_2$. We put $\Pi_1 = (\pi_1, \pi_2)$, $\Pi_2 = (\pi_2, \pi_3)$, $\widehat{\Sigma}_1 = \mathcal{F}_{sc}(\Pi_1)$, and $\widehat{\Sigma}_2 = \mathcal{F}_{sc}(\Pi_2)$. Let $\Sigma_1 \overset{Y_1}{\sim} \widehat{\Sigma}_1$ and $\Sigma_2 \overset{Y_2}{\sim} \widehat{\Sigma}_2$. By Proposition 2.3, we get $\Sigma_1 \cdot \Sigma_2 \sim \widehat{\Sigma}$ with the operator $P_{(31)}(Y_1, Y_2)$ realizing the similarity. It can easily be checked that $\widehat{L} = P_{(31)}(Y_1, Y_2)H_1$. Then, for $X = Y^{-1}P_{(31)}(Y_1, Y_2)$, we get $\Sigma \overset{X}{\sim} \Sigma_2 \cdot \Sigma_1$ and $L = XH_1$. $\qquad\square$

Besides, we have the following assertion.

Proposition 4.5. *Suppose the product of systems* $\Sigma_2 \cdot \Sigma_1 = \Sigma_2' \cdot \Sigma_1'$ *is simple,* $H_1 = H_1'$, *and* $\Theta_2\Theta_1 = \Theta_2'\Theta_1'$. *Then there exists* ψ *such that* ψ, $\psi^{-1} \in H^\infty(G_+, \mathcal{L}(\mathfrak{N}))$ *and* $\Sigma_1' \sim \Sigma_1'' = (T_1, M_1, N_1'')$, *where*

$$N_1''^* f_1 = -\frac{1}{2\pi i} \int_{\overline{C}} \psi(z)^* \, [N_1^*(T_1^* - \cdot)^{-1} f_1]_-(z)\, dz\,, \quad f_1 \in H_1\,.$$

Proof. Let $\Sigma = \Sigma_2 \cdot \Sigma_1 \overset{Y}{\sim} \widehat{\Sigma} = \mathcal{F}_{sc}(\theta)$ and $\Pi = (\pi_1, \pi_3) = \mathcal{F}_{mc}(\theta)$. Using the same notation as in the proof of Proposition 4.4, we obtain that the operators $P_{(31)}(Y_1, Y_2)$ and $P_{(31)}(Y_1', Y_2')$ realize the similarities $\Sigma_2 \cdot \Sigma_1 \sim \widehat{\Sigma}$ and $\Sigma_2' \cdot \Sigma_1' \sim \widehat{\Sigma}$, respectively. Since the system $\widehat{\Sigma} \sim \Sigma$ is simple, by Proposition 3.4(4), we get $P_{(31)}(Y_1, Y_2) = P_{(31)}(Y_1', Y_2')$ and therefore $P_{(31)}(Y_1, Y_2)H_1 = P_{(31)}(Y_1', Y_2')H_1'$. Then, by Proposition 4.3, there exists an operator-valued function ψ such that ψ, $\psi^{-1} \in H^\infty(G_+, \mathcal{L}(\mathfrak{N}_2))$, $\theta_1' = \psi^{-1}\theta_1$, and $\theta_2' = \theta_2\psi$. According to [9], $\Sigma_1' \sim \widehat{\Sigma}'' = \mathcal{F}_{sc}(\psi^{-1}\theta_1)$. Since $\Sigma_1' \sim \widehat{\Sigma}' = \mathcal{F}_{sc}(\theta_1')$, we get $\Sigma_1' \sim \Sigma_1''$. $\qquad\square$

Further, we shall say that a system $\Sigma \in \mathrm{Sys}$ possesses the property of uniqueness of characteristic function if there exists a unique characteristic function $\Theta \in \mathrm{Cfn}$ such that $\Sigma = \mathcal{F}_{cs}(\Theta)$. Recall (see the Introduction) the sufficient condition for this property: the transfer function $\Upsilon(z)$ of the system Σ is an operator-valued function of Nevanlinna class. For products of systems we have the following (non-trivial) fact: *suppose that a system $\Sigma = \Sigma_2\Sigma_1$ is simple, possesses the property of uniqueness, and $\rho(T_1) \cap G_+ \neq \emptyset$; then the system Σ_1 possesses the same property too.*

Proposition 4.6. *Suppose the product of systems $\Sigma_2 \cdot \Sigma_1 = \Sigma_2' \cdot \Sigma_1$ is simple and possesses the property of uniqueness. Suppose also $\rho(T_1) \cap G_+ \neq \emptyset$. Then $\Sigma_2 = \Sigma_2'$.*

Proof. Let $\Sigma = \Sigma_2 \cdot \Sigma_1 = \Sigma_2' \cdot \Sigma_1$ and $\theta = \theta_2\theta_1 = \theta_2'\theta_1'$ be the corresponding factorizations. Then $\theta_1 = \theta_1'$ (see the comments before the proposition). Since $\forall \; \lambda \in \rho(T_1) \cap G_+ \neq \emptyset \; \exists \; \theta_1(\lambda)^{-1}$, we get $\theta_2 = \theta_2'$. Then $\Sigma_2 \sim \mathcal{F}_{sc}(\theta_2)$, $\Sigma_2' \sim \mathcal{F}_{sc}(\theta_2)$ and therefore $\Sigma_2 \overset{X_2}{\sim} \Sigma_2'$. Taking this into account, we have $\Sigma \overset{I}{\sim} \Sigma$ and $\Sigma \overset{I \oplus X_2}{\sim} \Sigma$. By Proposition 3.4(4), we get $X_2 = I$. $\qquad\square$

5. Appendix

Proof of Lemma 1.3. Using Lemma 1.2, we have

1) $P_{(ij)}q_{k+} = P_{\pi_i \vee \cdots \vee \pi_j}(I - q_{j+})(I - q_{i-})q_{k+} = P_{\pi_i \vee \cdots \vee \pi_j}(I - q_{j+})q_{k+}$
$$= P_{\pi_i \vee \cdots \vee \pi_j}[(I - \pi_j\pi_j^{\dagger}) + q_{j-}]q_{k+} = P_{\pi_i \vee \cdots \vee \pi_j}(I - \pi_j\pi_j^{\dagger})\pi_k P_+\pi_k^{\dagger} = 0 ;$$

2) $q_{i-}P_{(jk)} = q_{i-}(I - q_{k+})(I - q_{j-})P_{\pi_j \vee \cdots \vee \pi_k} = q_{i-}(I - q_{j-})P_{\pi_j \vee \cdots \vee \pi_k}$
$$= q_{i-}[(I - \pi_j\pi_j^{\dagger}) + q_{j+}]P_{\pi_j \vee \cdots \vee \pi_k} = \pi_i P_-\pi_i^{\dagger}(I - \pi_j\pi_j^{\dagger})P_{\pi_j \vee \cdots \vee \pi_k} = 0 ;$$

3) $P_{(ij)}P_{(kl)} = P_{(ij)}(I - q_{l+})(I - q_{k-})P_{\pi_k \vee \cdots \vee \pi_l} = P_{(ij)}(I - q_{k-})P_{\pi_k \vee \cdots \vee \pi_l}$
$$= P_{(ij)}[(I - \pi_k\pi_k^{\dagger}) + q_{k+}]P_{\pi_k \vee \cdots \vee \pi_l}$$
$$= (I - q_{j+})(I - q_{i-})P_{\pi_i \vee \cdots \vee \pi_j}(I - \pi_k\pi_k^{\dagger})P_{\pi_k \vee \cdots \vee \pi_l} = 0 ;$$

4) $P_{(ik)}P_{(jk)} = P_{\pi_i \vee \cdots \vee \pi_k}(I - q_{k+})(I - q_{i-})P_{(jk)}$
$$= P_{\pi_i \vee \cdots \vee \pi_k}(I - q_{k+})P_{(jk)} = P_{(jk)} ;$$

5) $P_{(ij)}P_{(ik)} = P_{(ij)}(I - q_{k+})(I - q_{i-})P_{\pi_i \vee \cdots \vee \pi_k}$
$$= P_{(ij)}(I - q_{i-})P_{\pi_i \vee \cdots \vee \pi_k} = P_{(ij)} ;$$

6) $P_{(jk)}P_{(ij)} = P_{\pi_j \vee \cdots \vee \pi_k}(I - q_{k+})(I - q_{j-})(I - q_{j+})(I - q_{i-})P_{\pi_i \vee \cdots \vee \pi_j}$
$$= P_{\pi_j \vee \cdots \vee \pi_k}(I - q_{k+})(I - \pi_j\pi_j^{\dagger})(I - q_{i-})P_{\pi_i \vee \cdots \vee \pi_j}$$
$$= (I - q_{k+})P_{\pi_j \vee \cdots \vee \pi_k}(I - \pi_j\pi_j^{\dagger})P_{\pi_i \vee \cdots \vee \pi_j}(I - q_{i-}) = 0 . \qquad\square$$

Proof of Proposition 1.5. Beforehand note that though we systematically strive to deal only with nonorthogonal projections q_{i+} , sometimes we have to employ their orthogonal counterparts q_{i+}' . By [10], there exist isometries $\pi_i' \in \mathcal{L}(L^2(C, \mathfrak{N}_i), \mathcal{H})$

such that $\operatorname{Ran} \pi_i' = \operatorname{Ran} \pi_i$ and $\operatorname{Ran} q_{i+} = \pi_i' E_\alpha^2(G_+, \mathfrak{N}_i)$, where $E_\alpha^2(G_+, \mathfrak{N}_i)$ are Smirnov's spaces of character-automorphic functions (see [10, 11, 13] for the definition). Let $q_{i+}' = \pi_i' P_+' \pi_i'^*$. Then $\operatorname{Ran} q_{i+}' = \operatorname{Ran} q_{i+}$, $P_- \pi_i^\dagger \pi_i' P_+' = 0$, and $P_-' \pi_i'^* \pi_i P_+ = 0$, where P_+' is the orthoprojection onto $E_\alpha^2(G_+, \mathfrak{N}_i)$ and $P_-' = I - P_+'$. Define also the projections $q_{i-}' = \pi_i' P_-' \pi_i'^*$. Then we have

$$q_{i-}q_{j+}' = 0 \qquad \text{and} \qquad q_{i-}'q_{j+} = 0, \qquad i \geq j.$$

Indeed,

$$q_{i-}q_{j+}' = \pi_i P_- \pi_i^\dagger \pi_j' P_+' \pi_j'^* = \pi_i P_- \pi_i^\dagger \pi_j \pi_j^\dagger \pi_j' P_+' \pi_j'^*$$
$$= \pi_i P_- \pi_i^\dagger \pi_j P_+ \pi_j^\dagger \pi_j' P_+' \pi_j'^* + \pi_i P_- \pi_i^\dagger \pi_j P_- \pi_j^\dagger \pi_j' P_+' \pi_j'^* = 0 + 0 = 0.$$

By the same reason, $q_{i-}'q_{j+} = 0$. Using these identities and repeating mutatis mutandis proof of Lemma 1.3, we obtain

$$P_{(ij)}'q_{k+} = P_{(ij)}'q_{k+}' = 0; \qquad\qquad q_{i-}'P_{(jk)} = q_{i-}P_{(jk)}' = 0;$$
$$P_{(ij)}'P_{(kl)} = P_{(ij)}P_{(kl)}' = P_{(ij)}'P_{(kl)}' = 0, \qquad\qquad i \geq j \geq k \geq l.$$

Then, evidently,

$$P_{[m_i m_j]}P_{[m_k m_l]}' = P_{[m_i m_j]}'P_{[m_k m_l]} = P_{[m_i m_j]}'P_{[m_k m_l]}' = 0.$$

Since $\operatorname{Ran} q_{i+}' = \operatorname{Ran} q_{i+}$, we have $\mathcal{D}_{i+}' = \mathcal{D}_{i+}$. Evidently, $\mathcal{H}_{ij}' = \mathcal{H}_{ij}$. Further, let $f \in \mathcal{H}_{ij+} = \mathcal{H}_{ij} \cap \operatorname{Ker} q_{i-}$. Then $q_{i-}f = 0$, that is $\pi_i \pi_i^\dagger f = q_{i+}f$, and

$$q_{i-}'f = \pi_i' \pi_i'^* f - q_{i+}'f = \pi_i \pi_i^\dagger f - q_{i+}'f = q_{i+}f - q_{i+}'f \in \mathcal{D}_{i+}' = \mathcal{D}_{i+},$$

that is, $q_{i-}'f = q_{i+}'g$. Then, $q_{i-}'f = q_{i-}'^2 f = q_{i-}'q_{i+}'g = 0$ and therefore $\mathcal{H}_{ij+} \subset \mathcal{H}_{ij+}'$. For the same reason, $\mathcal{H}_{ij+}' \subset \mathcal{H}_{ij+}$. Thus we have

$$\mathcal{D}_{i+}' = \mathcal{D}_{i+}, \qquad \mathcal{H}_{ij}' = \mathcal{H}_{ij}, \qquad \mathcal{H}_{ij+}' = \mathcal{H}_{ij+}$$

and therefore

$$\mathcal{K}_{[m_i m_j]}' \subset \mathcal{H}_{m_i m_j +}, \qquad \mathcal{D}_{m_j +} \subset \operatorname{Ker} P_{[m_i m_j]}'.$$

We need to prove the following elementary lemmas, which are of interest in their own right.

Lemma (i). *Suppose M, N_+, N_- are subspaces of a Hilbert space and $N_+ \perp N_-$. Then $(N \vee M) \ominus N_- = ((N \vee M) \ominus N) \oplus N_+$, where $N = N_+ \oplus N_-$.*

Proof. Let $f \in (N \vee M) \ominus N_-$. Then $f \in (N \vee M)$, $f \perp N_-$. We have $f = f_N + f_N^\perp$, where $f_N \in N$, $f_N^\perp \in N^\perp$. Then $f_N = f - f_N^\perp \perp N_-$ and $f_N \in N_+$. Hence, $f = f_N^\perp + f_N \in ((N \vee M) \ominus N) \oplus N_+$.

Conversely, let $f \in N_+$. Then $f \in N$, $f \perp N_-$ and therefore $f \in (N \vee M) \ominus N_-$. Hence, $((N \vee M) \ominus N) \oplus N_+ \subset (N \vee M) \ominus N_-$. $\qquad\square$

Lemma (ii). *Suppose $P \in \mathcal{L}(\mathcal{H})$ is a projection; $\mathcal{D}_+, \mathcal{H}_+$ are subspaces of $\mathcal{H}$ such that $\mathcal{D}_+ \subset \mathcal{H}_+$, $\mathcal{K} = \operatorname{Ran} P \subset \mathcal{H}_+$ and $\mathcal{D}_+ \subset \operatorname{Ker} P$. Then the following conditions are equivalent:*

1) $\mathcal{H}_+ \cap \operatorname{Ker} P = \mathcal{D}_+$;
2) $\mathcal{H}_+ = \mathcal{K} \dotplus \mathcal{D}_+$;
3) $\operatorname{Ker}(P|\mathcal{H}_+) = \mathcal{D}_+$.

Proof. 1) $\implies$ 2) Let $f \in \mathcal{H}_+$. Then $f = f_1 + f_2$, where $f_1 = Pf \in \mathcal{K}$ and $f_2 = (I - P)f \in \operatorname{Ker} P$. Since $\mathcal{K} \subset \mathcal{H}_+$, we get $f_2 = f - Pf \in \mathcal{H}_+$ and therefore $f_2 \in \mathcal{H}_+ \cap \operatorname{Ker} P = \mathcal{D}_+$.

2) $\implies$ 3) It is clear that $\mathcal{D}_+ \subset \operatorname{Ker}(P|\mathcal{H}_+)$. Let $f \in \operatorname{Ker}(P|\mathcal{H}_+) \subset \mathcal{H}_+$. Then $f = f_1 + f_2$, where $f_1 \in \mathcal{K}$ and $f_2 \in \mathcal{D}_+$. Then $0 = Pf = P(f_1 + f_2) = f_1$ and therefore $f = f_2 \in \mathcal{D}_+$.

3) $\implies$ 1) It is clear that $\mathcal{D}_+ \subset \mathcal{H}_+ \cap \operatorname{Ker} P$. Let $f \in \mathcal{H}_+ \cap \operatorname{Ker} P$. Then $f \in \operatorname{Ker}(P|\mathcal{H}_+) = \mathcal{D}_+$. $\square$

Lemma (iii). *Suppose P_1 and P_2 are projections such that $\operatorname{Ker} P_1 = \operatorname{Ker} P_2$. Then $P_1 P_2 = P_1$ and $P_2 P_1 = P_2$.*

Proof. Since $\operatorname{Ran}(I - P_2) = \operatorname{Ker} P_2$, we get $P_1(I - P_2) = 0$. Hence, $P_1 P_2 = P_1$. $\square$

Corollary. *Suppose $P_1, P_2 \in \mathcal{L}(\mathcal{H})$ are projections; $\mathcal{D}_+, \mathcal{H}_+$ are subspaces of $\mathcal{H}$ such that $\mathcal{D}_+ \subset \mathcal{H}_+$, $\operatorname{Ran} P_1 \subset \mathcal{H}_+$, $\operatorname{Ran} P_2 \subset \mathcal{H}_+$, $\mathcal{D}_+ \subset \operatorname{Ker} P_1$, $\mathcal{D}_+ \subset \operatorname{Ker} P_2$ and $\operatorname{Ker}(P_1|\mathcal{H}_+) = \operatorname{Ker}(P_2|\mathcal{H}_+) = \mathcal{D}_+$. Then $P_1 P_2 P_1 = P_1$ and $P_2 P_1 P_2 = P_2$.*

Proof. It is clear that $P_1|\mathcal{H}_+$, $P_2|\mathcal{H}_+$ are projections. By Lemma (iii), we have $(P_1|\mathcal{H}_+)(P_2|\mathcal{H}_+) = P_1|\mathcal{H}_+$. Then $P_1 P_2 P_1 f = P_1 P_2 (P_1 f) = P_1(P_1 f) = P_1 f$. $\square$

Proof (of Proposition 1.5). First, we prove our assertion in the orthogonal context. Consider orthogonal projections

$$P'_{(ij)} = P_{\pi_i \vee \cdots \vee \pi_j}(I - q'_{j+})(I - q'_{i-}),$$

Since operators q'_{j+}, q'_{i-} are selfadjoint, we have $(q'_{j+} q'_{i-})^* = q'^*_{i-} q'^*_{j+} = q'_{i-} q'_{j+} = 0$ and hence

$$P'_{(ij)} = P_{\pi_i \vee \cdots \vee \pi_j}(I - q'_{j+} - q'_{i-}) = P_{\pi_i \vee \cdots \vee \pi_j} - q'_{j+} - q'_{i-}.$$

Define subspaces $N_{k\pm} := \operatorname{Ran} q'_{k\pm}$, $N_k := N_{k+} \oplus N_{k-} = \operatorname{Ran} \pi'_k$ $k = \overline{1, n}$. Then we have $P_{\pi_i \vee \cdots \vee \pi_j} = q'_{i-} + P'_{(ij)} + q'_{j+}$ and

$$\mathcal{H}_{ij} = N_{i-} \oplus \mathcal{K}'_{(ij)} \oplus N_{j+}, \qquad \mathcal{H}_{ij+} = \mathcal{K}'_{(ij)} \oplus N_{j+}, \qquad \mathcal{D}_{j+} = N_{j+}.$$

In particular, we get $N_k \vee N_{k+1} = N_{k+1-} \oplus \mathcal{K}'_{(k+1,k)} \oplus N_{k+}$ and therefore $N_{k+} \oplus \mathcal{K}'_{(k+1,k)} = (N_k \vee N_{k+1}) \ominus N_{k+1-}$. Applying the former identity and Lemma (i)

$i - j$ times, we have

$$
\begin{aligned}
N_{j+} &\oplus \mathcal{K}'_{(j+1,j)} \oplus \mathcal{K}'_{(j+2,j+1)} \oplus \cdots \oplus \mathcal{K}'_{(i,i-1)} \oplus N_{i-} \\
&= [(N_j \vee N_{j+1}) \ominus N_{j+1-}] \oplus \mathcal{K}'_{(j+2,j+1)} \oplus \cdots \oplus \mathcal{K}'_{(i,i-1)} \oplus N_{i-} \\
&= [(N_j \vee N_{j+1}) \ominus N_{j+1}] \oplus N_{j+1+} \oplus \mathcal{K}'_{(j+2,j+1)} \oplus \cdots \oplus \mathcal{K}'_{(i,i-1)} \oplus N_{i-} \\
&= [(N_j \vee N_{j+1}) \ominus N_{j+1}] \oplus [N_{j+1+} \oplus \mathcal{K}'_{(j+2,j+1)}] \oplus \cdots \oplus \mathcal{K}'_{(i,i-1)} \oplus N_{i-} \\
&= \cdots = [(N_j \vee N_{j+1}) \ominus N_{j+1}] \oplus [(N_{j+1} \vee N_{j+2}) \ominus N_{j+2}] \oplus \cdots \\
&\quad \oplus [(N_{i-1} \vee N_i) \ominus N_i] \oplus N_{i+} \oplus N_{i-} = [(N_j \vee N_{j+1}) \ominus N_{j+1}] \\
&\quad \oplus [(N_{j+1} \vee N_{j+2}) \ominus N_{j+2}] \oplus \cdots \oplus [(N_{i-1} \vee N_i) \ominus N_i] \oplus N_i \\
&= [(N_j \vee N_{j+1}) \ominus N_{j+1}] \oplus [(N_{j+1} \vee N_{j+2}) \ominus N_{j+2}] \oplus \cdots \oplus [(N_{i-1} \vee N_i)] \\
&= \cdots = N_j \vee N_{j+1} \vee \cdots \vee N_{i-1} \vee N_i = \mathcal{H}_{ij} .
\end{aligned}
$$

On the other hand, we have already shown $\mathcal{H}_{ij} = N_{i-} \oplus \mathcal{K}'_{(ij)} \oplus N_{j+}$. Therefore,

$$
\mathcal{K}'_{(ij)} = \mathcal{K}'_{(ii-1)} \oplus \cdots \oplus \mathcal{K}'_{(j+1j)} .
$$

Then we have $\mathcal{K}'_{[m_i m_j]} = \mathcal{K}'_{(m_i m_j)}$ and $\mathcal{K}'_{[m_i m_k]} = \mathcal{K}'_{[m_i m_j]} \oplus \mathcal{K}'_{[m_j m_k]}$. It is easy to check that $\mathcal{H}_{m_i m_j +} \cap \operatorname{Ker} P'_{[m_i m_j]} = \mathcal{D}_{m_j +}$.

The nonorthogonal case can be obtained by induction. In fact, we have already shown that

$$
\mathcal{H}_{m_{j+1} m_j +} \cap \operatorname{Ker} P_{[m_{j+1} m_j]} = \mathcal{H}_{m_{j+1} m_j +} \cap \operatorname{Ker} P_{(m_{j+1} m_j)} = \mathcal{D}_{m_j +} , \ j = \overline{1, n} .
$$

Let $i \geq j \geq k$. Assume that $\mathcal{H}_{m_i m_j +} \cap \operatorname{Ker} P_{[m_i m_j]} = \mathcal{D}_{m_j +}$ and $\mathcal{H}_{m_j m_k +} \cap \operatorname{Ker} P_{[m_j m_k]} = \mathcal{D}_{m_k +}$. Let $f \in \mathcal{H}_{m_i m_k +} \cap \operatorname{Ker} P_{[m_i m_k]}$. Using the recursion relation

$$
P_{[m_i m_k]} = P_{[m_j m_k]}(I - P_{[m_i m_j]}) + P_{[m_i m_j]}
$$

and properties of projections $P_{[\cdot, \cdot]}$, we have

$$
P_{[m_i m_j]} f = P_{[m_i m_j]}(P_{[m_j m_k]}(I - P_{[m_i m_j]}) + P_{[m_i m_j]}) f = P_{[m_i m_j]} P_{[m_i m_k]} f = 0 .
$$

Then, since $P_{[m_i m_k]} f = 0$ and $P_{[m_i m_j]} f = 0$, we also have $P_{[m_j m_k]} f = 0$. On the other hand, the vector f can be decomposed $f = f'_{ij} + f'_{jk} + g$, where $f'_{ij} \in \mathcal{K}'_{[m_i m_j]}$, $f'_{ij} \in \mathcal{K}'_{[m_i m_j]}$ and $g \in \mathcal{D}_{m_j +}$. Since $P_{[m_i m_j]} P'_{[m_j m_k]} = 0$, we have

$$
0 = P_{[m_i m_j]} f = P_{[m_i m_j]}(f'_{ij} + f'_{jk} + g) = P_{[m_i m_j]} f'_{ij} .
$$

By Lemma (ii), $\operatorname{Ker}(P_{[m_i m_j]} | \mathcal{H}_{m_i m_j +}) = \mathcal{D}_{m_j +}$. Then, by the corollary of Lemma (iii), we obtain $0 = P'_{[m_i m_j]} P_{[m_i m_j]} f'_{ij} = P'_{[m_i m_j]} P_{[m_i m_j]} P'_{[m_i m_j]} f'_{ij} = f'_{ij}$. Further, $0 = P_{[m_j m_k]} f = P_{[m_j m_k]}(f'_{jk} + g) = P_{[m_j m_k]} f'_{jk}$. As above, we get $0 = f'_{jk}$. Thus, we have $f = g \in \mathcal{D}_{m_j +}$ and therefore $\mathcal{H}_{m_i m_k +} \cap \operatorname{Ker} P_{[m_i m_k]} \subset \mathcal{D}_{m_j +}$. The inverse inclusion is obvious. $\qquad \square$

Proof of Proposition 2.1.

1) Let $\Sigma_1 \overset{X_1}{\sim} \Sigma_1'$, $\Sigma_2 \overset{X_2}{\sim} \Sigma_2'$. Then

$$M_2^{21} f_2 = -\frac{1}{2\pi i} \int_C \Theta_1^-(\zeta) \, [M_2(T_2 - \cdot)^{-1} f_2]_-(\zeta) \, d\zeta$$

$$= -\frac{1}{2\pi i} \int_C \Theta_1^-(\zeta) \, [M_2'(T_2' - \cdot)^{-1} X_2 f_2]_-(\zeta) \, d\zeta = M_2^{21'} X_2 f_2 \,.$$

Hence, $M_2^{21} = M_2^{21'} X_2$. Similarly, $X_1 N_1^{21} = N_1^{21'}$. Define $X_{21} := \begin{pmatrix} X_1 & 0 \\ 0 & X_2 \end{pmatrix}$.
Then, we get

$$T_{21}' X_{21} = \begin{pmatrix} T_1' & N_1' M_2' \\ 0 & T_2' \end{pmatrix} \begin{pmatrix} X_1 & 0 \\ 0 & X_2 \end{pmatrix} = \begin{pmatrix} T_1' X_1 & N_1' M_2' X_2 \\ 0 & T_2' X_2 \end{pmatrix}$$

$$= \begin{pmatrix} X_1 T_1 & X_1 N_1 M_2 \\ 0 & X_2 T_2 \end{pmatrix} = \begin{pmatrix} X_1 & 0 \\ 0 & X_2 \end{pmatrix} \begin{pmatrix} T_1 & N_1 M_2 \\ 0 & T_2 \end{pmatrix} = X_{21} T_{21} \,;$$

$$M_{21}' X_{21} = \begin{pmatrix} M_1' , & M_2^{21'} \end{pmatrix} \begin{pmatrix} X_1 & 0 \\ 0 & X_2 \end{pmatrix} = \begin{pmatrix} M_1' X_1 , & M_2^{21'} X_2 \end{pmatrix}$$

$$= \begin{pmatrix} M_1 , & M_2^{21} \end{pmatrix} = M_{21} \,;$$

$$N_{21}' = \begin{pmatrix} N_1^{21'} \\ N_2' \end{pmatrix} = \begin{pmatrix} X_1 N_1^{21'} \\ X_2 N_2' \end{pmatrix} = \begin{pmatrix} X_1 & 0 \\ 0 & X_2 \end{pmatrix} \begin{pmatrix} N_1^{21} \\ N_2 \end{pmatrix} = X_{21} N_{21}$$

and thus $\Sigma_2 \cdot \Sigma_1 \overset{X_{21}}{\sim} \Sigma_2' \cdot \Sigma_1'$.

2) Let $\Sigma_{21*} = (\Sigma_2 \cdot \Sigma_1)^*$. By straightforward calculation, we get

$$T_{*21} = \begin{pmatrix} T_1^* & 0 \\ M_2^* N_1^* & T_2^* \end{pmatrix}; \quad M_{*21} = \begin{pmatrix} N_1^{21*} , & N_2^* \end{pmatrix}; \quad N_{*21} = \begin{pmatrix} M_1^* \\ M_2^{21*} \end{pmatrix}.$$

On the other hand, let $\Sigma_{21*}' = \Sigma_1^* \cdot \Sigma_2^*$. Then we get

$$T_{*21}' = \begin{pmatrix} T_{*1} & 0 \\ N_{*2} M_{*1} & T_{*2} \end{pmatrix} = \begin{pmatrix} T_1^* & 0 \\ M_2^* N_1^* & T_2^* \end{pmatrix} = T_{*21}.$$

Since

$$M_{*1}^{21'} f_1 = -\frac{1}{2\pi i} \int_{\overline{C}} \Theta_{*2}^-(\zeta) \, [M_{*1}(T_{*1} - \cdot)^{-1} f_1]_-(\zeta) \, d\zeta$$

$$= -\frac{1}{2\pi i} \int_{\overline{C}} \Theta_{*2}^-(\zeta) \, [N_1^*(T_1^* - \cdot)^{-1} f_1]_-(\zeta) \, d\zeta = N_1^{21*} f_1 \,, \quad f_1 \in H_1 \,,$$

we have

$$M_{*21}' = \begin{pmatrix} M_{*1}^{21'} , & M_{*2} \end{pmatrix} = \begin{pmatrix} N_1^{21*'} , & N_2^* \end{pmatrix} = M_{*21} \,.$$

238   A. Tikhonov

Since

$$N_{*2}^{21'*} f_2 = -\frac{1}{2\pi i} \int_C \Theta_{**1}^{-}(\zeta)\, [N_{*2}^{*}(T_{*2}^{*} - \cdot)^{-1} f_2]_{-}(\zeta)\, d\zeta$$

$$= -\frac{1}{2\pi i} \int_C \Theta_1^{-}(\zeta)\, [M_2(T_2 - \cdot)^{-1} f_2]_{-}(\zeta)\, d\zeta = M_2^{21} f_2, \quad f_2 \in H_2\,,$$

we have

$$N'_{*21} = \begin{pmatrix} N_{*1} \\ N_{*2}^{21'} \end{pmatrix} = \begin{pmatrix} M_1^{*} \\ M_1^{21*} \end{pmatrix} = N_{*21}\,.$$

Thus we get $\Sigma_1^{*} \cdot \Sigma_2^{*} = \Sigma'_{21*} = \Sigma_{21*} = (\Sigma_2 \cdot \Sigma_1)^{*}$. $\qquad\qquad\square$

Proof of Proposition 2.2. For the sake of simplicity, consider the case $\Theta_{k-}^{-} \in H^{\infty}(G_{-}, \mathcal{L}(\mathfrak{N}_{k-}, \mathfrak{N}_{k+})))$, $k = 1, 2$ (in the general case we need to use expressions like $(M_{21}(T_{21} - \lambda)^{-1} N_{21} n, m)$). It can easily be shown that

$$(T_{21} - \lambda)^{-1} = \begin{pmatrix} (T_1 - \lambda)^{-1} & -(T_1 - \lambda)^{-1} N_1 M_2 (T_2 - \lambda)^{-1} \\ 0 & (T_2 - \lambda)^{-1} \end{pmatrix}.$$

Then, by straightforward computation, we obtain

$$M_{21}(T_{21} - \lambda)^{-1} N_{21} = M_1(T_1 - \lambda)^{-1} N_1^{21} - \Upsilon_1(\lambda)\Upsilon_2(\lambda) + M_2^{21}(T_2 - \lambda)^{-1} N_2\,.$$

Here, we have

$$M_2^{21}(T_2 - \lambda)^{-1} N_2$$

$$= -\frac{1}{2\pi i} \int_C \Theta_1^{-}(\zeta)\, [M_2(T_2 - \cdot)^{-1}(T_2 - \lambda)^{-1} N_2]_{-}(\zeta)\, d\zeta$$

$$= -\frac{1}{2\pi i} \int_C \frac{\Theta_1^{-}(\zeta)}{\zeta - \lambda}\, [M_2(T_2 - \cdot)^{-1} N_2 - M_2(T_2 - \lambda)^{-1} N_2]_{-}(\zeta)\, d\zeta$$

$$= -\frac{1}{2\pi i} \int_C \frac{\Theta_1^{-}(\zeta)(\Upsilon_2(\zeta)_{-} - \Upsilon_2(\lambda))}{\zeta - \lambda}\, d\zeta$$

$$= -\frac{1}{2\pi i} \int_C \frac{\Theta_{1+}^{-}(\zeta)(\Upsilon_2(\zeta)_{-} - \Upsilon_2(\lambda))}{\zeta - \lambda}\, d\zeta$$

$$= \frac{1}{2\pi i} \int_C \frac{\Theta_{1+}^{-}(\zeta)\Theta_{2-}^{-}(\zeta)}{\zeta - \lambda}\, d\zeta + \frac{1}{2\pi i} \int_C \frac{\Theta_{1+}^{-}(\zeta)\Upsilon_2(\lambda)}{\zeta - \lambda}\, d\zeta$$

$$= \frac{1}{2\pi i} \int_C \frac{\Theta_{1+}^{-}(\zeta)\Theta_{2-}^{-}(\zeta)}{\zeta - \lambda}\, d\zeta + \begin{cases} \Theta_{1+}^{-}(\lambda)(\Theta_{2+}^{-}(\lambda) - \Theta_2^{+}(\lambda)^{-1}), & \lambda \in G_{+} \\ 0, & \lambda \in G_{-} \end{cases}.$$

Similarly, we have

$$N_1^{21*}(T_1^* - \overline{\lambda})^{-1}M_1^*$$

$$= \frac{1}{2\pi i} \int_{\overline{C}} \frac{\Theta_{*2+}^-(\zeta)\Theta_{*1-}^-(\zeta)}{\zeta - \overline{\lambda}}\, d\zeta + \begin{cases} \Theta_{*2+}^-(\overline{\lambda})(\Theta_{*1+}^-(\overline{\lambda}) - \Theta_{*1}^+(\overline{\lambda})^{-1}), & \lambda \in G_+ \\ 0, & \lambda \in G_- \end{cases}$$

$$= -\frac{1}{2\pi i} \int_C \frac{\Theta_{*2+}^-(\overline{\zeta})\Theta_{*1-}^-(\overline{\zeta})}{\overline{\zeta} - \overline{\lambda}}\, d\overline{\zeta} + \begin{cases} \Theta_{*2+}^-(\overline{\lambda})(\Theta_{*1+}^-(\overline{\lambda}) - \Theta_{*1}^+(\overline{\lambda})^{-1}), & \lambda \in G_+ \\ 0, & \lambda \in G_- \end{cases}$$

Hence,

$$M_1(T_1 - \lambda)^{-1}N_1^{21}$$

$$= \frac{1}{2\pi i} \int_C \frac{\Theta_{*1-}^-(\overline{\zeta})^*\Theta_{*2+}^-(\overline{\zeta})^*}{\zeta - \lambda}\, d\zeta$$

$$+ \begin{cases} (\Theta_{*1+}^-(\overline{\lambda})^* - (\Theta_{*1}^+(\overline{\lambda})^*)^{-1})\,\Theta_{*2+}^-(\overline{\lambda})^*, & \lambda \in G_+ \\ 0, & \lambda \in G_- \end{cases}$$

$$= \frac{1}{2\pi i} \int_C \frac{\Theta_{1-}^-(\zeta)\Theta_{2+}^-(\zeta)}{\zeta - \lambda}\, d\zeta + \begin{cases} (\Theta_{1+}^-(\lambda) - \Theta_1^+(\lambda)^{-1})\,\Theta_{2+}^-(\lambda), & \lambda \in G_+ \\ 0, & \lambda \in G_- \end{cases}.$$

Consider the case when $\lambda \in G_-$. Then

$$M_{21}(T_{21} - \lambda)^{-1}N_{21}$$

$$= \frac{1}{2\pi i} \int_C \frac{\Theta_{1-}^-(\zeta)\Theta_{2+}^-(\zeta)}{\zeta - \lambda}\, d\zeta - \Theta_{1-}^-(\lambda)\Theta_{2-}^-(\lambda) + \frac{1}{2\pi i} \int_C \frac{\Theta_{1+}^-(\zeta)\Theta_{2-}^-(\zeta)}{\zeta - \lambda}\, d\zeta + 0$$

$$= \frac{1}{2\pi i} \int_C \frac{\Theta_{1-}^-(\zeta)\Theta_{2+}^-(\zeta)}{\zeta - \lambda}\, d\zeta + \frac{1}{2\pi i} \int_C \frac{\Theta_{1-}^-(\zeta)\Theta_{2-}^-(\zeta)}{\zeta - \lambda}\, d\zeta$$

$$+ \frac{1}{2\pi i} \int_C \frac{\Theta_{1+}^-(\zeta)\Theta_{2-}^-(\zeta)}{\zeta - \lambda}\, d\zeta + \frac{1}{2\pi i} \int_C \frac{\Theta_{1+}^-(\zeta)\Theta_{2+}^-(\zeta)}{\zeta - \lambda}\, d\zeta$$

$$= \frac{1}{2\pi i} \int_C \frac{(\Theta_{1+}^-(\zeta) + \Theta_{1-}^-(\zeta))(\Theta_{2+}^-(\zeta) + \Theta_{2-}^-(\zeta))}{\zeta - \lambda}\, d\zeta$$

$$= \frac{1}{2\pi i} \int_C \frac{\Theta_1^-(\zeta)\Theta_2^-(\zeta)}{\zeta - \lambda}\, d\zeta = \frac{1}{2\pi i} \int_C \frac{\Theta_{21}^-(\zeta)}{\zeta - \lambda}\, d\zeta = -\Theta_{21}^-(\lambda).$$

Consider the case when $\lambda \in G_+$. Then

$$M_{21}(T_{21} - \lambda)^{-1}N_{21}$$

$$= \frac{1}{2\pi i} \int_C \frac{\Theta_{1-}^-(\zeta)\Theta_{2+}^-(\zeta)}{\zeta - \lambda}\, d\zeta + (\Theta_{1+}^-(\lambda) - \Theta_1^+(\lambda)^{-1})\,\Theta_{2+}^-(\lambda)$$

$$- (\Theta_{1+}^-(\lambda) - \Theta_1^+(\lambda)^{-1})(\Theta_{2+}^-(\lambda) - \Theta_2^+(\lambda)^{-1})$$

$$+ \frac{1}{2\pi i} \int_C \frac{\Theta_{1+}^-(\zeta)\Theta_{2-}^-(\zeta)}{\zeta - \lambda} \, d\zeta + \Theta_{1+}^-(\lambda)(\Theta_{2+}^-(\lambda) - \Theta_2^+(\lambda)^{-1})$$

$$= \frac{1}{2\pi i} \int_C \frac{\Theta_{1-}^-(\zeta)\Theta_{2+}^-(\zeta) + \Theta_{1+}^-(\zeta)\Theta_{2-}^-(\zeta)}{\zeta - \lambda} \, d\zeta$$

$$+ \Theta_{1+}^-(\lambda)\Theta_{2+}^-(\lambda) - \Theta_1^+(\lambda)^{-1}\Theta_2^+(\lambda)^{-1}$$

$$= \frac{1}{2\pi i} \int_C \frac{\left(\Theta_1^-(\zeta) - \Theta_{1+}^-(\zeta)\right)\Theta_{2+}^-(\zeta) + \left(\Theta_1^-(\zeta) - \Theta_{1-}^-(\zeta)\right)\Theta_{2-}^-(\zeta)}{\zeta - \lambda} \, d\zeta$$

$$+ \Theta_{1+}^-(\lambda)\Theta_{2+}^-(\lambda) - \Theta_1^+(\lambda)^{-1}\Theta_2^+(\lambda)^{-1}$$

$$= \frac{1}{2\pi i} \int_C \frac{\Theta_1^-(\zeta)\Theta_2^-(\zeta)}{\zeta - \lambda} \, d\zeta$$

$$- \Theta_{1+}^-(\lambda)\Theta_{2+}^-(\lambda) + \Theta_{1+}^-(\lambda)\Theta_{2+}^-(\lambda) - \Theta_1^+(\lambda)^{-1}\Theta_2^+(\lambda)^{-1}$$

$$= \frac{1}{2\pi i} \int_C \frac{\Theta_{21}^-(\zeta)}{\zeta - \lambda} \, d\zeta - \Theta_{21}^+(\lambda)^{-1} = \Theta_{21}^-(\lambda) - \Theta_{21}^+(\lambda)^{-1}.$$

Thus, for $\lambda \in (G_+ \cup G_-) \cap \rho(T_{21})$, we obtain

$$M_{21}(T_{21} - \lambda)^{-1} N_{21} = \Upsilon_{21}(\lambda) = \begin{cases} \Theta_{21+}^-(\lambda) - \Theta_{21}^+(\lambda)^{-1}, & \lambda \in G_+ \cap \rho(T_{21}) \\ -\Theta_{21-}^-(\lambda), & \lambda \in G_- \end{cases}.$$

That is $\Upsilon_{21} = \mathcal{F}_{tc}(\Theta_2\Theta_1)$. $\square$

Computation of the operators Z^{-1} and $Z^{-1}Y$. Using Lemma 1.3 and the corollary of Lemma (iii), we have

$$(P_{(21)} + (P_{(32)} + P_{(43)})P_{(42)})Z$$

$$= (P_{(21)} + (P_{(32)} + P_{(43)})P_{(42)})P_{(41)}(P_{(21)}(I - P_{(43)}) + P_{(42)}(P_{(32)} + P_{(43)}))$$

$$= (P_{(21)}P_{(41)} + (P_{(32)} + P_{(43)})P_{(42)})(P_{(21)}(I - P_{(43)}) + P_{(42)}(P_{(32)} + P_{(43)}))$$

$$= P_{(21)}\underline{P_{(41)}P_{(21)}}(I - P_{(43)}) + P_{(21)}P_{(41)}P_{(42)}(P_{(32)} + P_{(43)})$$

$$+ (P_{(32)} + P_{(43)})\underline{P_{(42)}P_{(21)}}(I - P_{(43)}) + \underline{(P_{(32)} + P_{(43)})P_{(42)}(P_{(32)} + P_{(43)})}$$

$$= P_{(21)}(I - P_{(43)}) + P_{(21)}P_{(41)}P_{(42)}(P_{(32)} + P_{(43)}) + (P_{(32)} + P_{(43)})$$

$$= I + P_{(21)}P_{(41)}P_{(42)}(P_{(32)} + P_{(43)}).$$

Note that we will arrange our computations by underlining subexpressions that are in the focus of current step. Further, since

$$P_{(21)}P_{(41)}P_{(42)}$$

$$= (P_{(21)} + P_{(42)})P_{(41)}P_{(42)} - \underline{P_{(42)}P_{(41)}P_{(42)}}$$

$$= (P_{(21)} + P_{(42)})P_{(41)}P_{(42)} - P_{(42)}$$

$$= \underline{(P_{(21)} + P_{(42)})P_{(41)}(P_{(21)} + P_{(42)})} - (P_{(21)} + P_{(42)})P_{(41)}P_{(21)} - P_{(42)}$$

$$= (P_{(21)} + P_{(42)}) - (P_{(21)} + P_{(42)})P_{(41)}P_{(21)} - P_{(42)}$$
$$= P_{(21)} + P_{(42)} - P_{(21)} - P_{(42)} = 0\,,$$

we obtain $(P_{(21)} + (P_{(32)} + P_{(43)})P_{(42)})Z = I|_{\mathcal{K}_{(21)}\dotplus\mathcal{K}_{(32)}\dotplus\mathcal{K}_{(43)}}$. On the other hand, we have

$$Z(P_{(21)} + (P_{(32)} + P_{(43)})P_{(42)})$$
$$= P_{(41)}(P_{(21)}(I - P_{(43)}) + P_{(42)}(P_{(32)} + P_{(43)}))(P_{(21)} + (P_{(32)} + P_{(43)})P_{(42)})$$
$$= P_{(41)}(P_{(21)} + P_{(42)}(P_{(32)} + P_{(43)})P_{(42)}) = P_{(41)}(P_{(21)} + P_{(42)})|_{\mathcal{K}_{(41)}}$$
$$= P_{(41)}(P_{(21)} + P_{(42)})P_{(41)}|_{\mathcal{K}_{(41)}} = P_{(41)}|_{\mathcal{K}_{(41)}} = I|_{\mathcal{K}_{(41)}}\,.$$

We need to compute the operator $Z^{-1}Y$. We have

$$Z^{-1}Y = (P_{(21)} + (P_{(32)} + P_{(43)})P_{(42)})P_{(41)}(P_{(31)}(P_{(21)}(I - P_{(43)}) + P_{(32)}) + P_{(43)})$$
$$= (P_{(21)} + (P_{(32)} + P_{(43)})P_{(42)})(P_{(31)}(P_{(21)}(I - P_{(43)}) + P_{(32)}) + P_{(41)}P_{(43)})$$
$$= P_{(21)}P_{(31)}(P_{(21)}(I - P_{(43)}) + P_{(32)}) + P_{(21)}P_{(41)}P_{(43)}$$
$$\quad + (P_{(32)} + P_{(43)})P_{(42)}P_{(31)}(P_{(21)}(I - P_{(43)})$$
$$\quad + P_{(32)}) + (P_{(32)} + P_{(43)})P_{(42)}P_{(41)}P_{(43)}$$
$$= P_{(21)}(I - P_{(43)}) + P_{(21)}P_{(31)}P_{(32)} + P_{(21)}P_{(41)}P_{(43)}$$
$$\quad + P_{(32)}P_{(42)}P_{(31)}(P_{(21)}(I - P_{(43)}) + P_{(32)})$$
$$\quad + P_{(43)}P_{(31)}(P_{(21)}(I - P_{(43)}) + P_{(32)}) + (P_{(32)} + P_{(43)})P_{(42)}P_{(43)}$$
$$= P_{(21)}(I - P_{(43)}) + P_{(21)}P_{(31)}P_{(32)} + P_{(21)}P_{(41)}P_{(43)}$$
$$\quad + P_{(32)}P_{(42)}P_{(21)}(I - P_{(43)}) + P_{(32)}P_{(42)}P_{(31)}P_{(32)}$$
$$\quad + P_{(32)}P_{(42)}P_{(43)} + P_{(43)}\,.$$

Since

$$P_{(21)}P_{(31)}P_{(32)} = (P_{(21)} + P_{(32)})P_{(31)}(P_{(21)} + P_{(32)}) - P_{(21)}P_{(31)}P_{(21)}$$
$$\quad - P_{(32)}P_{(31)}P_{(21)} - P_{(32)}P_{(31)}P_{(32)}$$
$$= P_{(21)} + P_{(32)} - P_{(21)} - P_{(32)}P_{(21)} - P_{(32)} = 0$$

and

$$P_{(32)}P_{(42)}P_{(43)} = (P_{(32)} + P_{(43)})P_{(42)}(P_{(32)} + P_{(43)}) - P_{(32)}P_{(42)}P_{(32)}$$
$$\quad - P_{(43)}P_{(42)}P_{(32)} - P_{(43)}P_{(42)}P_{(43)}$$
$$= P_{(32)} + P_{(43)} - P_{(32)} - P_{(43)}P_{(32)} - P_{(43)} = 0\,,$$

we obtain

$$Z^{-1}Y = P_{(21)}(I - P_{(43)}) + P_{(21)}P_{(41)}P_{(43)} + P_{(32)}P_{(42)}P_{(31)}P_{(32)} + P_{(43)}\,.$$

Taking into account that $\operatorname{Ran} P_{(43)} \subset \mathcal{H}_{41+}$ and using Lemma (iii), we have

$$P_{(21)}P_{(41)}P_{(43)} = \underline{(P_{(21)} + P_{(42)})P_{(41)}}P_{(43)} - \underline{P_{(42)}P_{(41)}}P_{(43)}$$
$$= (P_{(21)} + P_{(42)})P_{(43)} - P_{(42)}P_{(43)} = P_{(21)}P_{(43)}\,.$$

Likewise, taking into account that $\operatorname{Ran} P_{(32)} \subset \mathcal{H}_{41+}$ and using Lemma (iii), we have

$$P_{(32)}P_{(42)}P_{(31)}P_{(32)} = P_{(32)}\underline{(P_{(21)} + P_{(42)})(P_{(31)} + P_{(43)})}P_{(32)}$$
$$= P_{(32)}(P_{(21)} + P_{(42)})P_{(32)}$$
$$= \underline{P_{(32)}P_{(21)}P_{(32)}} + P_{(32)}\underline{P_{(42)}P_{(32)}} = P_{(32)}P_{(32)} = P_{(32)}\,.$$

Therefore,

$$Z^{-1}Y = P_{(21)}(I - P_{(43)}) + P_{(21)}P_{(43)} + P_{(32)} + P_{(43)}$$
$$= (P_{(21)}(I - P_{(43)}) + P_{(32)} + P_{(43)}) + P_{(21)}P_{(43)}$$
$$= (I + P_{(21)}P_{(43)})|\mathcal{K}_{(21)}\dotplus\mathcal{K}_{(32)}\dotplus\mathcal{K}_{(43)}\,.$$

Note also that $Y^{-1}Z = (I - P_{(21)}P_{(43)})|\mathcal{K}_{(21)}\dotplus\mathcal{K}_{(32)}\dotplus\mathcal{K}_{(43)}$. Indeed, this follows from the identity $P_{(21)}P_{(43)}P_{(21)}P_{(43)} = 0$.

Proof of Lemma 3.1. 1) $\Longleftrightarrow$ 2) is obvious. To prove 2) $\Longleftrightarrow$ 3), we need the following

Lemma (iv). *One has* $\operatorname{clos} \operatorname{Ran}(I - V_2 V_2^*)V_1 = (\operatorname{Ran} V_1 \vee \operatorname{Ran} V_2) \ominus \operatorname{Ran} V_2$.

Proof. Let $f \in \operatorname{Ran}(I - V_2 V_2^*)V_1$. Then $f = (I - V_2 V_2^*)V_1 u \in \operatorname{Ran} V_1 \vee \operatorname{Ran} V_2$. On the other hand, $V_2^* f = V_2^*(I - V_2 V_2^*)V_1 u = (I - V_2^* V_2)V_2^* V_1 u = 0 \cdot V_2^* V_1 u = 0$, that is, $f \perp \operatorname{Ran} V_2$ and therefore $\operatorname{Ran}(I - V_2 V_2^*)V_1 \subset (\operatorname{Ran} V_1 \vee \operatorname{Ran} V_2) \ominus \operatorname{Ran} V_2$.

Conversely, let $f \in (\operatorname{Ran} V_1 \vee \operatorname{Ran} V_2) \ominus \operatorname{Ran} V_2$. Then we have $V_2^* f = 0$ and $f = \lim_{n \to \infty}(V_1 u_{1n} + V_2 u_{2n})$. Hence,

$$f = (I - V_2 V_2^*)f = \lim_{n \to \infty}\left((I - V_2 V_2^*)V_1 u_{1n} + (I - V_2 V_2^*)V_2 u_{2n}\right)$$
$$= (I - V_2 V_2^*)f = \lim_{n \to \infty}(I - V_2 V_2^*)V_1 u_{1n} \in \operatorname{clos} \operatorname{Ran}(I - V_2 V_2^*)V_1\,. \qquad \square$$

To complete the proof of Lemma 3.1 we need only to make use of the following observation

$$\operatorname{Ran}(I - V_2 V_2^*)V_1 \perp \operatorname{Ran}(I - V_2 V_2^*)V_3 \Longleftrightarrow V_3^*(I - V_2 V_2^*)V_1 = 0\,. \quad \square$$

Proof of Lemma 3.2. We need to make some preparations.

Lemma (v). *Assume that* $V_3^* V_1 - V_3^* V_2 V_2^* V_1 = 0$. *Then*

$$(\mathcal{E}_1 \vee \mathcal{E}_3) \ominus \mathcal{E}_3 \subset ((\mathcal{E}_1 \vee \mathcal{E}_2) \ominus \mathcal{E}_2) \oplus ((\mathcal{E}_3 \vee \mathcal{E}_2) \ominus \mathcal{E}_3)\,.$$

Proof. Using Lemma (iv) and the obvious identity

$$(I - V_3 V_3^*)V_1 = (I - V_2 V_2^*)V_1 + (I - V_3 V_3^*)V_2 V_2^* V_1\,,$$

we get

$$(\mathcal{E}_1 \vee \mathcal{E}_3) \ominus \mathcal{E}_3 \subset ((\mathcal{E}_1 \vee \mathcal{E}_2) \ominus \mathcal{E}_2) \vee ((\mathcal{E}_3 \vee \mathcal{E}_2) \ominus \mathcal{E}_3)\,.$$

By Lemma 3.1, $((\mathcal{E}_1 \vee \mathcal{E}_2) \ominus \mathcal{E}_2) \perp ((\mathcal{E}_3 \vee \mathcal{E}_2) \ominus \mathcal{E}_2)$. Then

$$((\mathcal{E}_1 \vee \mathcal{E}_2) \ominus \mathcal{E}_2) \perp ((\mathcal{E}_3 \vee \mathcal{E}_2) \ominus \mathcal{E}_2) \oplus \mathcal{E}_2 = \mathcal{E}_3 \vee \mathcal{E}_2$$

and therefore $((\mathcal{E}_1 \vee \mathcal{E}_2) \ominus \mathcal{E}_2) \perp ((\mathcal{E}_3 \vee \mathcal{E}_2) \ominus \mathcal{E}_3)$. $\qquad\qquad\square$

Remark. If we define the isometries $\tau_{jij} \colon \operatorname{clos} \operatorname{Ran} (I - V_j^* V_i V_i^* V_j)^{1/2} \to \mathcal{H}$ by the formula $\tau_{jij} (I - V_j^* V_i V_i^* V_j)^{1/2} = (I - V_i V_i^*) V_j$, we can rewrite the identity

$$(I - V_3 V_3^*) V_1 = (I - V_2 V_2^*) V_1 + (I - V_3 V_3^*) V_2 V_2^* V_1$$

in the form

$$\tau_{131} (I - A_{31}^* A_{31})^{1/2} = \tau_{121} (I - A_{21}^* A_{21})^{1/2} + \tau_{232} (I - A_{32}^* A_{32})^{1/2} A_{21}.$$

Note that $\operatorname{clos} \operatorname{Ran} \tau_{jij} = \operatorname{clos} \operatorname{Ran}(I - V_i V_i^*) V_j$. Then, by Lemma 3.1, the condition $V_3^* (I - V_2 V_2^*) V_1 = 0$ means $\tau_{232}^* \tau_{121} = 0$ and $\tau_{121}^* \tau_{232} = 0$. Therefore we have

$$Z\,(I - A_{31}^* A_{31})^{1/2} = \begin{pmatrix} (I - A_{21}^* A_{21})^{1/2} \\ (I - A_{32}^* A_{32})^{1/2} A_{21} \end{pmatrix} \tag{Z}$$

and the operator $Z = (\tau_{121}^* + \tau_{232}^*) \tau_{131}$ is an isometry. We need the following lemma established in [2].

Lemma (vi). *The following conditions are equivalent:*
1) $\operatorname{Ran} (I - A_{32}^* A_{32})^{1/2} \cap \operatorname{Ran} (I - A_{21} A_{21}^*)^{1/2} = \{0\}$;
2) $A_{21}^* (I - A_{32}^* A_{32})^{1/2} m + (I - A_{21}^* A_{21})^{1/2} n = 0,\ m \in \operatorname{clos} \operatorname{Ran} (I - V_2^* V_3 V_3^* V_2)^{1/2}$ *and* $n \in \operatorname{clos} \operatorname{Ran} (I - V_1^* V_2 V_2^* V_1)^{1/2} \implies m = 0,\, n = 0$;
3) *The operator* Z *defined by the condition* (Z) *is a unitary operator.*

Proof (of Lemma 3.2). 1) $\iff$ 2) By Lemma (vi), condition 1) is equivalent to the condition that the operator Z is unitary. Since under our assumptions Z is always isometrical, we can check only that $Z^* = \tau_{131}^* (\tau_{121} + \tau_{232})$ is an isometrical operator. The latter is equivalent to the condition $\operatorname{Ran} \tau_{121} \oplus \operatorname{Ran} \tau_{232} \subset \operatorname{Ran} \tau_{131}$. The inverse inclusion is Lemma (v).

2) $\implies$ 3) Since we have $(\mathcal{E}_1 \vee \mathcal{E}_3) \ominus \mathcal{E}_3 = ((\mathcal{E}_1 \vee \mathcal{E}_2) \ominus \mathcal{E}_2) \oplus ((\mathcal{E}_3 \vee \mathcal{E}_2) \ominus \mathcal{E}_3)$, we obtain

$$\begin{aligned}
\mathcal{E}_1 \vee \mathcal{E}_3 \ &= \mathcal{E}_3 \oplus ((\mathcal{E}_1 \vee \mathcal{E}_3) \ominus \mathcal{E}_3) - \mathcal{E}_3 \oplus ((\mathcal{E}_3 \vee \mathcal{E}_2) \ominus \mathcal{E}_3) \oplus ((\mathcal{E}_1 \vee \mathcal{E}_2) \ominus \mathcal{E}_2) \\
&= (\mathcal{E}_3 \vee \mathcal{E}_2) \oplus ((\mathcal{E}_1 \vee \mathcal{E}_2) \ominus \mathcal{E}_2).
\end{aligned}$$

Hence, $\mathcal{E}_2 \subset \mathcal{E}_1 \vee \mathcal{E}_3$.

3) $\implies$ 2) We have $\mathcal{E}_2 \subset \mathcal{E}_1 \vee \mathcal{E}_3$. Then

$$\mathcal{E}_3 \oplus ((\mathcal{E}_2 \vee \mathcal{E}_3) \ominus \mathcal{E}_3) = \mathcal{E}_2 \vee \mathcal{E}_3 \subset \mathcal{E}_1 \vee \mathcal{E}_3 = \mathcal{E}_3 \oplus ((\mathcal{E}_1 \vee \mathcal{E}_3) \ominus \mathcal{E}_3)$$

and therefore $(\mathcal{E}_2 \vee \mathcal{E}_3) \ominus \mathcal{E}_3 \subset (\mathcal{E}_1 \vee \mathcal{E}_3) \ominus \mathcal{E}_3$.

On the other hand, we have $((\mathcal{E}_1 \vee \mathcal{E}_2) \ominus \mathcal{E}_2) \perp \mathcal{E}_2)$ and $((\mathcal{E}_1 \vee \mathcal{E}_2) \ominus \mathcal{E}_2) \perp ((\mathcal{E}_3 \vee \mathcal{E}_2) \ominus \mathcal{E}_2)$. Hence, $((\mathcal{E}_1 \vee \mathcal{E}_2) \ominus \mathcal{E}_2) \perp \mathcal{E}_3 \vee \mathcal{E}_2$. Then, we get

$$\mathcal{E}_3 \oplus ((\mathcal{E}_1 \vee \mathcal{E}_2) \ominus \mathcal{E}_2) \subset \mathcal{E}_1 \vee \mathcal{E}_2 \vee \mathcal{E}_3 \subset \mathcal{E}_1 \vee \mathcal{E}_3 = \mathcal{E}_3 \oplus ((\mathcal{E}_1 \vee \mathcal{E}_3) \ominus \mathcal{E}_3)$$

and therefore $(\mathcal{E}_1 \vee \mathcal{E}_2) \ominus \mathcal{E}_2 \subset (\mathcal{E}_1 \vee \mathcal{E}_3) \ominus \mathcal{E}_3$. Thus, we obtain

$$((\mathcal{E}_1 \vee \mathcal{E}_2) \ominus \mathcal{E}_2) \oplus ((\mathcal{E}_2 \vee \mathcal{E}_3) \ominus \mathcal{E}_3) \subset (\mathcal{E}_1 \vee \mathcal{E}_3) \ominus \mathcal{E}_3.$$

The inverse inclusion is Lemma (v). $\qquad\square$

Proof of Proposition 3.4. 1) By [17], we have

$$-\widehat{M}(\widehat{T}-z)^{-1}f = \begin{cases} \Theta^+(z)^{-1}(\pi_-^\dagger f)(z) & , z \in \rho(\widehat{T}) \cap G_+ \\ (\pi_+^\dagger f)(z) & , z \in G_- \end{cases}$$

and therefore

$$\bigcap_{z \in \rho(T)} \operatorname{Ker} \widehat{M}(\widehat{T}-z)^{-1} = \{f \in \mathcal{K} : \pi_+^\dagger f = 0, \ \pi_-^\dagger f = 0\}$$
$$= \{f \in \mathcal{K} : f \perp \operatorname{Ran}\pi_+, \ f \perp \operatorname{Ran}\pi_-\} = \{0\}.$$

2) Let $\Sigma \overset{X}{\sim} \Sigma'$. Then $M(T-z)^{-1} = M'(T'-z)^{-1}X$ and the property follows straightforwardly from this identity.

3) is a direct consequence of properties 1) and 2).

4) It is sufficient to check that $\Sigma \overset{X}{\sim} \Sigma \Rightarrow X = I$. We have

$$M(T-z)^{-1}Xf = MX(T-z)^{-1}f = M(T-z)^{-1}f$$
$$\Longrightarrow Xf - f \in \bigcap_{z \in \rho(T)} \operatorname{Ker} M(T-z)^{-1} = \{0\} \Longrightarrow Xf = f.$$

5) It can easily be checked that $\rho(T) \cap \rho(T_1) \subset \rho(T_2)$. Then we have

$$(T_{21}-z)^{-1} = \begin{pmatrix} (T_1-z)^{-1} & -(T_1-z)^{-1}N_1 M_2(T_2-z)^{-1} \\ 0 & (T_2-z)^{-1} \end{pmatrix}.$$

and therefore $\forall f_1 \in H_1 \ M_{21}(T_{21}-z)^{-1}f_1 = M_1(T_1-z)^{-1}f_1$. Hence,

$$\bigcap_{z \in \rho(T)} \operatorname{Ker} M_1(T_1-z)^{-1} \subset \bigcap_{z \in \rho(T)} \operatorname{Ker} M_{21}(T_{21}-z)^{-1} = \{0\},$$

that is, the system Σ_1 is simple.

Further, by property 3), the system Σ^* is simple. Then, using the same arguments as above, it follows that the system Σ_2^* is simple. Hence the system Σ_2 is simple too. $\qquad\square$

Proposition 5.1. *Let $A_{21}, A_{42}, A_{31}, A_{43}$ be contractions. Suppose that factorizations $A_{42} \cdot A_{21}$, $A_{43} \cdot A_{31}$ are regular, $A_{42} \cdot A_{21} \prec A_{43} \cdot A_{31}$ and $A_{43} \cdot A_{31} \prec A_{42} \cdot A_{21}$. Then there exists a unitary operator U such that $A_{31} = UA_{21}$ and $A_{43} = A_{42}U^{-1}$.*

Proof. We shall make use of the following two lemmas.

Lemma (vii). *Suppose that $\|A\| \leq 1$ and $A|H_1 = I|H_1$. Then $A^*|H_1 = I|H_1$.*

Proof. We have $A = \begin{pmatrix} I & a_{12} \\ 0 & a_{22} \end{pmatrix}$. Then

$$0 \leq ((I - A^*A) \begin{pmatrix} f_1 \\ 0 \end{pmatrix}, \begin{pmatrix} f_1 \\ 0 \end{pmatrix}) = -(a_{12}^* f_1, a_{12}^* f_1) \leq 0.$$

Therefore, $a_{12} = 0$ and $A = \begin{pmatrix} I & 0 \\ 0 & a_{22} \end{pmatrix}$. $\qquad\square$

Lemma (viii). *Let* A_{21}, A_{32} *be contractions. Suppose that factorization* $A_{32} \cdot A_{21}$ *is regular. Then* $(\operatorname{Ran} A_{32}^* \vee \operatorname{Ran} A_{21})^\perp = \{0\}$.

Proof. Let $f \perp (\operatorname{Ran} A_{32}^* \vee \operatorname{Ran} A_{21})$. Then $f \in \operatorname{Ker} A_{32}$ and $f \in \operatorname{Ker} A_{21}^*$. Hence, $(I - A_{21}A_{21}^*)f = f$ and therefore $(I - A_{21}A_{21}^*)^{1/2}f = f$. Similarly, we have $(I - A_{32}^* A_{32})^{1/2}f = f$. Then $f \in \operatorname{Ran}(I - A_{32}^* A_{32})^{1/2} \cap \operatorname{Ran}(I - A_{21}A_{21}^*)^{1/2} = \{0\}$. $\qquad\square$

From the definition of the order relation $\prec$ we get that there exists contractions A_{32}, A_{23} such that $A_{42} = A_{43}A_{32}$, $A_{31} = A_{32}A_{21}$, $A_{43} = A_{42}A_{23}$, and $A_{32} = A_{23}A_{31}$. Let $A = A_{23}A_{32}$. Then we have $A_{21} = AA_{21}$ and $A_{42} = A_{42}A$ and therefore $A| \operatorname{Ran} A_{21} = I| \operatorname{Ran} A_{21}$ and $A^*| \operatorname{Ran} A_{42} = I| \operatorname{Ran} A_{42}$. By Lemma (vii), $A| \operatorname{Ran} A_{42} = I| \operatorname{Ran} A_{42}$. Finally, by Lemma (viii), we get $A = A|(\operatorname{Ran} A_{21} \vee \operatorname{Ran} A_{42}) = I|(\operatorname{Ran} A_{21} \vee \operatorname{Ran} A_{42}) = I$, that is, $A_{23}A_{32} = I$. Likewise, we get $A_{32}A_{23} = I$. Since A_{23} and A_{32} are contraction, they are unitary operators. It remains to put $U = A_{32}$. $\qquad\square$

Acknowledgement

Research for this article was supported by Central European University ("CEU") Special and Extension Programs. The opinions expressed herein are the author's own and do not necessarily express the views of CEU

References

[1] B. Szökefalvi-Nagy, C. Foiaş, *Harmonic analysis of operators on Hilbert space.* North-Holland, Amsterdam-London, 1970.

[2] M.S. Brodskiĭ, *Unitary operator colligations and their characteristic functions*, Uspekhi mat. nauk **33** (1978), no. 4, 141–168.

[3] P.L. Duren, *Theory of H^p spaces*, Pure Appl. Math., vol. 38, Academic Press, New York–London, 1970.

[4] N.K. Nikolski, *Operators, functions, and systems: an easy reading.* Vol. 1 Hardy, Hankel, and Toeplitz. Vol. 2 Model operators and systems, Math. Surveys and Monographs, 92, 93, AMS, Providence, RI, 2002.

[5] M.S. Livsic, *On a class of linear operators on Hilbert space*, Mat. Sb. **19** (1946), 239–260.

[6] A. Beurling, *On two problems concerning linear transformations in Hilbert space*, Acta Math. **81** (1949), 239–255.

[7] H. Bart, I. Gohberg, M.A. Kaashoek, *Minimal factorization of matrix and operator functions.* Operator Theory: Adv. and Appl., 1979.

[8] A.S. Tikhonov, *Functional model and duality of spectral components for operators with continuous spectrum on a curve.* Algebra i Analiz, **14** (2002), no. 4, 158–195.

[9] A.S. Tikhonov, *Free functional model related to simply-connected domains.* Operator Theory: Adv. and Appl., Vol. **154** (2004), 405–415.

[10] M.B. Abrahamse, R.G. Douglas, *A class of subnormal operators related to multiply connected domains,* Adv. in Math., **19** (1976), 106–148.

[11] J.A. Ball, *Operators of class C_{00} over multiply connected domains,* Michigan Math. J., **25** (1978), 183–195.

[12] B.S. Pavlov, S.I. Fedorov, *Group of shifts and harmonic analysis on Riemann surface of genus one,* Algebra i Analiz, **1** (1989), no. 2, 132–168.

[13] S.I. Fedorov, *On harmonic analysis in multiply connected domain and character-automorphic Hardy spaces.* Algebra i Analiz, **9** (1997), no. 2, 192–239.

[14] D.V. Yakubovich, *Linearly similar model of Sz.-Nagy-Foias type in a domain.* Algebra i Analiz, **15** (2003), no. 2, 180–227.

[15] S.M. Verduyn Lunel, D.V. Yakubovich, *A functional model approach to linear neutral functional differential equations,* Integr. Equ. Oper. Theory, **27** (1997), 347–378.

[16] N.K. Nikolski, V.I. Vasyunin, *Elements of spectral theory in terms of the free functional model. Part I: Basic constructions,* Holomorphic spaces (eds. Sh. Axler, J. McCarthy, D. Sarason), MSRI Publications **33** (1998), 211–302.

[17] A.S. Tikhonov, *Transfer functions for "curved" conservative systems.* Operator Theory: Adv. and Appl., Vol. 153 (2004), 255–264.

[18] S.S. Boiko, V.K. Dubovoi, *Unitary couplings and regular factorizations of operator functions in L^∞,* Dopovidi Nacional. Akad. Nauk Ukr. , (1997), no. 1, 41–44.

[19] A.S. Tikhonov, *Inner-outer factorization of J-contractive-valued functions.* Operator Theory: Adv. and Appl., Vol. **118** (2000), 405–415.

[20] W. Mlak, *Intertwining operators.* Studia Math., Vol. **43** (1972), 219–233.

[21] L. Kerchy, *On the factorization of operator-valued functions* Acta. Sci. Math. (Szeged), **69** (2003), 337–348.

Alexey Tikhonov
Department of Mathematics
Taurida National University
4, Yaltinskaya str.
Simferopol 95007 Crimea
Ukraine
e-mail: `tikhonov@club.cris.net`

List of Participants in OTAMP2004

1. **Vadim Adamyan**
   Odessa National University, Ukraine
   I.I. Mechnikov Odessa National University, 65026 Odessa, Ukraine
   vadamyan@paco.net
   Perturbation theory for a class of close selfadjoint extensions

2. **Walter van Assche**
   Katholieke Universiteit Leuven, Belgium
   Department of Mathematics Katholieke Universiteit Leuven
   Celestijnenlaan 200 B B-3001 Leuven Belgium
   walter@wis.kuleuven.ac.be
   Orthogonal polynomials with discrete spectrum and converging recurrence
   coefficients

3. **Yurij Berezansky**
   Institute of Mathematics of NASU
   3 Tereshchenkivs'ka Str., 01601 Kyiv, Ukraine
   berezan@mathber.carrier.kiev.ua
   Spectral theory of commutative Jacobi fields and its some applications

4. **Andrei Borisovich**
   Gdansk University, Poland
   Institute of Mathematics, University of Gdansk
   ul. Wita Stwosza 57, 80-952 Gdansk, POLAND
   andbor@math.univ.gda.pl
   Fredholm operators and bifurcations in Plateau problem

5. **Piotr Budzyński**
   Institute of Mathematics, Jagiellonian University, Poland
   Jagiellonian University, Department of Mathematics and Informatics,
   Institute of Mathematics, Reymonta 4, 30-059, Krakow, Poland
   piotr_budzynski@o2.pl

6. **Laura Cattaneo**
   Institute for Applied Mathematics, University of Bonn, Germany
   Institute for Applied Mathematics, Probability and Statistics Division,
   Wegelerstrasse 6, D-53115 Bonn, Germany
   cattaneo@wiener.iam.uni-bonn.de
   Mourre's inequality and embedded bound states

7. **Dariusz Cichon**
   Uniwersytet Jagiellonski, Instytut Matematyki, Poland
   Instytut matematyki UJ. Reymonta 4, 30059 Krakow, Poland
   cichon@im.uj.edu.pl
   Selfadjointness of non-Jacobi infinite matrices

8. **Petru Cojuhari**
   AGH University of Science and Technology, Department of Applied
   Mathematics, Poland
   Department of Applied Mathematics, AGH University of Science and
   Technology, Al. Mickiewicza 30, 30-059 Cracow, Poland
   cojuhari@uci.agh.edu.pl
   Spectral Analysis of Dirac Operators

9. **Monique Combescure**
   CNRS France
   IPNL, bat Paul Dirac 4 rue Enrico Fermi
   F-69622 Villeurbanne, France
   mcombe@ipnl.in2p3.fr
   Around the semiclassical behavior of "Quantum Fidelity"

10. **Horia Cornean**
    Aalborg Universitet, Danmark
    Institut for Matematiske Fag, Aalborg Universitet
    Fredrik Bajers Vej 7G, 9220 Aalborg, Danmark
    cornean@math.auc.dk
    One dimensional models of excitons in carbon nanotubes

11. **Serguei Denissov**
    California Institute of Technology
    Mathematics 253-37, Caltech, Pasadena, CA, 91125, USA
    denissov@caltech.edu
    On new developments in the spectral analysis of the Schrödinger and Dirac
    operators

12. **Gisele Ducati**
    Federal University of Parana, Brazil
    Departmento of Mathematics PO Box 19081 Curitiba (PR) - Brazil
    81531-970
    ducati@mat.ufpr.br
    Quaternionic potentials in non relativistic quantum mechanics

13. **Pavel Exner**
    Nuclear Physics Institute, Czech Academy of Sciences
    Department of Theoretical Physics, Nuclear Physics Institute, Academy of
    Sciences, 25068 Rez/Prague, Czechia
    exner@ujf.cas.cz
    Vertex coupling approximations in quantum graphs

14. **Rupert Frank**
    Royal Institute of Technology Stockholm, Sweden
    Royal Institute of Technology, Department of Mathematics,
    Lindstedtsvaegen 25, 10044 Stockholm, Sweden
    rupert@math.kth.se
    On the Laplacian on the halfplane with a periodic boundary condition

15. **Vladimir Geyler**
    Mordovian State University, Russia
    Department of Mathematics Mordovian State University 430000 Saransk,
    Russia
    geyler@mrsu.ru
    Selberg zeta-function and geometric scattering on compact manifolds

16. **Leonid Golinskii**
    Institute for Low Temperature Physics and Engineering, Ukraine
    Mathematics Division, Institute for Low Temperature Physics and
    Engineering, 47 Lenin ave., 61103, Kharkov, Ukraine
    golinskii@ilt.kharkov.ua
    Discrete spectrum of complex Jacobi matrices and Pavlov's theorems

17. **Ira Herbst**
    University of Virginia, Charlottesville, USA
    Department of Mathematics P. O. Box 400137 University of Virginia
    Charlottesville, VA 22904-4137 USA
    iwh@virginia.edu
    Absence of quantum states corresponding to unstable classical channels

18. **Rostyslav Hryniv**
Institut fur Angewandte Mathematik, Bonn (Germany)
Abt. fur Stochastik Institut fur Angewandte Mathematik Universität Bonn
Wegelerstr. 6 D-53115 Bonn Germany
rhryniv@wiener.iam.uni-bonn.de, rhryniv@iapmm.lviv.ua
Inverse spectral problems for Sturm-Liouville operators in impedance form

19. **Wataru Ichinose**
Shinshu University, Japan
Department of Mathmatical Sciences, Shinshu University, Asahi 3-1-1,
390-8621, Matsumoto, Japan
ichinose@math.shinshu-u.ac.jp
A mathematical theory of the Phase space Feynman path integral of the
functional

20. **Jan Janas**
Institute of Mathematics PAN, Poland
Branch in Cracow, Sw.Tomasz 30, 31-027, Cracow, Pland
najanas@cyf-kr.edu.pl

21. **Aref Jeribi**
Faculty of Sciences in Sfax Tunisia
Departement de Mathématiques, Faculté des sciences de Sfax
Route de Soukra Km 3.5 B.P. 802, 3018 Sfax Tunisie
Aref.Jeribi@fss.rnu.tn

22. **Witold Karwowski**
Institute of Physics, Zielona Gora University, Poland
Witold Karwowski ul. Gajowicka 100 m.3 53 407 Wroclaw, Poland
witoldkarwowski@go2.pl

23. **Amir Khosravi**
University for Teacher Education of IRAN
Faculty of Mathematical Sciences and Computer Engineering, University
For Teacher Education, 599 Taleghani Ave., Tehran, 15618, IRAN
khosravi_amir@yahoo.com
C_ρ^* -algebras and functional calculus homomorphism

24. **Alexander V. Kiselev**
Dublin Institute of Technology, Ireland
School of Mathematical Sciences Dublin Institute of Technology DIT Kevin
Street Dublin 8 IRELAND
alexander.kiselev@dit.ie
Weak annihilators for non-self-adjoint operators with almost Hermitian
spectrum

25. **Kamila Kliś**
Akademia Rolnicza, Krakow, Poland
Zakład Zastosowań Matematyki, Akademia Rolnicza, Al. Mickiewicza
24/28, 30-059 Krakow
rmklis@cyf-kr.edu.pl

26. **Frederic Klopp**
Université Paris-Nord, France
LAGA UMR 7539, Institut Galilée
Av. JB Clément, F-93430 Villetaneuse, France
klopp@math.univ-paris13.fr
Adiabatic quasi-periodic Schrödinger operators. Interactions between
spectral bands

27. **Sylwia Kondej**
Institute of Physics, University of Zielona Gora, Poland
University of Zielona Gora, Institute of Physics
ul. Szafrana 4a, 65246 Zielona Gora, Poland
kondej@ift.uni.wroc.pl
Schrödinger operators with singular perturbations: a resonance model

28. **Evgeni Korotyaev**
Humboldt -Univ. Germany
Humboldt-Universität zu Berlin Math. Institute
Rudower Chaussee 25, Johann von Neumann-Haus (I.313)
10099 Berlin, Germany
ek@mathematik.hu-berlin.de
Inverse problem for the discrete 1D Schrödinger operator with periodic
potential

29. **Volodymyr Koshmanenko**
Institute of Mathematics, Ukraine
Department of Mathematical Physics, Institute of Mathematics,
Tereshchenkivska str. 3, 01601 Kyiv, Ukraine
kosh@imath.kiev.ua
On singular perturbations given by Jacobi matrices

30. **Alexander Kozhevnikov**
University of Haifa, Israel
Department of Mathematics, University of Haifa, Haifa, 31905, Israel
kogevn@math.haifa.ac.il
On isomorphism of elliptic operators

31. **Stanislav Kupin**
University of Provence, France
CMI, UMR 6632 University of Provence 39, rue Joliot-Curie 13453
Marseille Cedex 13 France
kupin@cmi.univ-mrs.fr
Non-Szegö asymptotics for orthogonal polynomials on the unit circle

32. **Pavel Kurasov**
Dept. of Mathematics, Lund Institute of Technology
Dept of Mathematics, LTH, Solvegatan 18, Box 118, 221 00 Lund, Sweden
kurasov@maths.lth.se

33. **Heinz Langer**
TU Vienna, Austria
Institute of Analysis and Scientific Computing
Wiedner Hauptstrasse 8–10, A-1040 Wien, Austria
hlanger@mail.zserv.tuwien.ac.at
The generalized Schur algorithm and inverse spectral problems

34. **Ari Laptev**
Royal Institute of Technology in Stockholm, Sweden
Department of Mathematics, Royal Institute of Technology
100 44 Stockholm, Sweden
laptev@math.kth.se

35. **Annemarie Luger**
Vienna University of Technology, Austria
Department of Analysis and Scientific Computing
Wiedner Hauptstrasse 8–10, A-1040 Wien, Austria
aluger@mail.zserv.tuwien.ac.at
Reflectionless point interaction

36. **Witold Majdak**
The University of Mining and Metallurgy in Krakow, Poland
The University of Mining and Metallurgy
The Faculty of Applied Mathematics
al. Mickiewicza 30, 30-059, Krakow, Poland
majdak@wms.mat.agh.edu.pl (or: majdak@autograf.pl)

37. **Maria Malejki**
Faculty of Applied Mathematics
AGH University of Science and Technology, Poland
Faculty of Applied Mathematics, AGH University of Science and
Technology, al. Mickiewicza 30, 30-059 Krakow, Poland
malejki@uci.agh.edu.pl

38. **Marco Marletta**
    Cardiff University, United Kingdom
    School of Mathematics, Cardiff University
    PO Box 926, Cardiff CF24 4YH, United Kingdom
    Marco.Marletta@cs.cardiff.ac.uk
    Approximation of DN maps and spectra of Schrödinger operators on
    exterior domains

39. **Federico Menendez-Conde**
    Universidad Autonoma de Hidalgo, Mexico
    Centro de Investigacion en Matematicas Instituto de Ciencias Besicas e
    Ingeniera (ICBI) Carr. Pachuca-Tulancingo Km 4.5 Mineral de la Reforma,
    Hidalgo CP 42181
    fmclara@uaeh.reduaeh.mx

40. **Bernd Metzger**
    Fakultät für Mathematik, TU Chemnitz
    Technische Universität Chemnitz Fakultät für Mathematik 09107 Chemnitz
    b.metzger@mathematik.tu-chemnitz.de
    The Parabolic Anderson Model: The asymptotics of the statistical
    moments, Lifshitz tails and related topics

41. **Robert Minlos**
    Institute for Information Transmission Problems, Moscow, Russia
    Dobrushin Mathematical Lab., Inst. for Information Transmission Problems
    (RAS), Karetnyj pereulok, 19, Moscow, 101447 Russia
    minl@iitp.ru
    Two-particle bounded states of transfer-matrix for Gibbsian fields (high
    temperature regime)

42. **Boris Mityagin**
    the Ohio State University, Columbus, USA
    Department of Mathematics 231
    West 18th Ave Columbus, OH 43210 the USA
    mityagin.1@osu.edu
    Simple and double eigenvalues of the Hill operator with a two term potential

43. **Wojciech Młocek**
    Akademia Rolnicza w Krakowie
    Zakład Zastosowań Matematyki, al. Mickiewicza 24
    30-059 Krakow, Poland
    mlocek@ar.krakow.pl
    tel: (012) 662 40 20

44. **Marcin Moszynski**
Instytut Matematyki Stosowanej i Mechaniki, Wydział Matematyki
Informatyki i Mechaniki, Uniwersytet Warszawski
Uniwersytet Warszawski, Wydział Matematyki Informatyki i Mechaniki,
Instytut Matematyki Stosowanej i Mechaniki, ul. Banacha 2, 02-097,
Warszawa, Poland
mmoszyns@mimuw.edu.pl
On some unbounded tridiagonal matrices generating C_0 semigroups in l^p
spaces

45. **Yaroslav Mykytyuk**
Lviv National University, Ukraine
Department of mechanics and mathematics, Lviv National University, 1,
Universytetska st., 79000, Lviv, Ukraine
yamykytyuk@yahoo.com
An Inverse Scattering Problem for Sturm-Liouville Operators on Semiaxis

46. **Sergey Naboko**
St. Petersburg State University, Russia
Department of Mathematical Physics, Institute of Physics, NIIF,
St. Petersburg State University, ul.Ulianovskaya, 1, St. Petergoff,
St. Petersburg, Russia, 198904
naboko@math.su.se

47. **Hatem Najar**
I.P.E.I.Monastir Tunisia
Departement de mathematiques I.P.E.I.Monastir. 5000 Monastir Tunisie
hatem.najar@ipeim.rnu.tn

48. **Marlena Nowaczyk**
Matematik LTH, Sweden
Matematikcentrum LTH Box 118 221 00 Lund, Sweden
marlena@maths.lth.se
Inverse Scattering Problem for Quantum Graphs

49. **Konstantin Pankrashkin**
Institute of Mathematics, Humboldt-University of Berlin, Germany
Institute of Mathematics Humboldt-University of Berlin Rudower Chaussee
25 12489 Berlin Germany
const@mathematik.hu-berlin.de
Point perturbations as pseudopotentials

50. **Boris Pavlov**
    Dept. of Mathematics, University of Auckland, New Zealand
    Dept. of Mathematics, University of Auckland, Private Bag 92019,
    Auckland, New Zealand
    pavlov@math.auckland.ac.nz
    Transport properties of quantum networks

51. **Artem Pulemotov**
    Kyiv National T. Shevchenko Univ., Ukraine
    Department of Mathematics and Mechanics, Kyiv National T. Shevchenko
    Univ., 64 Volodymyrska str., 01033 Kyiv, Ukraine
    pulen@i.kiev.ua
    Subfields of a Jacobi Field

52. **Alexander Pushnitski**
    Loughborough University, U.K.
    Department of Mathematical Sciences, Loughborough University,
    Loughborough, LE11 2NL, United Kingdom
    a.b.pushnitski@lboro.ac.uk
    Trace formulae and high energy spectral asymptotics for the Landau
    Hamiltonian

53. **Roman Romanov**
    Cardiff University, UK, and St. Petersburg State University, Russia
    School of Computer Science Cardiff University, Cardiff Queen's Buildings,
    PO Box 916 Newport Road, Cardiff CF24 3XF UK
    r.v.romanov@cs.cf.ac.uk
    The instability of the absolutely continuous spectrum of nonself-adjoint
    ordinary differential operators under slowly decaying perturbations

54. **Hermann Schulz-Baldes**
    TU Berlin, Germany
    Inst. für Mathematik, Strasse des 17. Juni 136 D-10623 Berlin, Germany
    schuba@math.tu-berlin.de
    Weak disorder expansion for localization lengths of quasi-1D systems

55. **Roman Shterenberg**
    St. Petersburg State University, Russia
    Department of Mathematical Physics, Physical Faculty, St. Petersburg
    State University, ul. Ul'yanovskaya 1, 198504 St. Petersburg, Russia
    roman@rs3759.spb.edu
    Periodic magnetic Schrödinger operator with degenerate lower edge of the
    spectrum

56. **Luis Silva**
Instituto de Investigaciones en Matematicas Aplicadas y en Sistemas
(IIMAS) Universidad Nacional Autonoma de Mexico Mexico
Depto. de Metodos Matematicos y Numericos IIMAS-UNAM Apdo. postal
20-726 01000 Mexico, D.F.
silva@leibniz.iimas.unam.mx
Absence of accumulating points in the pure point spectrum of Jacobi
matrices

57. **Gerald Teschl**
University of Vienna, Austria
Department of Mathematics, University of Vienna, Nordbergstr. 15, A1090
Vienna, Austria
gerald.teschl@univie.ac.at
Scattering theory for Jacobi operators with quasi-periodic background

58. **Alessandro Teta**
Dipartimento di Matematica Pura e Applicata Università di L'Aquila Italy
Dipartimento di Matematica Pura e Applicata Università di L'Aquila
via Vetoio – Loc. Coppito 67100 L'Aquila Italy
teta@univaq.it
Analysis of decoherence in two-particle system

59. **Alexey Tikhonov**
Taurida National University, Ukraine
28 Kirov str., apt. 61 Simferopol, 95011 Crimea Ukraine
tikhonov@club.cris.net
Functional model for operators with spectrum on a curve and its
applications

60. **Françoise Truc**
Institut Fourier, France
Institut Fourier, Laboratoire de Mathématiques
BP 74, 38402 St. Martin d'Heres Cedex, France
trucfr@ujf-grenoble.fr
Remarks on the spectrum of the Neumann problem with magnetic field in
the half space

61. **Tomio Umeda**
Himeji Institute of Technology Japan
Department of Mathematics, Himeji Institute of Technology
Shosha, Himeji 671-2201 Japan
umeda@sci.himeji-tech.ac.jp

62. **Peter Yuditskii**
    Johannes Kepler University of Linz, Austria
    Institute for Analysis, Johannes Kepler University of Linz
    A-4040 Linz, Austria
    Petro.Yudytskiy@jku.at
    On generalized sum rules for Jacobi matrices

63. **Boris N. Zakhariev**
    Joint Institute for Nuclear Research
    Laboratory of theoretical Physics, JINR, Dubna, 141980, Russia
    zakharev@thsun1.jinr.ru
    New results in control of discrete, continuous abd band spectra of
    Shrödinger equation

64. **Grigory Zhislin**
    Doctor of phys. and math. sciences,
    Radiophysical Research Institute and
    Nizhny Novgorod State University, Russia
    Division of mathem.physics of Radiophysical Research Institut (NIRFI)
    25 Bolshaya Pecherskaya str. 603950, Nizhny Novgorod, Russia
    greg@nirfi.sci-nnov.ru
    Spectral properties of pseudorelativistic hamiltonians of atoms and positive
    ions with nuclei of infinite masses

65. **Lech Zielinski**
    Université du Littoral, France
    Centre Universitaire de la Mi-Voix, Laboratoire de Math. Pures et
    Appliqués, Université du Littoral
    50 rue F. Buisson B.P. 699, 62228 Calais, France
    Lech.Zielinski@lmpa.univ-littoral.fr
    On Semiclassical Spectral Asymptotics for Elliptic Operators with Critical
    Points

66. **Andrej Zlatos**
    University of Wisconsin-Madison
    Mathematics Department University of Wisconsin-Madison
    480 Lincoln Dr. Madison, WI 53706 USA
    andrej@math.wisc.edu
    Sum Rules for Jacobi Matrices and Divergent Lieb-Thirring Sums